Power-Aware Testing and Test Strategies for Low Power Devices

Patrick Girard · Nicola Nicolici · Xiaoqing Wen
Editors

Power-Aware Testing and Test Strategies for Low Power Devices

Springer

Editors
Patrick Girard
LIRMM / CNRS Laboratoire d'Informatique
de Robotique et de Microélectronique
de Montpellier
161 Rue Ada
34392 Montpellier
France
girard@lirmm.fr

Xiaoqing Wen
Department of Computer Science
and Electronics
Kyushu Institute of Technology
680-4 Kawazu
Iizuka
820-8502 Japan
wen@cse.kyutech.ac.jp

Nicola Nicolici
Department of Electrical
and Computer Engineering
McMaster University
1280 Main Street West
Hamilton ON L8S 4K1
Canada
nicola@ece.mcmaster.ca

ISBN 978-1-4419-0927-5 e-ISBN 978-1-4419-0928-2
DOI 10.1007/978-1-4419-0928-2
Springer New York Dordrecht Heidelberg London

Library of Congress Control Number: 2009930470

Printed on acid-free paper

Springer is part of Springer Science+Business Media (www.springer.com)

Summary and Objective of the Book

Power dissipation is becoming a critical parameter during manufacturing test as the device can consume much more power during test than during functional mode of operation. In the meantime, elaborate power management strategies, such as dynamic voltage scaling, clock gating or power gating techniques, are used today to control the power dissipation during functional operation. The usage of these strategies has various implications on manufacturing test, and power-aware test is therefore increasingly becoming a major consideration during design-for-test and test preparation for low-power devices. This book provides knowledge in this area. It is organized into three main parts. The first one gives necessary background and discusses issues arising from excessive power dissipation during test application. The second part provides comprehensive knowledge of structural and algorithmic solutions that can be used to alleviate such problems. The last part surveys low-power design techniques and shows how these low-power devices can be tested safely without affecting yield and reliability. EDA solutions for considering power during test and design-for-test are also described in the last chapter of the book.

About the Editors

Patrick Girard received a M.S. degree in Electrical Engineering and a Ph.D. degree in Microelectronics from the University of Montpellier, France, in 1988 and 1992 respectively. He is currently Research Director at CNRS (French National Center for Scientific Research), and works in the Microelectronics Department of the LIRMM (Laboratory of Informatics, Robotics and Microelectronics of Montpellier, France). Patrick Girard is the Vice-Chair of the European Test Technology Technical Council (ETTTC) of the IEEE Computer Society. He is currently the Editor-in-Chief of the *ASP Journal of Low Power Electronics* (*JOLPE*) and an Associate Editor of the *IEEE Transactions on VLSI Systems* and the *Journal of Electronic Testing – Theory and Applications* (*JETTA* – Springer). From 2005 to 2009, he was an Associate Editor of the *IEEE Transactions on Computers*. He has served as technical program committee member of the ACM/IEEE Design Automation Conference (DAC), ACM/IEEE Design Automation and Test in Europe (DATE), IEEE International Test Conference (ITC), IEEE International Conference on Computer Design (ICCD), IEEE International Conference on Design & Test of Integrated Systems (DTIS), IFIP International Conference on VLSI-SOC, IEEE VLSI Test Symposium (VTS), IEEE European Test Symposium (ETS), IEEE International On-Line Testing Symposium (IOLTS), IEEE Asian Test Symposium (ATS), ACM/IEEE International Symposium on Low Power Electronic Design (ISLPED), IEEE International Symposium on Electronic Design, Test & Applications (DELTA) and IEEE Workshop on Design and Diagnostics of Electronic Circuits and Systems (DDECS). He has served as Test Track Chair for DAC 2007, DATE 2007, DATE 2008, ICCD 2008 and DATE 2009. He has also served as Program Chair for DELTA 2006, DTIS 2006, DDECS 2007 and ETS 2008. Patrick Girard has been involved in several European research projects (ESPRIT III ATSEC, EUREKA MEDEA, MEDEA+ ASSOCIATE, IST MARLOW, MEDEA+ NanoTEST, CATRENE TOETS) and has managed industrial research contracts with major companies such as Infineon Technologies, Atmel, STMicroelectronics, etc. His research interests include the various aspects of digital testing and memory testing, with special emphasis on DfT, BIST, diagnosis, delay testing and power-aware testing. He has supervised 22 PhD dissertations and has published 6 books or book chapters, 34 journal papers, and more than 110 conference and symposium papers on these fields. He received the Best Paper Award at ETS 2004 and at DDECS 2005. Patrick Girard is a Senior Member of IEEE.

Nicola Nicolici is an Associate Professor in the Department of Electrical and Computer Engineering at McMaster University, Canada. He received a Dipl. Ing. degree in Computer Engineering from the "Politehnica" University of Timisoara, Romania in 1997, and a Ph.D. in Electronics and Computer Science from the University of Southampton, UK in 2000. His research interests are in the area of computer-aided design and test. He has coauthored over 70 research papers and one book in this area and he received the IEEE TTTC Beausang Award for the Best Student Paper at the IEEE International Test Conference in 2000 and the Best Paper Award at the IEEE/ACM Design Automation and Test in Europe Conference in 2004. He serves on technical program committees of several conferences in the fields such as IEEE/ACM Design Automation Conference, IEEE/ACM Design Automation and Test in Europe and the IEEE European Test Symposium. He was the guest coeditor for a special issue on Silicon Debug and Diagnosis, published by the *IET Proceedings on Computers and Digital Techniques* (*IET-CDT*) in November 2007, and a special issue on Low Power Test published by the *Journal of Electronic Testing – Theory and Applications* (*JETTA*) in August 2008. He currently serves on the editorial board of *IET-CDT*, *JETTA* and *Integration, the VLSI Journal*. He is a member of the ACM SIGDA and the IEEE Computer and Circuits and Systems Societies.

Xiaoqing Wen is a Professor and Chair of the Department of Creative Informatics, Kyushu Institute of Technology, Japan. He received a B.S. degree in Computer Science and Technology from Tsinghua University, China, in 1986, a M.S. degree in Information Engineering from Hiroshima University, Japan, in 1990, and a Ph.D. degree in Applied Physics from Osaka University, Japan, in 1993. He was an Assistant Professor at Akita University, Japan, from 1993 to 1997, and a Visiting Researcher at University of Wisconsin-Madison from October 1995 to March 1996. He joined SynTest Technologies, Inc. (Sunnyvale, CA) in 1998 and served as its Chief Technology Officer (CTO) until 2003. In 2004, he joined the Faculty of Kyushu Institute of Technology, Japan, as an Associate Professor, and became a full Professor in 2007. His research interests include low-power test generation, fault diagnosis, logic BIST, test compression and design for testability. He has published more than 100 journal and conference papers, and co-edited the book *VLSI Test Principles and Architectures: Design for Testability* (Morgan Kaufmann, 2006). He received the ISPJ Research Promotion Award in 1993, the Best Paper Award at the 7th IEEE Workshop on RTL and High Level Testing, and the IEICE-ISS Excellent Paper Award in 2008. He currently holds 15 US patents, with 22 more filed US/Japan patent applications. He has served on program committees for numerous IEEE-sponsored technical events, and was the Program Committee Co-Chair of the 16th IEEE Asian Test Symposium and the 8th IEEE Workshop on RTL and High Level Testing. He is an Associate Editor of *IPSJ Transactions on System LSI Design Methodology*, a member of ITC Asian Subcommittee, and a technical consultant for the Association of Southeast Asian Nations (ASEAN). He is a senior member of IEEE, a member of IEICE and a member of REAJ.

Preface

Power consumption has already become a critical issue that must be taken into consideration when developing and implementing modern integrated circuits and systems. In many application markets enabled by portable devices, the limited energy supply, which is available in-field by means of either energy scavenging or rechargeable batteries, is defining key product features, such as the form factor or the "talk time" for mobile phones. For example, the power budget for today's smartphones is in the range of 2–4 W. When excluding the power required for operating the user display and radios, there is approximately 1 W available for almost 100 billion operations per second needed by the digital workload. The continuous user demand to further expand the functionality of portable devices, without affecting the key product features, imposes tight constraints on the design technology with power optimization as its focal point. Besides, power impacts the cost of fabrication, packaging and cooling, as well as the system's reliability and its maintenance cost. With the proliferation of cloud computing and data centers, the power requirements for this elaborate computing infrastructure may exceed 100 MW in the foreseeable future. Therefore, even a modest 10% reduction in the average power consumed by each circuit in a server farm can visibly reduce the operational expenses.

Given the above trends, in the past two decades, we have witnessed extensive research and development in the area of low-power circuits and systems. A large pool of circuit techniques, design methodologies and tool flows have been invented by exploiting redundancies in the implementation space and manipulating design parameters, such as scaling the supply voltage, dynamically adjusting the operating frequency at runtime, gating clocks to match the circuit activity to the application workload; or modifying the software and/or the processor architecture, only to mention a few well-established solutions. In addition to the above, according to Moore's Law, there was a steady shift to finer semiconductor process geometries, which also enables a power reduction for each device, if the same level of functionality is preserved. Nonetheless, as the number of transistors integrated onto a silicon die has increased, and because the number of pins available for screening the devices for fabrication defects is not scaling at the same rate, the cost of manufacturing test is on the rise. This problem has been exacerbated by both the new types of defect mechanisms that have been found in advanced process technologies and the test constraints unique to low-power devices.

This book provides a comprehensive coverage of the established interrelationships between low-power design and manufacturing test of integrated circuits. It deals with both power-aware testing that addresses excessive, and potentially damaging, power surges that occur only during test, and the unique test challenges posed by the power-management techniques. The material takes the reader from the fundamental principles to the advanced concepts in the field and all the known directions of work are explored, ranging from low-power automatic test pattern generation and power-aware design-for-test to the core test strategies for low-power devices.

Chapter 1 provides the background material on manufacturing test of very large-scale integrated (VLSI) circuits. Fundamental concepts such as defects, fault models and coverage, manufacturing yield, defect level, logic testing and memory testing are introduced. The basic algorithms and methods developed to make VLSI test tractable, such as automatic test pattern generation (ATPG), design-for-test (DFT), built-in self-test (BIST) and test data compression, are also described in this chapter.

Sudden changes in power consumption affect supply voltage levels, while variations in chip temperature, caused by excessive power, can affect both reliability and timing. These are only two power-related issues that affect the testing process. Chapter 2 elaborates this relationship between power consumption and VLSI test in detail and it presents basic concepts such as power droop, power delivery and on-chip thermal gradients. How power impacts test throughput and yield loss, both of which influence test economics, is also discussed.

ATPG algorithms have been researched for over four decades and they have become ubiquitous in test tools. However, the unique challenges posed by power consumption during test introduce a new dimension to the ATPG problem. The key advantage of using an ATPG-based approach to reduce test power lies in its algorithmic nature, where the added effort is placed during test preparation and there is no area or performance overhead on-chip. Chapter 3 details the recent developments in low-power ATPG, including also advanced concepts such as low-power test compaction, low-power filling of don't care values in test patterns, low-power test vector ordering during scan testing and low-power algorithms for memory testing.

Chapter 4 discusses low-power DFT techniques. Employing scan chains to improve the controllability and observability of state elements is the most common DFT technique. However, the power consumed during the shift cycle in the combinational logic blocks brings no value to the testing process. Hence, gating the outputs of scan cells during shift can be employed to eliminate the useless circuit activity. DFT techniques for low-power test facilitate or can be combined with other test power reduction techniques such as test planning and low-power ATPG. Therefore, more advanced DFT techniques for test-power reduction, such as test clock gating, low-power scan cell design, scan chain partitioning and ordering of scan cells, among others, are discussed in this chapter.

Due to the steady growth in the design complexity and the variety of fabrication defects in modern process technologies, both the number of bits per test pattern and the number of test patterns are increasing, thus causing a test data volume problem. BIST and test data compression are the two main approaches employed to deal with the excessive test data volume. Chapter 5 describes several low-power BIST and test data compression techniques, by pointing out their advantages and disadvantages in terms of area, performance and power. The discussion is focused on both entropy coding techniques, as well as compression based on linear-feedback shift registers, which are commonly used as pseudorandom sequence generators in BIST.

A modular approach based on the design reuse philosophy has been widely adopted for system-on-chip (SOC) designs. Testing SOCs can also be done in a modular fashion. An important advantage of using a modular test strategy is the ability to develop test plans that are power-aware. Chapter 6 contains an introduction to core-based testing, followed by a discussion on test power modeling and test plan scheduling. The advantages of a modular approach are also highlighted for SOCs with multiple clock domains and when monitoring the steady-state current.

In the previous chapters, the focus has been placed on algorithms and techniques for low-power testing, with the main objective on avoiding overstressing and overheating the devices. In the second part of the book, the focus shifts toward low-power devices and the unique test challenges posed by the power-management structures. Chapter 7 provides an overview of the adopted design techniques for the static and dynamic power reduction. For a better understanding of the following chapters, it discusses the impact of low-power design techniques on test and it covers the test implications of the post-silicon adaptation approaches for power reduction.

Using multiple supply voltages (also called multi-voltage or multi-Vdd) enables dynamic voltage scaling, which is often employed in practice to match the circuit speed and power at runtime to the application workload. Chapter 8 discusses testing strategies for multi-voltage designs. There are fabrication defects which have Vdd dependency and hence they can be activated at some but not at all the power supply settings. Due to cost considerations, the aim is to screen the fabrication defects by avoiding repetitive tests at several Vdd settings. To adequately address this unique test challenge, new techniques for defect modeling, test generation and DFT have been investigated in recent years. Some of these techniques, such as test point insertion, elevated Vdd testing and low-cost scan for multi-voltage design, are presented in depth in this chapter.

Another widely adopted low-power technique is to gate off clocks to logic blocks that are not doing any useful computations in the present state. Chapter 9 discusses DFT approaches that deal with gated clocks, such that the ATPG algorithms can be reused and correctly interpret the operation of gated clocks to avoid any loss in test coverage. Besides, as elaborated in this chapter, clock-gating logic, which is inserted for lowering power during the functional operation, can also be leveraged to reduce the switching activity during both test application and test data loading/offloading through scan.

Power management techniques used in low-power circuits commonly rely on special low-power cells such as level shifters, state-retention registers and isolation cells. To ensure that the defect level is not affected by not screening these cells adequately for fabrication defects, in addition to testing logic and memory blocks, the power-management structures also need to be tested in a structured way. Chapter 10 discusses different methods for testing low-power cells and for validating the integrity of power distribution networks in low-power devices.

Unlike the previous chapters, the last one is focused neither on techniques for reducing power during test nor on testing the power-management structures present in low-power devices. Rather its focus is on explaining the unique challenges posed by integrating these techniques in electronic design automation (EDA) tool flows. For example, on the one hand, DFT insertion tools must be power-aware so that scan is robustly implemented across different clock and voltage domains. On the other hand, the power overhead of the DFT logic must be minimal. Hence, as elaborated in Chapter 11, EDA tool users must be provided with the choice to reach trade-offs, since DFT and low power can present conflicting constraints and implementation costs.

Low-power design and manufacturing test are well-researched and published topics in the broad field of VLSI circuits and systems. However, given the technology trends from the past decade and the stronger interrelation between power and test, this book provides the first comprehensive reference material on power-aware testing and testing low-power devices. It covers the fundamentals, the established techniques that have been adopted in practice and the latest research in the field. It is hoped that this book will have a two-pronged effect: to provide the reader with the fundamental material to be used as a reference and to motivate further innovation in the field. Therefore, it can become a valuable asset for a diverse group of readers: VLSI design and test professionals, EDA tool developers, academics that are planning to develop or to bring their course material up to date and, most importantly, students who are entering in the VLSI and EDA fields.

Montpellier, France — *Patrick Girard*
Hamilton, Canada — *Nicola Nicolici*
Iizuka, Japan — *Xiaoqing Wen*

Contents

Contributors

Bashir M. Al-Hashimi University of Southampton, Southampton, UK

Swarup Bhunia Case Western Reserve University, Cleveland, OH, USA

Krishnendu Chakrabarty Duke University, Dhuram, NC, USA

Krishna Chakravadhanula Cadence Design Systems Inc., Endicott, NY, USA

Sandeep Kumar Goel LSI Corporation, Milpitas, CA, USA

Mokhtar Hirech Synopsys Inc., Mountain View, CA, USA

Mark Kassab Mentor Graphics Corp., Wilsonville, OR, USA

Brion Keller Cadence Design Systems Inc., Endicott, NY, USA

Saqib Khursheed University of Southampton, Southampton, UK

Sandip Kundu University of Massachusetts, Amherst, MA, USA

Erik Larsson Linköping University, Linköping, Sweden

C.P. Ravikumar Texas Instruments Inc., Bangalore, India

Kaushik Roy Purdue University, West Lafayette, IN, USA

Alodeep Sanyal University of Massachusetts, Amherst, MA, USA

Charles E. Stroud Auburn University, Auburn, AL, USA

Mohammad Tehranipoor University of Connecticut, Storrs, CT, USA

Laung-Terng Wang SynTest Technologies Inc., Sunnyvale, CA, USA

Seongmoon Wang NEC Labs, Princeton, NJ, USA

Xiaoqing Wen Kyushu Institute of Technology, Iizuka, Japan

Hans-Joachim Wunderlich University of Stuttgart, Stuttgart, Germany

Christian G. Zoellin University of Stuttgart, Stuttgart, Germany

Chapter 1
Fundamentals of VLSI Testing

Laung-Terng Wang and Charles E. Stroud

Abstract *Very-large-scale integration* (VLSI) testing encompasses all spectrums of test methods and structures embedded in a *system-on-chip* (SOC) to ensure the quality of manufactured devices during manufacturing test. The test methods typically include fault simulation and test generation, so that quality test patterns can be supplied to each device. The test structures often employ specific *design for testability* (DFT) techniques, such as scan design and *built-in self-test* (BIST), to test the digital logic portions of the device. To provide readers with basic understanding of the most recent DFT advances in logic testing, memory testing, and SOC testing for low-power device applications, this chapter covers a number of fundamental test methods and DFT structures to facilitate testing of modern SOC circuits. These methods and structures are required to improve the product quality and reduce the defect level and test cost of the manufactured devices, while at the same time simplifying the test, debug, and diagnosis tasks.

1.1 Introduction

Logic testing involves the process of testing the digital logic portion of a *circuit under test* (CUT). The digital logic can be reconfigured in the test mode (TM) to include test logic to improve the testability and test quality of circuit. Logic testing typically consists of applying a set of test stimuli to the inputs of the digital logic while analyzing the output responses. Both input test stimuli and output response analysis can be generated and performed externally or inside the chip. Circuits that produce the correct output responses for all input stimuli pass the test and are considered to be fault-free. Those circuits that fail to produce a correct response at any point during the test sequence are assumed to be faulty.

L.-T. Wang (✉)
SynTest Technologies Inc., Sunnyvale, CA, USA
e-mail: wang@syntest.com

C.E. Stroud
Auburn University, Auburn, AL, USA

P. Girard et al. (eds.), *Power-Aware Testing and Test Strategies for Low Power Devices*,
DOI 10.1007/978-1-4419-0928-2_1,

A circuit *defect* may lead to a *fault* causing an *error* that can result in a system *failure*. Two major defect mechanisms can cause the digital design to malfunction: manufacturing defects and soft errors. *Manufacturing defects* are physical (circuit) defects introduced during manufacturing that cause the design to fail to function properly in the device, on the *printed circuit board* (PCB), or in the system or field. These manufacturing defects can result in *static faults*, such as stuck-at faults, or *timing faults*, such as delay faults. A general consensus, known as the *rule of ten*, says that the cost of detecting a faulty device increases by an order of magnitude as we move through each stage of manufacturing, from device level to board level, to system level, and finally, to system operation in the field (Wang et al. 2006). *Soft errors*, which are also referred to as *single event upsets* (SEUs), are *transient faults* induced by environmental conditions such as *α-particle radiation* that cause a fault-free circuit to malfunction during operation (May and Woods 1979, Baumann 2005). The probability of the occurrence of soft errors increases as feature sizes decrease. For example, the probability of SEUs increased by a factor of more than 21 when moving from a feature size of 0.6 to 0.35 μm with an accompanying decrease in supply voltage, VDD, from 5 to 3.3 V (Ohlsson et al. 1998). Transient faults are nonrepeatable, temporary faults, and thus, they cannot be detected during manufacturing. Defect mechanisms must be tolerated in the system to enhance device reliability and yield, reduce defect level and test costs, as well as to improve system reliability and availability. Such robust methods and circuit structures to tolerate soft errors at the device, circuit, or system level are generally referred to as *design for reliability* (DFR).

Some percentage of the manufactured devices is expected to be faulty due to manufacturing defects. The *yield* of a manufacturing process is defined as the percentage of acceptable parts among all parts that are fabricated:

$$\text{Yield} = \frac{\text{Number of acceptable parts}}{\text{Total number of parts fabricated}}$$

Two types of yield loss are catastrophic and parametric. Catastrophic yield loss is caused by random defects, and parametric yield loss is caused by process variations. Automation of and improvements in an IC fabrication process line drastically reduce the particle density that creates random defects over time; consequently, parametric variations caused by process fluctuations become the dominant sources of yield losses.

Methods to reduce the effects of process variations during fabrication are generally referred to as *design for yield enhancement* (DFY) (Wang et al. 2007). The circuit implementation methods to avoid random defects are generally referred to as *design for manufacturability* (DFM). Broadly speaking, any DFM method is helpful for increasing manufacturing yield, and thus, it can also be considered a DFY method. Manufacturing yield relates to the failure rate λ. The *bathtub curve* shown in Fig. 1.1 is a typical device or system failure chart indicating how early failures, wearout failures, and random failures contribute to the overall device or system failures.

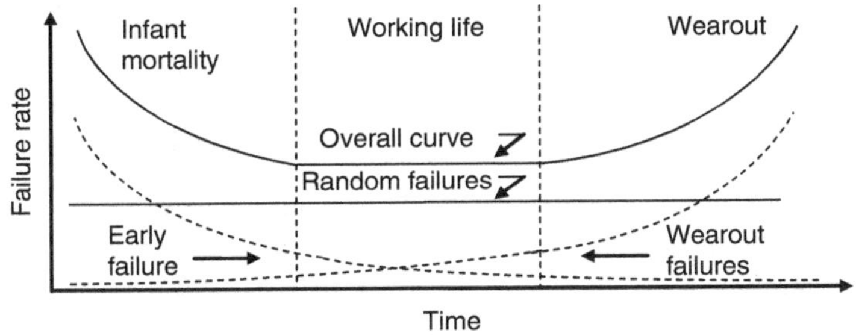

Fig. 1.1 Bathtub curve

The *infant mortality* period (with decreasing failure rate) occurs when a product is in its early production stage. Failures that occur in this period are mostly attributable to poor process or design quality, which leads to poor product quality. The product should not be shipped during this period to avoid massive field returns. The *working life* period (with constant failure rate) represents the product's "working life." Failures during this period tend to occur randomly. The *wearout* period (with increasing failure rate) indicates the "end-of-life" of the product. Failures during this period are caused by age defects, such as metal fatigue, hot carriers, electromigration, dielectric breakdown. For electronic products, this period is of less concern because end users often replace electronic products before the devices reach their respective wearout periods.

When ICs are tested, the following two undesirable situations may occur:

1. A good device fails the test and is declared as faulty.
2. A faulty device passes the test and is declared to be a good part.

These two outcomes are often due to a poorly designed test or the lack of *design for testability* (DFT). As a result of the first case, some good devices will be erroneously marked as faulty by the test. This outcome may be caused by IR-drop, for example. Overkill and hence yield loss will be induced in this case. As a result of the second case, even if all devices pass an acceptance test, some faulty devices may still be found in the manufactured electronic system. This outcome is mainly caused by insufficient test patterns for screening the faulty devices. The ratio of field-rejected parts to all parts passing quality assurance testing is referred to as the *reject rate*, also called the *defect level*

$$\text{Reject rate} = \frac{\text{Number of faulty parts passing final test}}{\text{Number of parts passing final test}}$$

For a given device, defect level DL is a function of process yield Y and fault coverage FC (McCluskey and Buelow 1988):

$$\text{DL} = 1 - Y^{(1-\text{FC})}$$

where *fault coverage* is defined as:

$$\text{Fault coverage} = \frac{\text{Number of detected faults}}{\text{Total number of faults}}$$

The defect level provides an indication of the overall quality of the testing process (Williams and Brown 1981). Generally speaking, a defect level of 500 *parts per million* (ppm) may be considered to be acceptable, whereas 100 ppm or less represents high quality. The goal of *six sigma* manufacturing, which is also referred to as *zero defect*, is 3.4 ppm or less.

For example, assume the process yield is 50%, and the fault coverage for a device is 90% for the given test set. By the above equation, we obtain a DL $= 1 - 0.5^{(1-0.9)} = 0.067$, which means that 6.7% of shipped parts will be defective, or the defect level of the product is 67,000 ppm. However, if a DL of 100 ppm is required for the same process yield of 50%, then the fault coverage required to achieve the target ppm level is FC $= 1-(\log(1-\text{DL})/\log(Y)) = 0.99986$. Because it could be extremely difficult, if not impossible, to generate tests that have 99.986% fault coverage, improvements over process yield might become mandatory to meet such a stringent *DL* goal.

As advances in the manufacturing technology have led to complex very-large-scale integration (VLSI) designs, it has become a requirement that DFT features be incorporated in the digital logic. The most popular DFT techniques in use today for testing digital logic include *scan design* and *scan-based logic built-in self-test* (BIST) (Williams and Parker 1983, McCluskey 1986). Both techniques have proven to be effective in producing testable VLSI designs. Additionally, *test compression*, which is a DFT technique that supplements scan, is growing in importance for further reducing test data volume and test application time during manufacturing test (Touba 2006, Wang et al. 2006, Kapur et al. 2008).

The test logic is typically inserted at the *register-transfer level* (RTL) or gate level prior to physical design to ensure the quality of the fabricated devices. Once the designer verifies the combined digital logic and test logic, test patterns are generated for the design to ensure that it meets the manufactured device test requirements. The test requirements the product must meet are often specified in terms of *DL* and *manufacturing yield*, test cost, and whether it is necessary to perform self-test and diagnosis. As the test requirements primarily target manufacturing defects rather than soft errors, which would require *online fault detection and correction* (Wang et al. 2007), one needs to decide the fault models that should be considered.

The logic test process now consists of: (1) defining the targeted fault models for *DL* and manufacturing yield considerations, (2) deciding what types of DFT features should be incorporated in the design to meet the test requirements, (3) generating and fault-grading test patterns to calculate the final fault coverage, and (4) conducting manufacturing test to screen bad chips from shipping to customers and performing *failure mode analysis* (FMA) when the chips do not achieve desired *DL* or yield requirements.

In general, functional test patterns and structural test patterns can be used as manufacturing tests. Applying all possible input test patterns to an n-input combinational logic circuit illustrates the basic idea of *functional testing* where every entry in the truth table for the combinational logic circuit is tested to determine whether it produces the correct response. In practice, functional testing is considered by many designers and test engineers as testing the circuit as thoroughly as possible in a system-like mode of operation. A more practical approach is *structural testing* where test patterns are selected based on the circuit structural information and a set of *fault models*. Structural testing saves test time and improves test efficiency because the total number of test patterns is decreased by targeting specific faults that would result from defects in the manufactured device. Unfortunately, structural testing cannot guarantee detection of all possible manufacturing defects because the test patterns are generated based on specific fault models.

1.2 Fault Models

The diversity of defects makes it difficult to generate tests for real defects. Fault models are necessary for generating and evaluating a set of test patterns. Generally, a good fault model should satisfy two criteria: (1) it should accurately reflect the behavior of the defects and (2) it should be computationally efficient in terms of the time required for fault simulation and test generation (Stroud 2002). Many fault models have been proposed but, unfortunately, no single fault model accurately reflects the behavior of all possible defects that can occur. As a result, a combination of different fault models is often used in the generation and evaluation of test patterns. Some well-known and commonly used fault models for general sequential logic include the following:

1. *Gate-level stuck-at fault model*: The stuck-at fault is a logical fault model that has been used successfully for decades. A stuck-at fault transforms the correct value on the faulty signal line to appear to be stuck-at a constant logic value, either logic 0 or 1, referred to as *stuck-at-0* (SA0) or *stuck-at-1* (SA1), respectively. This model is commonly referred to as the *line stuck-at fault model*, where any line can be SA0 or SA1. It is also referred to as the gate-level stuck-at fault model where any input or output of any gate can be SA0 or SA1 (Wadsack 1978).
2. *Transistor-level stuck fault model*: At the switch level, a transistor can be *stuck-off* or *stuck-on*, which are also referred to as *stuck-open* or *stuck-short*, respectively. The line stuck-at fault model cannot accurately reflect the behavior of stuck-off and stuck-on transistor faults in *complementary metal oxide semiconductor* (CMOS) logic circuits because of the dual transistors used to construct the load and driver circuits in CMOS logic gates. A stuck-open transistor fault in a CMOS combinational logic gate can cause the gate to behave like a dynamic level-sensitive latch. Thus, a stuck-open fault in a CMOS combinational circuit requires a sequence of two vectors. The first vector *sensitizes* the fault by establishing the opposite logic value to that of the fault-free circuit at the faulty node and the second vector propagates the faulty circuit value to a point of observability. Stuck-short faults can produce a conducting path between power

(V_{DD}) and ground (V_{SS}) and may be detected by monitoring the power supply current, I_{DDQ}, during steady-state operation. This technique of monitoring the steady-state power supply current to detect transistor stuck-short faults is called I_{DDQ} *testing* (Bushnell and Agrawal 2000).

3. *Bridging fault models*: Defects can also include opens and shorts in the wires that interconnect the transistors that form the circuit. Opens tend to behave like line stuck-at faults; however, a *resistive open* does not behave the same as a transistor or line stuck-at fault; instead, it can affect the propagation delay of the signal path. A short between two wires is commonly referred to as a *bridging fault.* The case of a wire being shorted to V_{DD} or V_{SS} is equivalent to the line stuck-at fault model; however, when two signal wires are shorted together, bridging fault models are needed. The three most commonly used bridging fault models are illustrated in Fig. 1.2. The first bridging fault model proposed was the *wired-AND/wired-OR* bridging fault model. This model was originally developed for

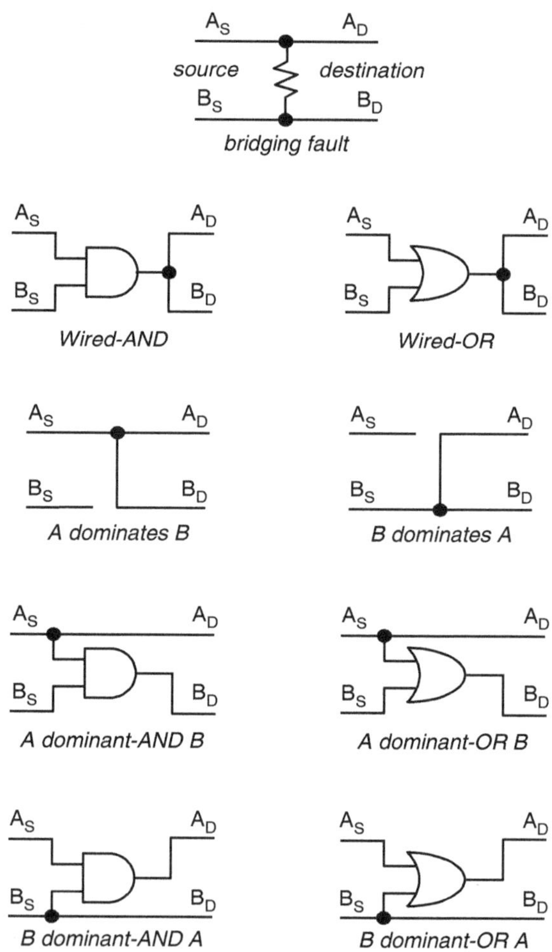

Fig. 1.2 Bridging fault models

bipolar technology and does not accurately reflect the behavior of those bridging faults typically found in CMOS devices. Therefore, the *dominant bridging fault* model was proposed for CMOS. In this model, one driver is assumed to dominate the logic value on the two shorted nets; however, the dominant bridging fault model does not accurately reflect the behavior of a resistive short in some cases. The most recent bridging fault model *four-way* bridging fault model, also known as the *dominant-AND/dominant-OR* bridging fault model, assumes that one driver dominates the logic value of the shorted nets for one logic value only (Emmert et al. 2000).

4. *Delay fault models*: Resistive opens and shorts in wires as well as parameter variations in transistors can cause excessive delays such that the total propagation delay falls outside the specified limit. *Delay faults* have become more prevalent with decreasing feature sizes, and different delay fault models are available. In *gate-delay fault* and *transition fault* models, a delay fault occurs when the time interval for a transition through a single gate exceeds its specified range. The *path-delay fault* model, on the contrary, considers the cumulative propagation delay along any signal path through the circuit. The *small delay defect* model takes into consideration the timing delays associated with the fault sites and propagation paths from the layout (Sato et al. 2005).

As tests generated for one fault model can potentially detect faults of other models, identifying a good order of fault models to target during test generation can help reduce the number of test vectors and, in turn, test time. A common practice is to target delay faults first, followed by gate-level stuck-at faults, bridging faults, and finally, transistor-level stuck faults.

1.3 Design for Testability

To test a given circuit, we need to control and observe the logic values of internal nodes. Unfortunately, some nodes in sequential circuits can be difficult to control and/or observe. DFT techniques have been proposed to improve the *controllability* and *observability* of internal nodes. These techniques generally fall into one of the following four categories: (1) *ad hoc DFT* methods, (2) *scan design*, (3) BIST, and (4) test compression.

1.3.1 Ad Hoc Methods

Ad hoc methods were the first DFT techniques introduced in the 1970s (Abramovici et al. 1990) to target only those portions of the circuit that were difficult to test. Circuitry typically referred to as *test points* was added to improve the observability and/or controllability of internal nodes (Wang et al. 2006). Figure 1.3 shows an example of observation point insertion for a logic circuit with three low-observability

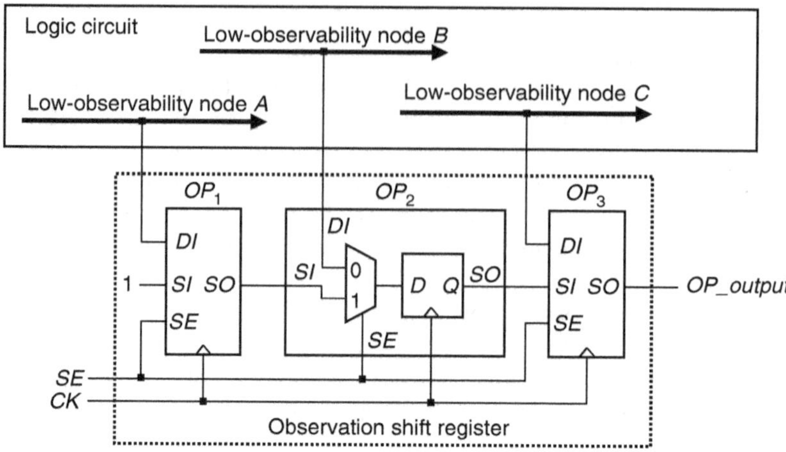

Fig. 1.3 Observation point insertion

nodes. OP_2 shows the structure of an observation point, which is composed of a multiplexer (MUX) and a D flip-flop. A low-observability node is connected to the 0 port of the MUX in an observation point, and all observation points are serially connected into an observation shift register using the one port of the MUX. An *SE* signal is used for MUX port selection. When *SE* is set to 0 and the clock *CK* is applied, the logic values of the low-observability nodes are captured into the D flip-flops. When *SE* is set to 1, the D flip-flops within OP_1, OP_2, and OP_3 operate as a shift register, allowing us to observe the captured logic values through *OP_output* during sequential clock cycles. As a result, the observability of internal nodes is greatly improved.

Figure 1.4 shows an example of control point insertion for a logic circuit with three low-controllability nodes. CP_2 shows the structure of a control point, which is composed of a MUX and a D flip-flop. The original connection at a low-controllability node is cut, and a MUX is inserted between the source and destination ends. During normal operation, *TM* is set to 0 such that the value from the source end drives the destination end through the 0 port of the MUX. During test, *TM* is set to 1 such that the value from the D flip-flop drives the destination end through the one port of the MUX. The D-flip-flops in OP_1, OP_2, and OP_3 are designed to form a shift register so that the required value can be shifted into the flip-flops using *CP_input* and used to control the destination ends of low-controllability nodes. As a result, the controllability of the circuit nodes is dramatically improved. However, this results in an additional delay to the logic path. Hence, care must be taken not to insert control points on a critical timing path. Furthermore, it is preferable to add a scan point, which is a combination of a control point and an observation point, instead of a control point, since this allows us to observe the source end as well.

Testability analysis is often used to measure the testability of a circuit by calculating the *controllability* and *observability* of each signal line in the circuit, where *controllability* reflects the difficulty of setting a signal line to a required logic value

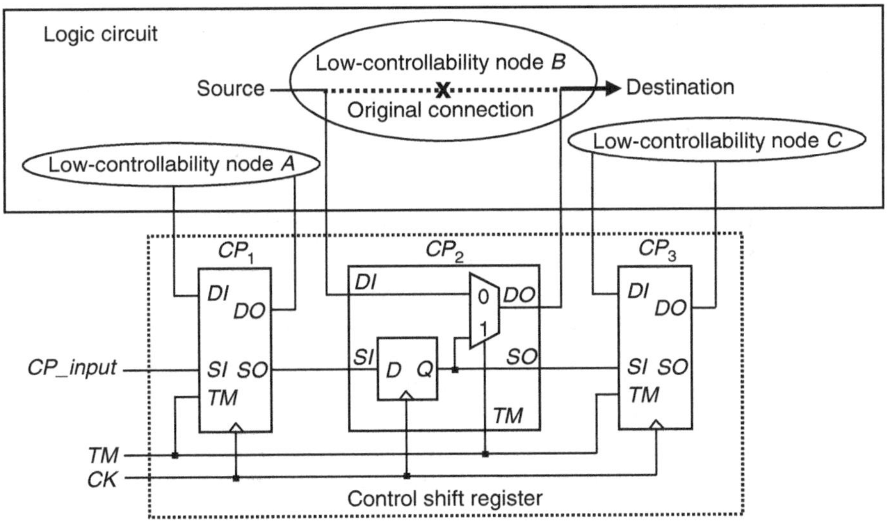

Fig. 1.4 Control point insertion

from primary inputs, and *observability* reflects the difficulty of propagating the logic value of the signal line to primary outputs. Since the 1970s, many testability analysis techniques have been proposed. The *Sandia Controllability/Observability Analysis Program* (SCOAP) (Goldstein and Thigpen 1980) was the first topology-based program for testability analysis applications. Enhancements based on SCOAP have also been developed and used to aid in test point selection (Wang and Law 1985).

Traditionally, a circuit's gate-level topological information is used for testability analysis. Depending on the target application, deterministic and/or random testability measures are calculated (Wang et al. 2006). In general, *topology-based testability analysis* (such as SCOAP) or probability-based testability analysis (Parker and McCluskey 1975, Savir et al. 1984) is computationally efficient but can produce inaccurate results for circuits that contain many reconvergent fanouts. *Simulation-based testability analysis* such as the *STAtistical Fault ANalysis* (STAFAN) algorithm (Jain and Agrawal 1985), however, can generate more accurate estimates by simulating the circuit behavior using deterministic, random, or pseudorandom test patterns, but these may require long simulation times.

Although attempts to use ad hoc methods have substantially improved the testability of a design and reduced the complexity of sequential *automatic test pattern generation* (ATPG), their end results were far from satisfactory; it was still difficult to reach a satisfactory level of fault coverage, say more than 90%, for large designs. Even with these testability aids, deriving functional patterns by hand or generating test patterns for a sequential circuit is a much more difficult problem than generating test patterns for a combinational circuit (Fujiwara and Toida 1982, Jha and Gupta 2003).

1.3.2 Scan Design

Currently, *scan design* is the most widely used structured DFT approach. It is implemented by connecting selected storage elements of a design into one or more shift registers, which are called *scan chains*, to provide them with external access. This task is accomplished by replacing each of the selected storage elements with *scan cells*, each of which has one additional *scan input* (SI) port and one shared/additional *scan output* (SO) port. By connecting the SO port of one scan cell to the SI port of the next scan cell, one or more scan chains are created.

The scan-inserted design operates in three modes: *normal mode*, *shift mode*, and *capture mode*. Circuit operations with associated clock cycles conducted in these three modes are referred to as normal operation, shift operation, and capture operation, respectively. In normal mode, all test signals are turned off, and the scan design operates in the original functional configuration. In both shift and capture modes, a *TM* signal is often used to turn on all test-related functions in compliance with scan design rules. A set of *scan design rules* that can be found in Cheung and Wang (1997) and Wang et al. (2006) is necessary to simplify the test, debug, and diagnosis tasks, improve fault coverage, and guarantee the safe operation of the device under test. These circuit modes and operations are distinguished using additional test signals or test clocks.

The fundamental scan architectures include the following (Wang et al. 2006): (1) *muxed-D scan design*, where storage elements are converted into muxed-D scan cells, (2) *clocked-scan design*, where storage elements are converted into clocked-scan cells, and (3) *level-sensitive scan design (LSSD) scan design*, where storage elements are converted into LSSD *shift register latches* (Eichelberger and Williams 1978). The scan cell designs are illustrated in Fig. 1.5. The major difference among the three types of scan cell designs are that (1) the muxed-D scan cell uses a scan enable signal *SE* to select data input *DI* or scan input *SI* connected to the previous scan cell output *SO*, (2) the clocked-scan cell uses a data clock *DCK* and a scan clock *SCK* to select *DI* and *SI*, respectively; and (3) the LSSD *shift register latches* uses two nonoverlapping clocks *C* and *B* to select data input *D* and two nonoverlapping scan clocks *A* and *B* to select scan input *I*.

The basic idea used to create a scan design is to reconfigure each *flip-flop* (FF) or latch in the sequential circuit such that it becomes either a *scan flip-flop* (SFF) or a scan latch, which is often called scan cell. Consider the example of a muxed-D scan design illustrated in Fig. 1.6, where each FF has been reconfigured as an SFF as shown in Fig. 1.5a. The scan cells (SFFs) are connected in series to form a shift register, or scan chain, that has direct access to a primary input (*scan in*) and a primary output (*scan out*). During the shift operation, when the *scan enable* is set to 1, the scan chain is used to shift in a test pattern through the primary input *scan in.* After the test pattern is shifted into the scan cells, it is applied to the combinational logic. The circuit is then configured in capture mode, by setting the *scan enable* to 0, for one clock cycle. The response of the combinational logic with respect to the test pattern is then captured in the scan cells. The scan chain is then configured in scan mode again to shift out the response captured in the scan cells for observation.

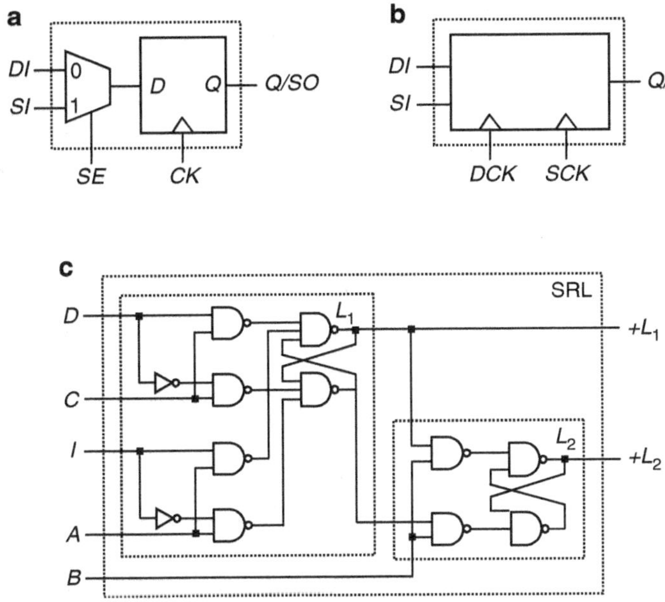

Fig. 1.5 Scan cell designs including muxed-D scan cell (**a**), clocked-scan cell (**b**), and polarity-hold shift register latch (**c**)

While shifting out the response, the next test pattern can be shifted into the scan cells concurrently. As a result, scan design reduces the problem of testing sequential logic to that of testing combinational logic and, thereby, it facilitates the use of combinational ATPG for testing sequential circuits.

Although scan design has provided many benefits for manufacturing test, traditional test schemes using ATPG software to target single faults have become expensive and ineffective because (1) the increase in test data volume for testing multimillion-gate designs could exceed the tester memory and (2) sufficiently high fault coverage levels for these deep submicron or nanometer VLSI designs are difficult to sustain from the chip level to the board and system levels.

1.3.3 Built-In Self-Test

BIST seeks to alleviate the aforementioned test problems by incorporating circuits that generate test patterns and analyze the output responses of the CUT. As illustrated in Fig. 1.7, a *test pattern generator* (TPG) is used to automatically supply the internally generated test patterns to the CUT and an *output response analyzer* (ORA) is used to compact the output responses from the CUT (Stroud 2002). The TPG and ORA are either embedded in the chip or elsewhere on the same board where the chip resides.

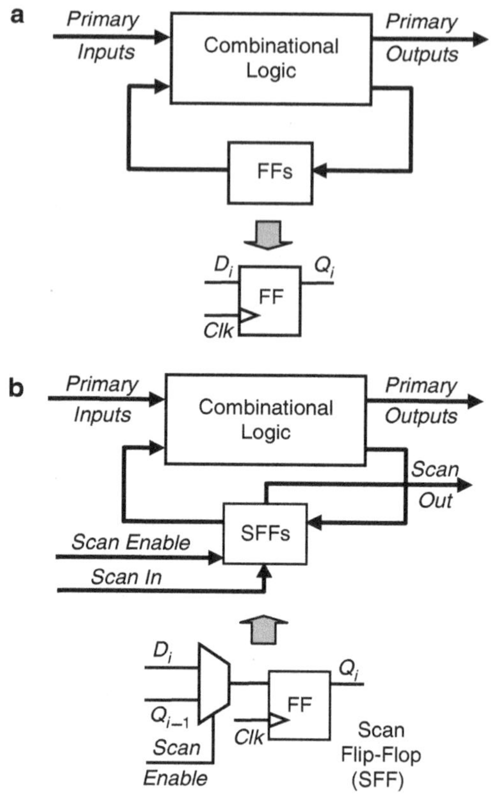

Fig. 1.6 Transforming a sequential circuit (**a**) to a scan design (**b**)

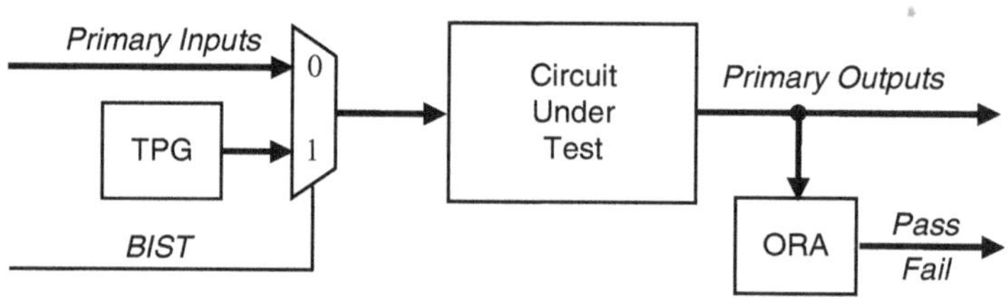

Fig. 1.7 Simple BIST architecture

BIST architectures can be classified into two categories: (1) those that use *test-per-scan BIST* and (2) those that use *test-per-clock BIST*. Test-per-scan BIST takes advantage of already built-in scan chains of the scan design and applies a test pattern to the CUT after a shift operation is completed; hence, the hardware overhead is low. Test-per-clock BIST, however, applies a test pattern to the CUT and captures its test response every system clock cycle; hence, the scheme can execute tests much faster than the test-per-scan BIST scheme, but usually at the expense of more hardware overhead. A number of test-per-clock BIST schemes have been implemented for general sequential logic, including *built-in logic block observer*

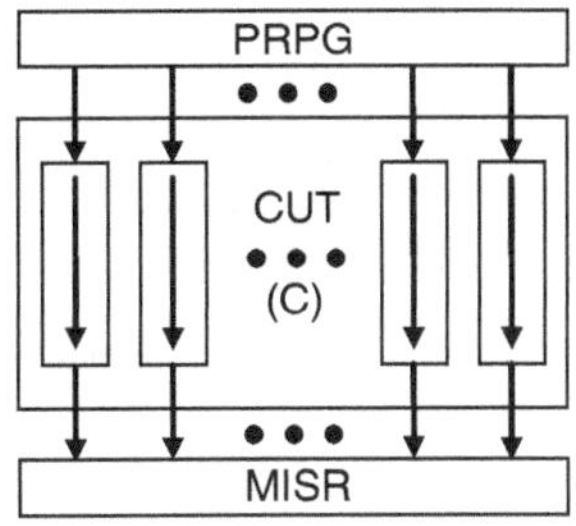

Fig. 1.8 Self-testing using MISR and parallel SRSG (STUMPS)

(BILBO) (Könemann et al. 1979) and circular BIST (Stroud 1988). However, the wide acceptance of scan design for manufacturing test led to the dominance of scan-based *logic BIST*. BIST architectures for regular structures, such as large memories, are test-per-clock due to the algorithmic nature and large number of test patterns that must be applied, as will be discussed later in this chapter.

A test-per-scan BIST design was presented in (Bardell and McAnney 1982). This design, which is shown in Fig. 1.8, uses a *pseudorandom pattern generator* (PRPG), which is also known as a parallel *shift register sequence generator* (SRSG), as the TPG and a *multiple-input signature register* (MISR) as the ORA. Typically, both PRPG and MISR are constructed from *linear feedback shift registers* (LFSRs) composed of D flip-flops and XOR gates. The pseudorandom patterns are generated by the PRPG and shifted into the scan chains embedded in the CUT. The system clocks are then triggered for one cycle and the test responses captured in the scan cells are then shifted to the MISR for compaction. New test patterns are shifted in at the same time when test responses are being shifted out. This BIST architecture using the test-per-scan BIST scheme is referred to as *self-testing using MISR and parallel SRSG* (STUMPS) (Bardell and McAnney 1982). Because of the ease of integration with the traditional scan architecture, *STUMPS* is the only BIST architecture widely used in industry today. Analysis and experimental data showed that applying multiple capture cycles after each scan sequence, instead of just one capture cycle as suggested in the original test-per-scan BIST scheme, helps to produce patterns with a different profile (other than pseudorandom patterns) in the scan cells and helps improve BIST quality (Tsai et al. 1999).

1.3.4 Test Compression

A supplement to scan design, test compression is commonly used to reduce the amount of test data – both input stimuli and output responses – that must be stored on the *automatic test equipment* (ATE) (Touba 2006). Reductions in test data volume and test application time by 10× or more can be achieved. This result is typically accomplished by including a decompressor before the m scan chain inputs of the CUT to decompress the compressed input stimuli and also adding a compactor after the m scan chain outputs of the CUT to compact the output responses, as illustrated

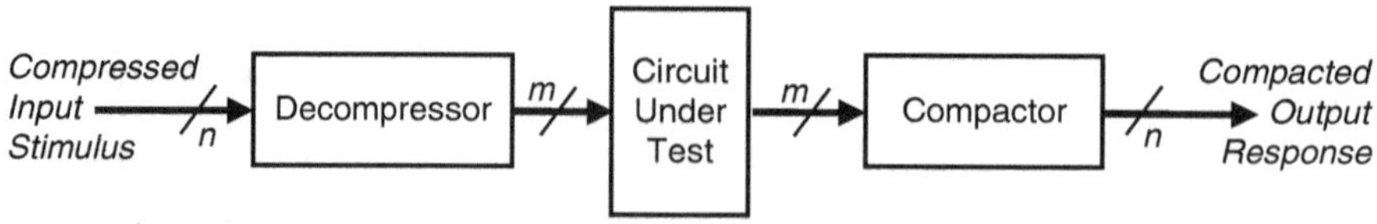

Fig. 1.9 Test compression architecture

in Fig. 1.9. The compressed input stimulus and compacted output response are each connected to n tester channels on the ATE, where $n < m$ and n is typically at least $10\times$ less than m. Modern test synthesis tools can now directly incorporate these test compression features into either an RTL design or a gate-level design.

Define a Test cube as a deterministic test pattern in which the bits that are not assigned values by the ATPG procedure are left as don't cares (Xs). ATPG procedures normally perform *random-fill* in which all the Xs in the test cubes are filled randomly with 0s and 1s to create fully specified test vectors. For test stimulus compression, *random-fill* is not performed during ATPG so the Xs make the test cubes much easier to compress than the fully specified test vectors.

Many schemes for compressing test cubes have been proposed (Touba 2006). They can be broadly classified into the three categories (Wang et al. 2006):

1. *Code-based schemes.* These schemes use data compression codes to encode the test cubes, such as *(fixed-to-fixed) dictionary code* (Reddy et al. 2002), *(fixed-to-variable) Huffman code* (Jas et al. 2003), *(variable-to-fixed) run-length code* (Jas and Touba 1998), and *(variable-to-variable) Golomb code* (Chandra and Chakrabarty 2001).
2. *Linear-decompression-based schemes.* These schemes decompress the data using only linear operations (e.g., LFSRs and XOR networks) (Rajski et al. 2004).
3. *Broadcast-scan-based schemes.* These schemes use pure combinational logic to decompress the data, such as *broadcast scan* (Lee et al. 1999) and *Illinois scan* (Hamzaoglu and Patel 1999) for broadcasting the same data to multiple scan chains, *adaptive scan* using MUXs (Sitchinava et al. 2004, Wang et al. 2008b, c), *XOR compression* using XOR gates (Konemann et al. 2003), and *virtual scan* using combinational logic (Wang et al. 2004, 2008b, c, 2009).

The industry does not currently favor *code-based schemes* because of their high implementation costs. The main difference between *linear-decompression-based schemes* and *broadcast-scan-based schemes* is the manner in which the ATPG engine is used. For designs using *linear-decompression-based schemes*, test compression is achieved in two distinct steps. During the first step, conventional ATPG is used to generate sparse ATPG patterns or test cubes, and *dynamic compaction* is performed in a nonaggressive manner that leaves unspecified bit locations in each test cube as Xs. This task is accomplished by not aggressively performing the *random-fill* operation on the test cubes, which is used to increase coverage of individual patterns, and hence reduce the total pattern count. During the second step, a system of linear equations that describes the hardware mapping from the external scan input ports to the internal scan chain inputs are solved in order to map each test cube into a compressed stimulus that can be applied externally. If a mapping is not found, then a new attempt at generating a new test cube is required.

For designs using *broadcast-scan-based schemes*, only a single step is required to perform test compression (Wang et al. 2008c). This result is achieved by embedding the constraints introduced by the decompressor into the ATPG tool, such that the tool operates with much more restricted constraints. Hence, whereas in conventional ATPG, each individual scan cell can be set to 0 or 1 independently, for *broadcast-scan-based schemes*, the values to which related scan cells can be set are constrained. Thus, a limitation of this solution is that in some cases, the constraints among scan cells can preclude some faults from being tested. These faults are typically tested as part of a subsequent top-up ATPG process, if required, similar to using *linear-decompression-based schemes*.

In response compaction, a major issue that arises is dealing with unknown values (Xs) that may arise in circuit response due to the use of uninitialized memory elements, bus contention, floating tri-states, etc. Solutions to mask or tolerate Xs have used either *space compactors* (Wang et al. 2004, Mitra and Kim 2004, Wohl et al. 2007) or *time compactors* (Rajski et al. 2004, Touba 2007). Currently, space compactors have a greater acceptance rate in the industry because of the simplicity of design. A hybrid solution, reported in Chao et al. (2007), combines space and time compaction and achieves a better compaction rate with a low fault coverage loss and unknown blocking capability.

1.4 Logic Testing

The objective of VLSI testing is to minimize the number of defective chips, resulting from imperfect manufacturing processes, shipped to customers. The thoroughness of the testing process strongly depends on the quality of test patterns. Thus, a quality assessment to test patterns, which are developed either manually or automatically, is necessary in order to determine if the desired level of product quality can be achieved.

Fault simulation is the process for such quality assessment. The tasks involved in fault simulation are illustrated in Fig. 1.10. First, a set of target faults, which is referred to as the *fault list*, inside the CUT, is enumerated. Often *fault collapsing* is applied to identify equivalent faults so that only one fault in each equivalent fault set is included in the fault list. This process reduces the number of faults in the fault list and thus the fault simulation time. The size of a collapsed fault set is typically about 40% of the original fault set. Next, for each fault from the fault set, the CUT is simulated in the presence of the fault. The output responses with respect to the given input stimuli are then compared with the expected fault-free responses to determine whether the fault can be detected by the given input stimuli. For fault simulation, the CUT is typically synthesized down to a gate-level design that is referred to as a *circuit netlist*.

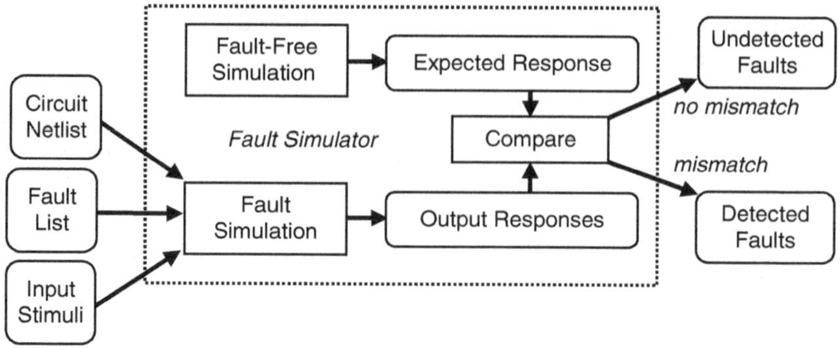

Fig. 1.10 Fault simulation

For simulation-based design verification, determining if sufficient design verification has been achieved is a difficult task for the designer. While the ultimate criterion for such determination is whether the design works in the system, fault simulation can be used to serve the purpose of providing a rough quantitative measure of the thoroughness of design verification vectors early in the design process (Stroud 2002). Fault simulation can also be used to identify portions of a design that require additional verification.

Test development consists of selecting specific test patterns based on circuit structural information and a set of fault models. Traditionally, functional test patterns are manually created by designers to verify the correctness of a design and these test patterns are fault-graded for their effectiveness. As this *functional testing* approach requires much more-than-needed test patterns to reach the desired fault coverage, a *structural testing* approach is usually employed for generating test patterns to save test time and improve test efficiency by targeting specific faults that would result from defects in the manufactured circuit. The fault models provide a quantitative measure of the fault detection capabilities for a given set of test patterns for the targeted fault model. This measure is called *fault coverage*, as defined in Sect. 1.1.

Any input pattern or sequence of input patterns that produces a different output response for a faulty circuit from that of the fault-free circuit is a test pattern or sequence of test patterns for detecting the fault. Therefore, the goal of ATPG is to find a set of test patterns that detects all targeted faults in the CUT. Because a test pattern or a sequence of test patterns generated by ATPG can often detect other faults as well, fault simulation is typically used right after a test pattern or a sequence of test patterns is generated to identify additional faults detected by the pattern(s). These detected faults can then be removed from further ATPG consideration. This strategy helps reduce total ATPG time as well as total test length.

Historically, ATPG has focused on *faults* derived from the gate-level fault model. ATPG for a given target fault consists of two steps: *fault activation* and *propagation. Fault activation* establishes a signal value at the fault site opposite to the value produced by the fault. *Fault propagation* propagates the fault effect forward

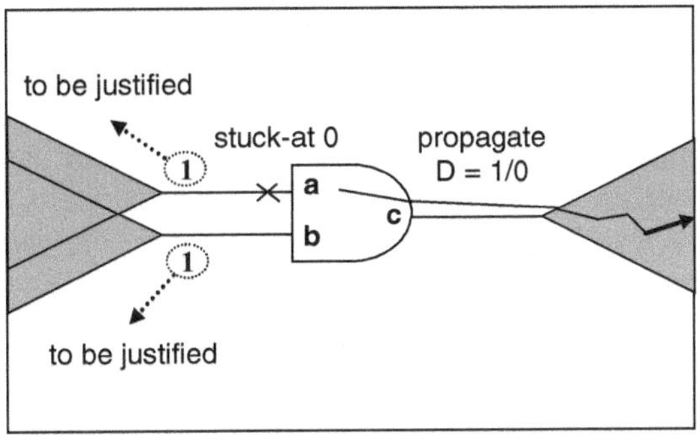

Fig. 1.11 Fault activation and propagation for *a* stuck-at 0

by sensitizing a path from the fault site to a primary output. The ATPG objective is to find an input test pattern or sequence that, when applied to the circuit, distinguishes between the fault-free and the faulty circuit responses in the presence of the target fault.

In the example in Fig. 1.11, the target fault is the stuck-at-0 fault, denoted as SA0, at the input line a of the AND gate. In order to activate this fault, the test pattern must produce a logic valuc 1 at line a. That is, for the fault-free circuit, line a has logic value 1 when the test pattern is applied. For the faulty circuit, line a has logic value 0. The symbol $D = 1/0$ is used to denote the situation. D needs to be *propagated* through a path sensitized to one of the primary outputs. For D to be propagated from line a to line c, input line b has to be set to logic 1 that is the *noncontrolling logic value* for an AND gate. Once b is set to the noncontrolling value, line c will have the same logic value that line a has.

In general, only a subset of primary inputs (e.g., directly or indirectly connected to the shadow areas shown in Fig. 1.11) are required to be set to 0s or 1s in order to activate and propagate a fault. Normally, ATPG procedures perform random-fill in which all the Xs in the test cubes are filled randomly with 0s and 1s to create fully specified test patterns. However, for *low-power testing*, random-fill is often not performed during ATPG so that the resulting test set consists of incompletely specified test cubes. These Xs can then be assigned for the purpose of reducing the *shift power* and *capture power* (Girard 2002, Wen et al. 2006).

The ATPG process involves simultaneous justification of the logic value 1 at lines a and b and the propagation of the fault difference D to a primary output. In a typical circuit with reconvergent fanouts, the process involves a search for decisions to assign logic values at primary inputs and at internal signal lines to accomplish both justification and propagation. The ATPG problem is an NP-complete problem (Ibarra and Sahni 1975, Fujiwara and Toida 1982). Hence, all known algorithms have an exponential worst-case run time.

The basic ATPG process described in conjunction with the example of Fig. 1.11 is called the *D-algorithm* (Roth 1966) and fits into an ATPG category known as combinational ATPG (Cheng and Wang 2006). The more advanced ATPG tasks, on the contrary, target faults beyond traditional gate-level fault models, which include crosstalk, bridging faults, and delay faults.

For example, during delay fault ATPG, a 0-to-1 transition (for detecting a slow-to-rise delay fault) or a 1-to-0 transition (for detecting a slow-to-fall delay fault) is required to launch the delay fault at the fault site followed by a capture clock pulse to capture the output test response for analysis. Two basic capture-clocking schemes are commonly used to test delay faults in a clock domain: (1) *skewed-load* (also called *launch-on-shift* [LOS]) and (2) *double-capture* (also called *launch-on-capture* [LOC] or *broad-side*) (Wang et al. 2006). Skewed-load uses the last shift clock pulse followed immediately by a capture clock pulse to launch the transition and capture the output test response, respectively. Double-capture uses two consecutive capture clock pulses to launch the transition and capture the output test response, respectively. In both schemes, the second capture clock pulse must be running at the domain's operating frequency or at-speed. The difference is that skewed-load requires the domain's scan enable signal *SE* to switch its value between the launch and capture clock pulses, making *SE* act as a clock signal. Figure 1.12 shows sample waveforms using the basic skewed-load and double-capture at-speed test schemes.

Both delay fault test schemes are in sharp contrast with stuck-at fault testing where only a 0 (for detecting a stuck-at-1 fault) or a 1 (for detecting a stuck-at-0 fault) is required to activate the fault site before a capture clock pulse is applied to the clock domain to capture the output test response. This means that an ordered pair of test vectors is required for detecting a delay fault, whereas only one test vector is required for detecting a stuck-at fault.

Because of the poor scalability of sequential ATPG and the immaturity of RTL-based ATPG solutions, generating high-quality tests for large designs can only rely on combinational ATPG techniques. To date, combinational ATPG tools can effectively and efficiently generate tests for detecting stuck-at faults, delay faults, and bridging faults. The tools can also be configured to generate low-power test patterns. ATPG effectiveness is measured by the fault coverage achieved for the fault model and the number of generated patterns, which collectively affect the test application

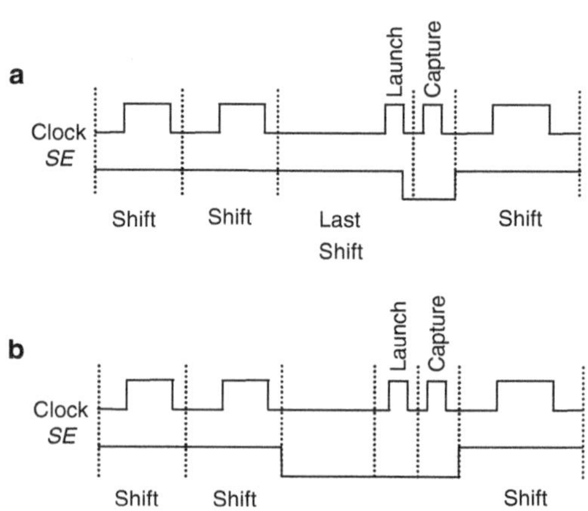

Fig. 1.12 Basic at-speed delay fault test schemes including skewed-load (*a.k.a.* launch-on-shift) (**a**), and double-capture (*a.k.a.* launch-on-capture) (**b**)

time. ATPG efficiency is influenced by the fault model under consideration, the type of CUT, the level of abstraction used to represent the CUT (gate, register-transfer, transistor), and the required test quality.

Once the resulting fault coverage is satisfactory, the tester, referred to as ATE, applies the functional test vectors and structural test patterns to each of the fabricated circuits and compares the output responses with the expected responses obtained from the simulation of the fault-free circuit.

The first type of testing performed during the manufacturing process examines the devices fabricated on the wafer to identify defective dies. Only those chips that pass the wafer-level test are packaged. The packaged devices are retested to eliminate those defective dies that are missed by the wafer-level test or those devices that may have been damaged during the packaging process or are put into defective packages. Additional testing is used to assure the final quality standards are being met prior to shipping the chips to customers. This final testing procedure includes the measurement of parameters such as input/output timing specifications, voltage, and current. In addition, *burn-in* or *stress testing* is often performed in which chips are subject to high temperatures and supply voltages. The purpose of burn-in testing is to accelerate the effect of defects that could lead to the infant mortality as shown in Fig. 1.1. FMA is typically used at all stages of manufacturing test to identify improvements to processes that can result in an increase in the number of defect-free electronic devices and systems produced (Amerasekera and Campbell 1987, Gizopoulos 2006, Wang et al. 2006, Wang et al. 2007).

In the case of a VLSI device, the chip may be discarded, or it may be investigated by using FMA for yield enhancement. In the case of a PCB, FMA may be performed for yield enhancement, or the board may undergo further testing for fault location and repair.

1.5 Memory Testing

Manufacturing defects can be of a wide variety and manifest themselves as faults that are not covered by the specific fault models for digital circuits discussed thus far. This is particularly true in the case of densely packed memories. The classical fault models associated with *random access memories* (RAMs), whose basic architecture is illustrated in Fig. 1.13, including:

1. *Cell stuck-at fault* caused when a cell is stuck-at-0 or stuck-at-1. To detect these faults, one must test each cell with both logic 0 and logic 1 values.
2. *Address decoder fault* caused when a wrong address is selected due to a fault in the address decoder. To detect these faults, one must test all addresses with unique data.
3. *Data line fault* caused when input and output data registers (or bit and bit-bar lines, where bit-bar is the complement of the bit line) have a fault that prevents correct data from being written into or read from a cell in the array. To detect

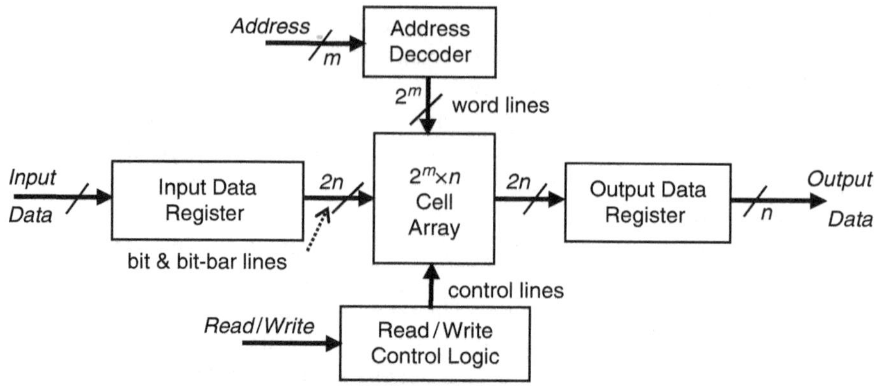

Fig. 1.13 Basic RAM architecture

these faults, one must pass both logic 0 and logic 1 values through every data input and output.

4. *Read/write fault* caused when a fault on the read/write control lines or in the read/write control logic prevents a read or write operation in the cell array. To detect these faults, one must write and read all cells.

In high-density RAMs, there are many faults that do not behave like the classical fault models given above. For example, the contents of a cell or the ability of a memory cell to change can be influenced by the contents of its neighboring cells or changes in its neighboring cells. These fault models include:

1. *Transition fault* caused when a cell cannot undergo a 0-to-1 or 1-to-0 transition. These faults can be detected by testing both transitions.
2. *Data retention fault* caused when a cell loses its contents after a certain period of time. These faults can be detected by reading data after a period of no activity.
3. *Destructive read fault* caused when a read operation changes the contents of a cell. These faults are sometimes referred to as *read disturb faults*. These faults can be detected by performing multiple read operations of the cells in the array. Those read operations should be performed for both logic 0 and logic 1 values.
4. *Pattern sensitivity fault* caused when the contents of a given cell are affected by the contents of other cells. A bridging fault between cells would be one example of a fault causing pattern sensitivity. These faults can be detected by surrounding the cell under test with specific logic values in adjacent cells and verifying that the contents of the cell under test are not changed.
5. *Coupling fault* caused when the contents of a given cell are affected by operations on other cells. There are several types of coupling faults:

   a. *Inversion coupling fault* caused when a transition in one cell inverts the contents of another cell.
   b. *Idempotent coupling fault* caused when a transition in one cell forces a constant value (0 or 1) in another cell.

Table 1.1 RAM test algorithms

Test Algorithm	March Test Sequence
MATS	⇕(w0); ⇕(r0, w1); ⇕(r1)
MATS+	⇕(w0); ↑(r0, w1); ↓(r1,w0)
MATS++	⇕(w0); ↑(r0, w1); ↓(r1,w0,r0)
March X	⇕(w0); ↑(r0, w1); ↓(r1, w0); ↑(r0)
March Y	⇕(w0); ↑(r0, w1,r1); ↓(r1, w0, r0); ⇕(r0)
March C−	⇕(w0); ↑(r0, w1); ↑(r1, w0); ↓(r0, w1); ↓(r1, w0,); ⇕(r0)
March LR	⇕(w0); ↓(r0, w1); ↑(r1, w0, r0, w1); ↑(r1, w0); ↑(r0, w1, r1, w0); ↑(r0)
March LR with BDS	⇕(w00); ↓(r00, w11); ↑(r11, w00, r00, w11); ↑(r11, w00); ↑(r00, w11, r11, w00); ↑(r00, w01, w10, r10); ↑(r10, w01, r01); ↑(r01)
March S2pf−	⇕(w0:n); ↑(r0:r0, r0:-, w1:r0); ↑(r1:r1, r1:-, w0:r1); ↓(r0:r0, r0:-, w1:r0); ↓(r1:r1, r1:-, w0:r1); ↓(r0);
March D2pf	⇕(w0:n); $\uparrow_{c=0}^{C-1}$ ($\uparrow_{r=0}^{R-1}$ ($w1_{r,c} : r0_{r+1,c}, r1_{r,c} : w1_{r-1,c}, w0_{r,c} : r1_{r-1,c}, r0_{r,c} : w0_{r+1,c}$)); $\uparrow_{c=0}^{C-1}$ ($\uparrow_{r=0}^{R-1}$ ($w1_{r,c} : r0_{r,c-1}, r1_{r,c} : w1_{r,c-1}, w0_{r,c} : r1_{r,c+1}, r0_{r,c} : w0_{r,c+1}$));

Notation: w0 = write 0 (or all 0s); r1 = read 1 (or all 1s)
portA:portB ↑= address up; ↓= address down; ⇕= address either way

Coupling faults and pattern sensitivity faults may seem similar, but the basic difference is that the pattern sensitivity fault is a function of static contents of neighboring cells while coupling faults result from transitions in other cells due to write/read operations. Pattern sensitivity faults are sometimes referred to as *state coupling faults*. Therefore, when testing memories it is necessary to add tests for pattern sensitivity and coupling faults in addition to those not caused by cell adjacency such as stuck-at faults. Extensive work has been done for memory testing and many memory test algorithms have been proposed (van de Goor 1991, Wang et al. 2006). Some of these RAM test algorithms are summarized in Table 1.1 with the notation for the algorithms given at the bottom of this table.

The simplest test algorithm is the *modified algorithmic test sequence* (*MATS*) that consists of three March elements and is of length $4N$, where N is the number of address locations. In the first March element, all memory cells in the RAM are written to logic 0. During the second March element, each address is first read, with logic 0 being the expected result, and then written to logic 1. In the final March element, each address is read with an expected result of logic 1. While the *MATS* algorithm detects cell stuck-at faults, it does not detect all address decoder faults or transition faults. For example, memory cells are tested for a 0-to-1 transition but are not tested for a 1-to-0 transition with *MATS*. Address decoder faults are not detected since the RAM can be addressed in any direction, either ascending or descending.

Most test algorithms incorporate a specified address order for the March elements to detect address decoder faults. For example, *MATS+* (a $5N$ test with three March elements) begins by writing all memory cells to logic 0 with addressing in either ascending or descending order. However, addressing is in ascending order during the second March element and in descending order during the third March element. While *MATS+* detects address decoder faults, it does not detect all transition faults since there is no read after the final write operation. Algorithms that detect transition faults include *MATS++* (a $6N$ test with three March elements), *March X* (a $6N$ test with four March elements), *March Y* (an $8N$ test with four March elements), and *March C−* (a $10N$ test with six March elements). In general, longer test algorithms detect a broader range of fault models and, hence, more defects in manufactured memories. For example, coupling fault detection improves with each of these test algorithms as we move from *MATS ++* to *March C−*.

One of the most efficient RAM test algorithms, in terms of test time and fault detection capability, currently in use is the *March LR* algorithm (van de Goor et al. 1996). This algorithm consists of six March elements and has a test time of $14N$. In addition to classical stuck-at faults and transition faults, this algorithm is capable of detecting data retention faults, destructive data faults, pattern sensitivity faults, intraword coupling faults, as well as bridging faults in the RAM.

The test algorithms discussed thus far are specified for bit-oriented memories, where each address location corresponds to a single memory cell. For word-oriented memories, a *background data sequence* (BDS) must be added to detect pattern sensitivity and coupling faults within each word of the memory (Hamdioui 2004). The *March LR* with BDS given in Table 1.1 is for a RAM with two-bit words, but in general the number of background data sequences, N_{BDS}, is given by:

$$N_{\mathrm{BDS}} = \lceil \log_2 K \rceil + 1$$

where K is the number of bits per word. For example, the four-pair BDS for a byte-wide RAM is {00000000/11111111, 00001111/11110000, 00110011/11001100, and 01010101/10101010}.

For RAMs that support true dual-port operations, with two separate address and data ports that can access any location in the memory, the March test algorithm is applied to each port in turn to test the single port operations. This is followed by application of the *March S2pf-* and *March D2pf* algorithms to test the dual port operations (Hamdioui 2004). In these two test algorithms, the simultaneous operations on port A and port B are separated by a semicolon (where a '–' indicates no operation on that port). The subscripts in the *March D2pf* algorithm indicate the physical relationship to the current row and column coordinates r and c, respectively.

In summary, the fault models and defect mechanisms of memories tend to be much more complicated than general sequential logic. Hence, the test algorithms tend to be more complicated as well. Furthermore, as technology continues to advance and design rules continue to shrink, new defect and failure mechanisms are observed that, in turn, require new fault models and test algorithms. A good example of this is new dynamic fault behaviors that can result from resistive opens in different parts of SRAMs (core-cells, address decoders, write drivers, sense amplifiers, etc.), which are currently among the most difficult faults to detect (Ney et al. 2007). However, since the expected output response is known for each operation, the test algorithms are well suited for BIST implementation (Wang et al. 2006).

1.6 System-On-Chip Testing

Most DFT, DFM, DFY, and DFR methods that have been proposed in the literature are mainly used to improve the manufactured device quality and to extend the availability of the system once the manufactured devices are used in the field. When an SOC fails as a chip, on a board, or in the system, the ability to find the root causes of the failure in a timely manner becomes critical. In this section, we briefly discuss the IEEE 1500 standard [IEEE 1500–2005] and other techniques that can reduce the overall test time and ease silicon test and debug. For detailed descriptions, refer to the key references cited in this section.

The IEEE 1500 standard is effective as it supports various *test access mechanisms* (TAMs) for the testing of core-based designs within an SOC. These mechanisms constitute hardware architecture and leverage the *core test language* (CTL), which is a subset of the IEEE 1450.6 standard [IEEE 1450.6–2001] to facilitate communication between core designers and core integrators. The primary structure is a "*wrapper*" surrounding the boundary (I/O signals) of each core that facilitates the isolation or access of the core from its SOC environment. The wrapper standardizes the test interface of the core so as to allow *core reuse* and *test reuse*. It should be noted that the IEEE 1500 standard only provides mechanisms for core-based test; the actual test patterns can be generated using any method.

An overall architecture of an SOC with N cores, each wrapped by an IEEE 1500-compliant wrapper, is shown in Fig. 1.14. The *wrapper serial port* (WSP) is a set of I/O signals of the wrapper for serial operations, which consists of the *wrapper serial input* (WSI), the *wrapper serial output* (WSO), and several *wrapper serial*

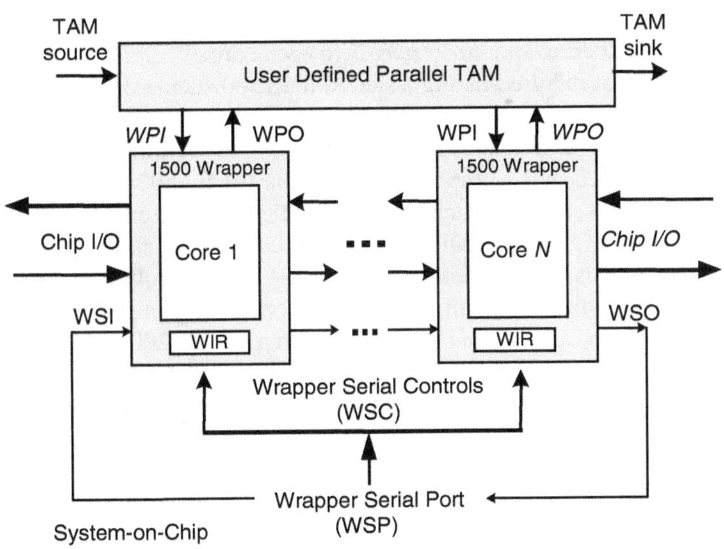

Fig. 1.14 IEEE 1500-compliant SOC architecture

control (WSC) signals. Each wrapper has a *wrapper instruction register* (WIR) to store the instruction to be executed in the corresponding core, which controls operations in the wrapper including accessing the *wrapper boundary register* (WBR), the *wrapper bypass register* (WBY), or other user-defined function registers. The WBR consists of *wrapper boundary cells* (WBCs) that can be as simple as a single storage device (for observation only), similar to the *boundary-scan cell* (BSC) used in the IEEE 1149.1 boundary-scan standard [IEEE 1149.1–2001], or a complex cell with multiple storage devices on its shift path.

The WSP supports the serial *test mode* (TM) similar to that in the boundary-scan architecture, but without using a *test access port* (TAP) controller. This means that the *wrapper serial control* (WSC) signals defined in the IEEE 1500 standard can be directly applied to the cores, hence providing more test flexibility. For example, delay testing that requires a sequence of test patterns to be consecutively applied to a core can be supported by the IEEE 1500 standard [IEEE 1500–2005] (Wang et al. 2007). In addition to the serial TM, the IEEE 1500 standard also provides an optional parallel TM with a user-defined, parallel *test access mechanism* (TAM). Each core can have its own *wrapper parallel input* (WPI), *wrapper parallel output* (WPO), and *wrapper parallel control* (WPC) signals. A user-defined parallel TAM (Marinissen et al. 2002, Xu and Nicolici 2005, Zorian and Yessayan 2005) can transport test signals from the TAM-source (either inside or outside the chip) to the cores through WPC and WPI and from the cores to the TAM-sink through WPO in a parallel manner, and hence can greatly reduce the total amount of test time.

A variety of architectures can be implemented in the TAM for providing parallel access to control and test signals (both input and output) via the *wrapper parallel port* (WPP) (Wang et al. 2006, 2007). Some of these architectures are illustrated in Fig. 1.15 that include *multiplexed access* where the cores time-share the test control and data ports, *daisy-chained access* where the output of one core is connected to the input of the next core, and *direct access* to each core.

Although it is not required or suggested in the 1500 standard, a chip with 1500-wrapped cores may use the same four mandatory pins – *test data in* (TDI), *test data out* (TDO), *test clock* (TCK), and *test mode select* (TMS) – as the IEEE 1149.1 standard for chip interface so the primary access to the IEEE 1500-compliant architecture is via boundary scan. An on-chip test controller with the capability of the TAP controller in the IEEE 1149.1 standard can be used to generate the *wrapper serial control* (WSC) signals for each core. This on-chip test controller concept can also be used to deal with the testing of hierarchical cores in a complex system (Cheng et al. 2004, Goel et al. 2004, Lee et al. 2005, Wang et al. 2006, 2008a). Because each core can be accessed via the on-chip test controller, the IEEE 1500 standard provides a nice mechanism for core-based test and to isolate errors down to the core level, greatly reducing silicon debug and diagnosis efforts (Wang et al. 2009).

One challenge faced with the increasing design size and complexity of SOCs is the total amount of test time and test data volume required to test all embedded cores in the SOC during manufacturing test. If the overall test time and test data volume are not managed properly, the increase in test cost may diminish most benefits of core-based designs. Test strategies and techniques for TAM optimization,

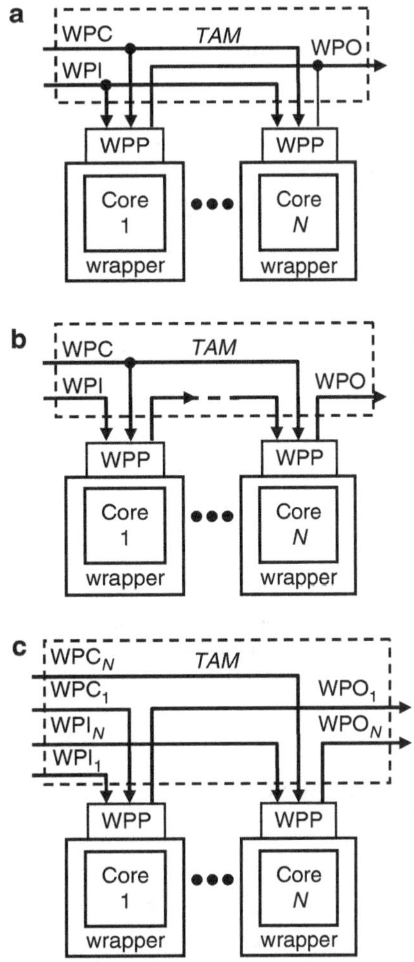

Fig. 1.15 Example user-defined parallel TAM architectures including multiplexed (**a**), daisy-chain (**b**), and direct access (**c**) TAMs

test scheduling and test resource partitioning, as well as their impacts on power consumption are important considerations. For more information on SOC test and optimization, the reader is referred to Iyengar et al. (2002), Iyengar et al. (2003), Nahvi and Ivanov (2004), Pande et al. (2005), Xu and Nicolici (2005), Saleh et al. (2006), Bahukudumbi and Chakrabarty (2007), and Wang et al. (2007).

1.7 Summary and Conclusions

This chapter provides an overview of the fundamentals of VLSI testing as an area of both theoretical and great practical significance. The concepts of defect, fault, error, failure, fault coverage, manufacturing yield, defect level, and bathtub curve were discussed along the typical yet popular fault models of interest in measuring VLSI circuit testability and in developing test patterns. The industry-widely used DFT structures along with test methods for logic testing, memory testing, and *system-on-chip* (SOC) testing were described to further ensure VLSI product quality and reduce test cost. Why do we need specific test structures and test methods for low-power VLSI applications? What are the prevailing concepts and techniques for low-power testing? While a few of the fundamental issues were briefly reviewed in this chapter, a more detailed discussion of these questions can be found in the remaining chapters of this book.

Acknowledgments The authors would like to thank Professor Xiaoqing Wen of Kyushu Institute of Technology for providing a portion of the material in the Scan Design section, Professor Kwang-Ting (Tim) Cheng of the University of California at San Barbara for providing a portion of the material in the Logic Testing section, and Professor Kuen-Jong Lee of National Cheng Kung University for providing a portion of the material in the System-on-Chip Testing section. The authors also would like to thank Professor Wen-Ben Jone of University of Cincinnati, Professor Nur A. Touba of the University of Texas at Austin, Professor Michael S. Hsiao of Virginia Tech, and the three coeditors of this book for reviewing the chapter and providing very helpful comments. The authors drew material from their prior work in the Logic Testing article in *Wiley Encyclopedia of Computer Science and Engineering* (2008) published by John Wiley & Sons. Material was also drawn from the three DFT and EDA textbooks published by Morgan Kaufmann: *VLSI Test Principles and Architectures: Design for Testability* (2006), *System-on-Chip Test Architectures: Nanometer Design for Testability* (2007), and *Electronic Design Automation: Synthesis, Verification, and Test* (2009).

References

M. Abramovici, M. A. Breuer, and A. D. Friedman, *Digital Systems Testing and Testable Design*, IEEE Press, Piscataway, NJ, 1990.

E. A. Amerasekera and D. S. Campbell, *Failure Mechanisms in Semiconductor Devices*, John Wiley & Sons, London, 1987.

S. Bahukudumbi and K. Chakrabarty, "Wafer-Level Modular Testing of Core-Based SOCs," *IEEE Trans. on VLSI Systems*, vol. 15, no. 10, pp. 1144–1154, Oct. 2007.

P. H. Bardell and W. H. McAnney, "Self-Testing of Multiple Logic Modules," in *Proc. of the International Test Conf.*, Nov. 1982, pp. 200–204.

M. L. Bushnell and V. D. Agrawal, *Essentials of Electronic Testing for Digital, Memory & Mixed-Signal VLSI Circuits*, Springer, Boston, 2000.

A. Chandra and K. Chakrabarty, "System-on-a-Chip Test-Data Compression and Decompression Architectures Based on Golomb Codes," *IEEE Trans. on Computer-Aided Design*, vol. 20, no. 3, pp. 355–368, Mar. 2001.

M. Chao, K.-T. Cheng, S. Wang, S. Chakradhar, and W. Wei, "A Hybrid Scheme for Compacting Test Responses with Unknown Values," in *Proc. of the International Conf. on Computer-Aided Design*, Nov. 2007, pp. 513–519.

B. Cheung and L.-T. Wang, "The Seven Deadly Sins of Scan-Based Designs," Integrated System Design, Aug. 1997. (http://www.eetimes.com/editorial/1997/test9708.html).

K.-T. Cheng and L.-C. Wang, "Chapter 22: Automatic Test Pattern Generation," in *EDA for IC System Design, Verification, and Testing*, L. Scheffer, L. Lavagno, and G. Martin, editors, CRC Press, Boca Raton, FL, 2006.

K. L. Cheng, J. R. Huang, C. W. Wang, C. Y. Lo, L. M. Denq, C. T. Huang, C. W. Wu, S. W. Hung, and J. Y. Lee, "An SOC Test Integration Platform and Its Industrial Realization," in *Proc. of the International Test Conf.*, Oct. 2004, pp. 1213–1222.

E. B. Eichelberger and T. W. Williams, "A Logic Design Structure for LSI Testability," *Journal of Design Automation and Fault-Tolerant Computing*, vol. 2, no. 2, pp. 165–178, Feb. 1978.

J. M. Emmert, C. E. Stroud, and J. R. Bailey, "A New Bridging Fault Model for More Accurate Fault Behavior," in *Proc. of the Design Automation Conf.*, Sep. 2000, pp. 481–485.

H. Fujiwara and S. Toida, "The Complexity of Fault Detection Problems for Combinational Logic Circuits," *IEEE Trans. on Computers*, vol. 31, no. 6, pp. 555–560, Jun. 1982.

P. Girard, "Survey of Low-Power Testing of VLSI Circuits," *IEEE Design & Test of Computers*, vol. 19, no. 3, pp. 82–92, May-Jun. 2002.

D. Gizopoulos, editor, *Advances in Electronic Testing: Challenges and Methodologies*, Morgan Kaufmann, San Francisco, 2006.

S. K. Goel, K. Chiu, E. J. Marinissen, T. Nguyen, and S. Oostdijk, "Test Infrastructure Design for the Nexperia Home Platform PNX8550 System Chip," in *Proc. of the Design, Automation, and Test in Europe Conf.*, Feb. 2004, pp. 108–113.

L. H. Goldstein and E. L. Thigpen, "SCOAP: Sandia Controllability/Observability Analysis Program," in *Proc. of the Design Automation Conf.*, Jun. 1980, pp. 190–196.

S. Hamdioui, *Testing Static Random Access Memories*, Springer, Boston, 2004.

I. Hamzaoglu and J. H. Patel, "Reducing Test Application Time for Full Scan Embedded Cores," in *Proc. of the Fault-Tolerant Computing Symp.*, Jul. 1999, pp. 260–267.

P. H. Ibarra and S. K. Sahni, "Polynomially Complete Fault Detection Problems," *IEEE Trans. on Computers*, vol. C-24, no. 3, pp. 242–249, Mar. 1975.

IEEE Std. 1149.1–2001, *IEEE Standard Test Access Port and Boundary Scan Architecture*, IEEE Press, New York, 2001.

IEEE Std. 1450.6–2001, *Core Test Language (CTL)*, IEEE Press, New York, 2001.

IEEE Std. 1500–2005, *IEEE Standard for Embedded Core Test*, IEEE Press, New York, 2005.

V. Iyengar, K. Chakrabarty, and E. J. Marinissen, "Test Wrapper and Test Access Mechanism Co-Optimization for System-on-a-Chip," *Journal of Electronic Testing: Theory and Applications, Special Issue on Low Power Testing*, vol. 18, pp. 213–230, Apr. 2002.

V. Iyengar, K. Chakrabarty, and E. J. Marinissen, "Test Access Mechanism Optimization, Test Scheduling and Tester Data Volume Reduction for System-on-Chip," *IEEE Trans. on Computers*, vol. 52, no. 12, pp. 1619–1632, Dec. 2003.

S. K. Jain and V. D. Agrawal, "Statistical Fault Analysis," *IEEE Design & Test of Computers*, vol. 2, no. 2, pp. 38–44, Feb. 1985.

A. Jas and N. A. Touba, "Test Vector Compression via Cyclical Scan Chains and Its Application to Testing Core-Based Designs," in *Proc. of the International Test Conf.*, Oct. 1998, pp. 458–464.

A. Jas, J. Ghosh-Dastidar, M. Ng, and N. A. Touba, "An Efficient Test Vector Compression Scheme Using Selective Huffman Coding," *IEEE Trans. on Computer-Aided Design*, vol. 22, no. 6, pp. 797–806, Jun. 2003.

N. K. Jha and S. K. Gupta, *Testing of Digital Systems*, Cambridge University Press, London, 2003.

R. Kapur, S. Mitra, and T. W. Williams, "Historical Perspective on Scan Compression," *IEEE Design & Test of Computers*, vol. 25 no. 2, pp. 114–120, Mar.-Apr. 2008.

B. Könemann, J. Mucha, and G. Zwiehoff, "Built-In Logic Block Observation Techniques," in *Proc. of the International Test Conf.*, Oct. 1979, pp. 37–41.

B. Könemann, C. Barnhart, and B. Keller, "Real-Time Decoder for Scan Test Patterns," United States Patent No. 6,611,933, Aug. 26, 2003.

K.-J. Lee, J.-J. Chen, and C.-H. Huang, "Broadcasting Test Patterns to Multiple Circuits," *IEEE Trans. on Computer-Aided Design*, vol. 18, no. 12, pp. 1793–1802, Dec. 1999.

K.-J. Lee, C.-Y. Chu, and Y.-T. Hong, "An Embedded Processor Based SOC Test Platform," in *Proc. of the International Symp. on Circuits and Systems*, 3, May 2005, pp. 2983–2986.

E. Marinissen, R. Kapur, M. Lousberg, T. McLaurin, M. Ricchetti, and Y. Zorian, "On IEEE P1500's Standard for Embedded Core Test," *Journal of Electronic Testing: Theory and Applications, Special Issue on Low Power Testing*, vol. 18, no. 4, pp. 365–383, Aug. 2002.

T. C. May and M. H. Woods, "Alpha-Particle-Induced Soft Errors in Dynamic Memories," *IEEE Trans. on Electron Devices*, vol. ED-26, no. 1, pp. 2–9, Jan. 1979.

E. J. McCluskey, *Logic Design Principles: With Emphasis on Testable Semicustom Circuits*, Prentice-Hall, Englewood Cliffs, NJ, 1986.

E. J. McCluskey and F. Buelow, "IC Quality and Test Transparency," in *Proc. of the International Test Conf.*, Sep. 1988, pp. 295–301.

S. Mitra and K. S. Kim, "X-Compact: An Efficient Response Compaction Technique," *IEEE Trans. on Computer-Aided Design*, vol. 23, no. 3, pp. 421–432, Mar. 2004.

A. Ney, P. Girard, C. Landrault, S. Pravossoudovitch, A. Virazel, and M. Bastian, "Dynamic Two-Cell Incorrect Read Fault due to Resistive-Open Defects in the Sense Amplifiers of SRAMs", in *Proc. of European Test Symp.*, May 2007, pp. 97–104.

M. Nahvi and A. Ivanov, "Indirect Test Architecture for SoC Testing," *IEEE Trans. on Computer-Aided Design*, vol. 23, no. 7, pp. 1128–1142, Jul. 2004.

M. Ohlsson, P. Dyreklev, K. Johansson, and P. Alfke, "Neutron Single Event Upsets in SRAM-Based FPGAs," in *Proc. of the Nuclear and Space Radiation Effects Conf.*, Jul. 1998, pp. 177–180.

P. Pande, C. Crecu, A. Ivanov, R. Saleh, and G. de Micheli, "Design, Synthesis and Test of Networks on Chip: Challenges and Solutions," *IEEE Design & Test of Computers*, vol. 22, no. 5, pp. 404–413, Sep.-Oct. 2005.

K. P. Parker and E. J. McCluskey, "Probability Treatment of General Combinational Networks," *IEEE Trans. on Computers*, vol. 24, no. 6, pp. 668–670, Jun. 1975.

J. Rajski, J. Tyszer, M. Kassab, and N. Mukherjee, "Embedded Deterministic Test," *IEEE Trans. on Computer-Aided Design*, vol. 23, no. 5, pp. 776–792 May 2004.

S. M. Reddy, K. Miyase, S. Kajihara, and I. Pomeranz, "On Test Data Volume Reduction for Multiple Scan Chain Designs," in *Proc. of the VLSI Test Symp.*, Apr. 2002, pp. 103–108.

J. P. Roth, "Diagnosis of Automata Failure: A Calculus & A Method," *IBM Journal of Research and Development*, vol. 10, no. 4, pp. 278–291, Apr. 1966.

R. Saleh, S. Wilton, S. Mirabbasi, A. Hu, M. Greenstreet, G. Lemieux, P. Pande, C. Grecu, and A. Ivanov, "System on Chip: Reuse and Integration," *Proceedings of the IEEE*, vol. 94, no. 6, pp. 1050–1069, Jun. 2006.

Y. Sato, S. Hamada, T. Maeda, A. Takatori, Y. Nozuyama, and S. Kajihara, "Invisible Delay Quality – SDQM Model Lights Up What Could Not Be Seen," in *Proc. of the International Test Conf.*, Nov. 2005, Paper 47.1.

J. Savir, G. S. Ditlow, and P. H. Bardell, "Random Pattern Testability," *IEEE Trans. on Computer*, vol. C-3, no. 1, pp. 79–90, Jan. 1984.

SIA, "The International Technology Roadmap for Semiconductors: 2007 Update," Semiconductor Industry Association, San Jose, CA, http://public.itrs.net, 2007.

N. Sitchinava, S. Samaranayake, R. Kapur, E. Gizdarski, F. Neuveux, and T. W. Williams, "Changing the Scan Enable During Shift," in *Proc. of the VLSI Test Symp.*, Apr. 2004, pp. 73–78.

C. E. Stroud, "An Automated Built-In Self-Test Approach for General Sequential Logic Synthesis," in *Proc. of the Design Automation Conf.*, Jun. 1988, pp. 3–8.

C. E. Stroud, *A Designer's Guide to Built-In Self-Test*, Springer, Boston, 2002.

N. A. Touba, "Survey of Test Vector Compression Techniques," *IEEE Design & Test of Computers*, vol. 23, no. 4, pp. 294–303, Jul.-Aug. 2006.

N. A. Touba, "X-Canceling MISR – An X-Tolerant Methodology for Compacting Output Responses with Unknowns Using a MISR," in *Proc. of the International Test Conf.*, Oct. 2007, Paper 6.2.

H.-C. Tsai, K.-T. Cheng, and S. Bhawmik, "Improving the Test Quality for Scan-Based BIST Using a General Test Application Scheme," in *Proc. of the Design Automation Conf.*, Jun. 1999, pp. 748–753.

A. van de Goor, *Testing Semiconductor Memories: Theory and Practice*, John Wiley & Sons, London, 1991.

A. van de Goor, G. Gaydadjiev, V. Jarmolik, and V. Mikitjuk, "March LR: A Test for Realistic Linked Faults," in *Proc. of the VLSI Test Symp.*, Apr. 1996, pp. 272–280.

R. Wadsack, "Fault Modeling and Logic Simulation for CMOS and NMOS Integrated Circuits," *The Bell System Technical Journal*, vol. 57, no. 5, pp. 1449–1474, May 1978.

L.-T. Wang and E. Law, "An Enhanced Daisy Testability Analyzer (DTA)," in *Proc. of the Design Automation Conf.*, Oct. 1985, pp. 223–229.

L.-T. Wang, X. Wen, H. Furukawa, F.-S. Hsu, S. H. Lin, S. W. Tsai, K. S. Abdel-Hafez, and S. Wu, "VirtualScan: A New Compressed Scan Technology for Test Cost Reduction," in *Proc. of the International Test Conf.*, Oct. 2004, pp. 916–925.

L.-T. Wang, C.-W. Wu, and X. Wen, editors, *VLSI Test Principles and Architectures: Design for Testability*, Morgan Kaufmann, San Francisco, 2006.

L.-T. Wang, C. E. Stroud, and N. A. Touba, editors, *System-on-Chip Test Architectures: Nanometer Design for Testability*, Morgan Kaufmann, San Francisco, 2007.

L.-T. Wang, C. E. Stroud, and K.-T. Cheng, "Logic Testing," in *Wiley Encyclopedia of Computer Science and Engineering*, B. W. Wah (ed.), John Wiley & Sons, Hoboken, NJ, 2008a.

L.-T. Wang, X. Wen, S. Wu, Z. Wang, Z. Jiang, B. Sheu, and X. Gu, "VirtualScan: Test Compression Technology Using Combinational Logic and One-Pass ATPG," *IEEE Design & Test of Computers*, vol. 25, no. 2, pp. 122–130, Mar.-Apr. 2008b.

L.-T. Wang, B. Sheu, Z. Jiang, Z. Wang, and S. Wu, "Method and Apparatus for Broadcasting Test Patterns in a Scan Based Integrated Circuit," United States Patent No. 7,412,637, Aug. 12, 2008c.

L.-T. Wang, R. Apte, S. Wu, B. Sheu, K.-J. Lee, X. Wen, W.-B. Jone, J. Guo, W.-S. Wang, H.-J. Chao, J. Liu, Y. Niu, Y.-C. Sung, C.-C. Wang, and F. Li, "Turbo1500: Core-Based Design for Test and Diagnosis," *IEEE Design & Test of Computers*, vol. 26, no. 1, pp. 26–35, Jan.-Feb. 2009.

X. Wen, S. Kajihara, K. Miyase, T. Suzuki, K. K. Saluja, L.-T. Wang, K. S. Abdel-Hafez, and K. Kinoshita, "A New ATPG Method for Efficient Capture Power Reduction During Scan Testing," in *Proc. of the VLSI Test Symp.*, May 2006, pp. 58–63.

T. W. Williams and N. C. Brown, "Defect Level as a Function of Fault Coverage," *IEEE Trans. on Computers*, vol. 30, no. 12, pp. 987–988, Dec. 1981.

T. W. Williams and K. Parker, "Design for Testability – A Survey," *Proceedings of the IEEE*, vol. 71, no. 1, pp. 98–112, Jan. 1983.

P. Wohl, J. A. Waicukauski, and S. Ramnath, "Fully X-Tolerant Combinational Scan Compression," in *Proc. of the International Test Conf.*, Oct. 2007, Paper 6.1.

Q. Xu and N. Nicolici, "Resource-Constrained System-on-a-Chip Test: A Survey," *IEE Proceedings – Computers and Digital Techniques*, vol. 152, no. 1, pp. 67–81, Jan. 2005.

Y. Zorian and A. Yessayan, "IEEE 1500 Utilization in SOC Test and Design," in *Proc. of the International Test Conf.*, Nov. 2005, pp. 1203–1212.

Chapter 2
Power Issues During Test

Sandip Kundu and Alodeep Sanyal

Abstract An unintended consequence of technology scaling has increased power consumption in a chip. Without specialized solutions, level of power consumption and rate of change of power consumption is even greater during test. Power delivery during test is somewhat limited by mechanical and electrical constraints. This chapter introduces the basic concepts related to power and energy and describes typical manufacturing test flow and associated constraints with power delivery. It also describes various types of power droop mechanisms, thermal issues, and how they interfere with the test process. Test economics issues, such as throughput and yield loss, are also discussed to further develop the low-power test problem statement.

2.1 Introduction

Continuous scaling of the feature size of complementary metal oxide semiconductor (CMOS) technology has resulted in exponential growth in transistor densities, enabling more functionality to be placed on a silicon die. The growth in transistor density has been accompanied with linear reduction in the supply voltage that has not been adequate in keeping power densities from rising. Elevated power densities lead to a two pronged problem: (1) supplying adequate power for circuit operation and (2) a heat flux from resulting dissipation. The power delivery issue can lead to supply integrity problems, whereas the heat flux issue affects packaging at chip, module, and system levels. In several situations, the form factor dictates a thermal envelope. Many modern systems from mobile to high-performance computers implement power management to address both energy and thermal envelope issues (Nicolici and Wen 2007).

S. Kundu (✉) and A. Sanyal
University of Massachusetts, Amherst, MA, USA
e-mail: kundu@ecs.umass.edu

P. Girard et al. (eds.), *Power-Aware Testing and Test Strategies for Low Power Devices*,
DOI 10.1007/978-1-4419-0928-2_2, © Springer Science+Business Media, LLC 2010

Power issues are not confined to functional operation of devices only. They also manifest during testing. First, power consumption may rise during testing (Dabholkar et al. 1998, Girard 2002, Nicolici and Wen 2007):

- Typical power management schemes are disabled during testing leading to increased power consumption.
  - Clock gating is turned off to improve observability of internal nodes during testing.
  - Dynamic frequency scaling is turned off during test either because the system clock is bypassed or because the *phase locked loop* (PLL) suffers from a relocking time overhead during which no meaningful test can be conducted.
  - Dynamic voltage scaling is usually avoided due to time constants in stabilizing supply voltage.
- Switching activity may be higher during testing.
  - Because of *automatic test pattern generation* (ATPG) complexity, testing is predominantly done structurally. Structural testing tends to produce more toggling than functional patterns because the goal of (structural) testing is to activate as many nodes as possible in the shortest test time, which is not the case during functional mode. Another reason is that the design-for-testability (DFT) (e.g., scan) circuitry is intensively used and stresses the circuit-under-test (CUT) much more than during functional mode.
  - Test compaction leads to higher switching activity due to parallel fault activation and propagation in a circuit.
  - Multiple cores in a *system-on-a-chip* (SoC) are tested in parallel to reduce test application time, which inherently lead to significant rise in switching activity.

Second, power availability and quality may be limited during testing:

- Longer connectors from *tester power supply* (TPS) to probe-card often result in higher inductance on the power delivery path. This may lead to voltage drop during test power cycling.
- During wafer sort test, all power pins may not be connected to the TPS, resulting in reduced power availability.
- Current limiters placed on TPS to prevent burn-out due to short-circuit current may interfere with both availability and quality of supply voltage during power surges that may result from testing.
- Reduced power availability may impact performance and in some cases may lead to loss of correct logic state of the device resulting in *manufacturing yield loss*.

Finally, there may be a reliability aspect of power to be considered during testing:

- Bus contention problem: during structural testing, nonfunctional vectors may cause illegal circuit operation such as creating a path from V_{DD} to ground with short circuit power dissipation.

- Memory contention problem: this occurs in a multiported memory, where simultaneous writes with conflicting data may take place to the same address, typically by nonfunctional patterns applied during structural testing.
- Bus and memory contention problems may cause short-circuit and permanent damage to the device. Therefore, it is important to conduct electrical verification of test vectors from a circuit operation point of view before they are applied from a tester.

In the rest of the chapter, we will explore these issues in greater depth. Subsequent sections in this chapter provide introduction to basic concepts related to power and energy, describe typical manufacturing test flow with staggered, multifaceted test objectives to provide a context for power and thermal issues during test. The discussion of test issues regarding power is contextualized with constraints arising out of test instrument, environment, test patterns, and test economics.

2.2 Power and Energy Basics

There are two major components of power dissipation in a CMOS circuit (Rabaey et al. 1996, Weste and Eshraghian 1988):

- Static dissipation due to leakage current or other currents drawn continuously from the power supply.
- Dynamic dissipation due to
  - charging and discharging of load capacitances and
  - short-circuit current.

The following two subsections discuss briefly the individual power components with the aid of a CMOS inverter.

2.2.1 *Static Dissipation*

The static (or steady-state) power dissipation of a circuit is given by the following expression:

$$P_{\text{stat}} = \sum_{i=1}^{n} I_{\text{stat}_i} \cdot V_{\text{DD}} \tag{2.1}$$

where I_{stat} is the current that flows between the supply rails in the absence of switching activity and i is the index of a gate in a circuit consisting of n gates.

Ideally, the static current of the CMOS inverter is equal to zero, as the positive and negative metal oxide semiconductor (PMOS and NMOS) devices are never ON simultaneously in the steady-state operation. However, there are some leakage currents that cause static power dissipation. The sources of leakage current

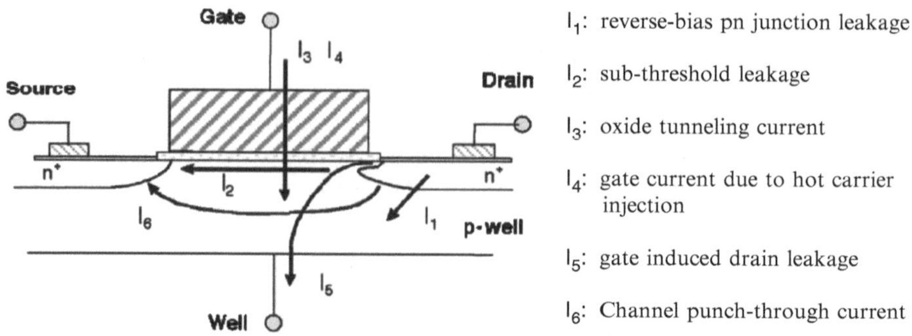

Fig. 2.1 Summary of leakage current mechanisms of deep-submicron transistors [figure adopted from Roy et al. (2003)]

for a CMOS inverter are indicated in Fig. 2.1. Major leakage contributors are (1) reverse-biased leakage current (I_1) between source and drain diffusion regions and the substrate, (2) sub-threshold conduction current (I_2) between source and drain, and (3) pattern-dependent leakage (I_3) across gate oxide (Roy et al. 2003). The other two sources of leakage which are often taken into consideration (especially as we move into nanometer design) are (1) gate-induced drain leakage (GIDL) and (2) drain-induced barrier lowering. In the following text, we describe them with some details.

2.2.1.1 Reverse-Biased pn Junction Leakage Current

Drain and source to well junctions are typically reverse-biased, causing pn junction leakage current (I_1). A reverse-biased pn junction leakage has two main components: (1) minority carrier diffusion/drift near the edge of the depletion region and (2) electron–hole pair generation in the depletion region of the reverse-biased junction (Pierret 1996). If both n and p regions are heavily doped as is the case for nanoscale CMOS devices, the depletion width is smaller and the electric field across depletion region is higher. Under this condition ($E > 1\,\text{MV/cm}$), direct band-to-band tunneling (BTBT) of electrons from the valence band of the p region to the conduction band of the n region becomes significant. In nanoscale CMOS circuits, BTBT leakage current dominates the pn junction leakage. For tunneling to occur, the total voltage drop across the junction has to be more than the band gap (Roy et al. 2003). Logical bias conditions for I_{BTBT} are shown in Fig. 2.2.

2.2.1.2 Sub-threshold Leakage Current

Sub-threshold current is the most dominant among all sources of leakage. It is caused by minority carriers drifting across the channel from drain to source due to the presence of weak inversion layer when the transistor is operating in cut-off

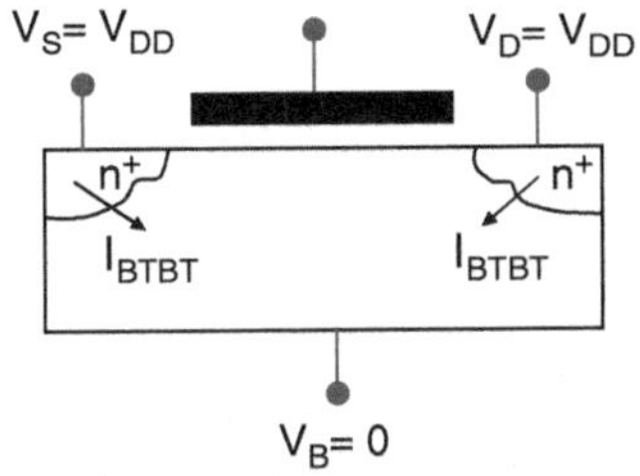

Fig. 2.2 Illustration of band-to-band tunneling (*BTBT*) leakage in a negative metal oxide semiconductor (*NMOS*) transistor

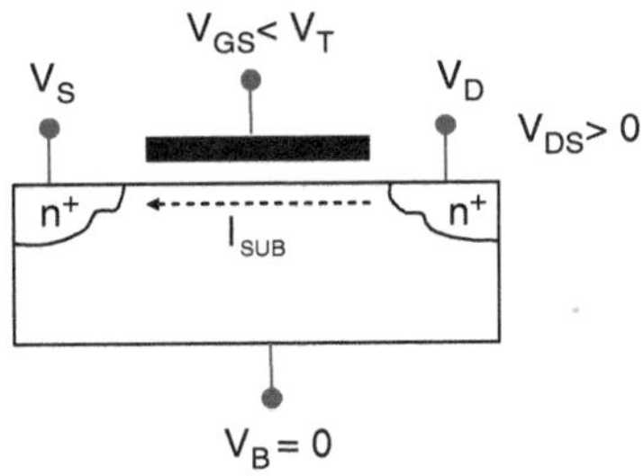

Fig. 2.3 Illustration of sub-threshold leakage in a negative metal oxide semiconductor (*NMOS*) transistor

region ($V_{GS} < V_t$). The minority carrier concentration rises exponentially with gate voltage V_G. The plot of log (I_2) versus V_G is a linear curve with typical slopes of 60–80 mV per decade. Sub-threshold leakage current depends on the channel doping concentration, channel length, threshold voltage V_t, and the temperature. In Fig. 2.3, the bias condition for sub-threshold current (I_{SUB}) on an NMOS device has been illustrated.

2.2.1.3 Gate Leakage Current

Reduction of gate oxide thickness results in an increase in the electric field across the oxide. The high electric field coupled with low oxide thickness results in tunneling of electrons from substrate to gate and also from gate-to-substrate through the gate oxide, resulting in the gate oxide tunneling current. The mechanism of tunneling between substrate and gate polysilicon can be primarily divided into two parts, namely: (1) *Fowler-Nordheim tunneling* and (2) *direct tunneling*. In the case of *Fowler-Nordheim* tunneling, electrons tunnel through a triangular potential barrier, whereas in the case of direct tunneling, electrons tunnel through a trapezoidal potential barrier. The tunneling probability of an electron depends on the thickness of the barrier, the barrier height, and the structure of the barrier (Roy et al. 2003).

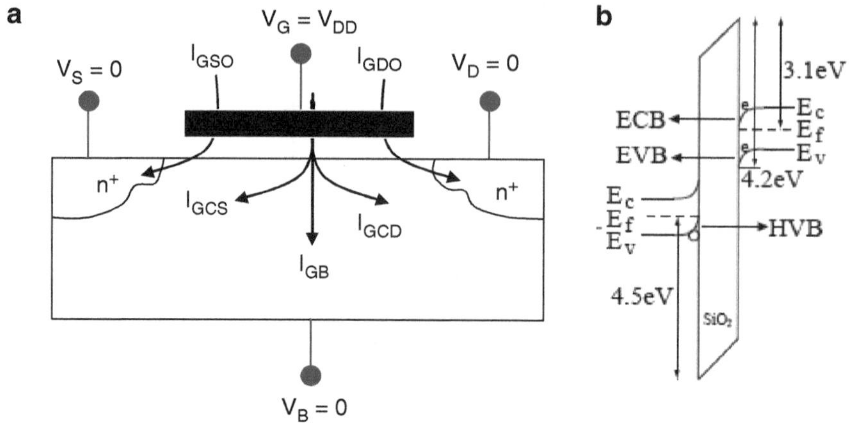

Fig. 2.4 (**a**) Illustration of gate leakage in a negative metal oxide semiconductor (*NMOS*) device and (**b**) the tunneling mechanism in band diagram [figure adopted from Drazdziulis and Larsson-Edefors (2003)]

The gate tunneling current can be divided into five major components, namely, parasitic leakage current through gate-to-source/drain extension overlap region (I_{GSO} and I_{GDO}); gate-to-inverted channel current (I_{GC}), part of which goes to the source (I_{GCS}) and the rest goes to the drain (I_{GCD}) (Hu et al. 2000); and the gate-to-substrate leakage current (I_{GB}). I_{GSO} and I_{GDO} are parasitic leakage currents that pass through gate-to-source/drain extension overlap region. I_{GDO} in off-state ($V_G = 0$) NMOS device is also known as *edge directed tunneling current* (EDL) (Yang et al. 2001) and is higher than its on-state counterpart. PMOS devices have less gate leakage compared with NMOS devices as holes have higher barrier of 4.5 eV compared with 3.1 eV for electron. Total gate leakage current is given as:

$$I_2 = I_{GSO} + I_{GDO} + I_{GCS} + I_{GCD} + I_{GB} \quad (2.2)$$

A bias condition at which I_{GC}, I_{GB}, and I_{EDL} occur is shown in Fig. 2.4.

2.2.1.4 Gate-Induced Drain Leakage Current

GIDL is due to high field effect in the drain junction of an MOS transistor. When the gate is biased to form an accumulation layer at the silicon surface, the silicon surface under the gate has almost same potential as the p-type substrate. Because of the presence of accumulated holes at the surface, the surface behaves like a p region more heavily doped than the substrate. This causes the depletion layer at the surface to be much narrower than elsewhere. The narrowing of the depletion layer at or near the surface causes field crowding or an increase in the local electric field, thereby enhancing the high field effects near that region. Large negative gate bias increases

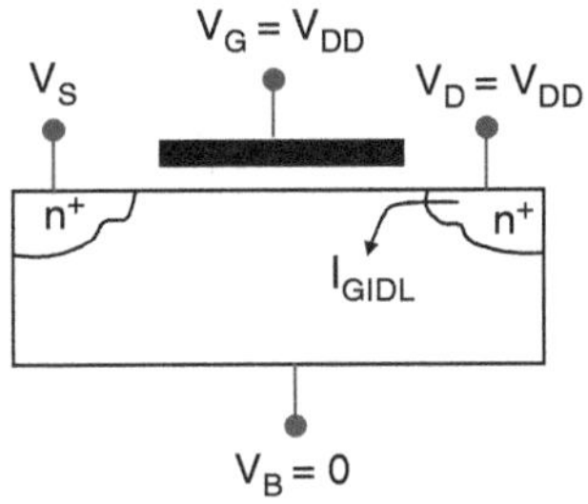

Fig. 2.5 Illustration of gate-induced drain leakage (*GIDL*) leakage in a negative metal oxide semiconductor (*NMOS*) transistor

field crowding further and peak field also increases, and the possibility of tunneling via near-surface traps also increases (Taur and Ning 1998). As a result of all these effects, minority carriers are emitted in the drain region underneath the gate. Since the substrate is at a lower potential for minority carriers, the minority carriers that have been accumulated or formed at the drain depletion region underneath the gate are swept laterally to the substrate, completing a path for the GIDL (Roy et al. 2003).

GIDL current is gaining importance as we move deeper into nanometer technologies. In Fig. 2.5, I_{GIDL} is illustrated for an NMOS device. Explanation of I_{GIDL} in PMOS can be similarly described.

2.2.2 *Dynamic Dissipation*

2.2.2.1 Dynamic Dissipation Due to Charging and Discharging of Load Capacitors

For a CMOS inverter, the dynamic power is dissipated mainly due to charging and discharging of the load capacitance (lumped as C_{L} as shown in Fig. 2.6). When the input to the inverter is switched to logic state 0 (Fig. 2.6a), the PMOS is turned ON and the NMOS is turned OFF. This establishes a resistive DC path from power supply rail to the inverter output and the load capacitor C_{L} starts charging, whereas the inverter output voltage rises from 0 to V_{DD}. During this charging phase, a certain amount of energy is drawn from the power supply. Part of this energy is dissipated in the PMOS device which acts as a resistor, whereas the remainder is stored on the load capacitor C_{L}. During the high-to-low transition (Fig. 2.6b), the NMOS is turned ON and the PMOS is turned OFF, which establishes a resistive DC path from the inverter output to the Ground rail. During this phase, the capacitor C_{L} is discharged, and the stored energy is dissipated in the NMOS transistor (Cirit 1987, Rabaey et al. 1996, Weste and Eshraghian 1988).

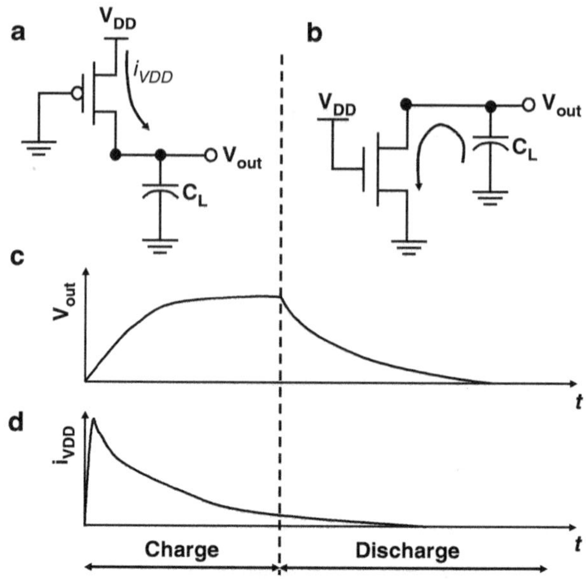

Fig. 2.6 Equivalent circuit during the (**a**) low-to-high transition, (**b**) high-to-low transition, (**c**) output voltages, and (**d**) supply current during corresponding charging and discharging phases of C_L [figure adopted from Rabaey et al. (1996)]

A precise measure for this energy consumption can be derived. Let us first consider the low-to-high transition. We start with a simplifying assumption that the NMOS and PMOS devices have zero *rise* and *fall times*, or in other words, the NMOS and PMOS devices are never ON simultaneously. Under this assumption, the equivalent circuits for charging and discharging of the load capacitor as shown in Fig. 2.6a,b are valid. The expressions for the energy E_{VDD}, taken from the supply during the transition, as well as the energy E_C, stored on the load capacitor at the end of the transition, can be derived by integrating the instantaneous power over the period of interest (Rabaey et al. 1996):

$$E_{VDD} = \int_0^{\infty} i_{VDD}(t) V_{DD}\, dt = V_{DD} \int_0^{\infty} C_L \frac{dv_{out}}{dt} dt = C_L V_{DD} \int_0^{V_{DD}} dv_{out} = C_L V_{DD}^2 \tag{2.3}$$

and

$$E_C = \int_0^{\infty} i_{VDD}(t) v_{out} dt = \int_0^{\infty} C_L \frac{dv_{out}}{dt} v_{out} dt = C_L \int_0^{V_{DD}} v_{out} dv_{out} = \frac{1}{2} C_L V_{DD}^2 \tag{2.4}$$

The corresponding waveforms of $v_{out}(t)$ and $i_{VDD}(t)$ are depicted in Fig. 2.6c,d, respectively. From (2.3) and (2.4), we may infer that only half of the energy supplied by the power source is stored on C_L. The other half is dissipated as heat by the

PMOS transistor that acts as a resistor. During the discharge phase, the charge is removed from the load capacitor, and its energy gets dissipated as heat in the NMOS transistor forming a resistive path to the Ground. In summary, each switching cycle (consisting of an L $\rightarrow$ H and an H $\rightarrow$ L transition) takes a fixed amount of energy, equal to $C_L V_{DD}^2$. In order to compute the power consumption, we have to take into account how often the device is switched. If the inverter is switched on and off during a given time period, the power consumption is given by

$$P_d = C_L V_{DD}^2 f_{0\rightarrow 1} \tag{2.5}$$

where $f_{0\rightarrow 1}$ represents the number of rising transitions at the inverter output per second.

2.2.2.2 Dynamic Dissipation Due to Short-Circuit Current

Even though under the simplifying assumption of zero rise and fall times for NMOS and PMOS devices for static CMOS logic gates, there exists no direct current path between the power and ground rails, a more realistic timing model for CMOS technology reveals that the input switching is gradual and not abrupt. Consequently, during switching of input, the PMOS and NMOS devices remain ON simultaneously for a finite period. The current associated with this DC current between supply rails is known as short-circuit current (I_{sc}) (Veendrick 1984, Vemuri and Scheinberg 1994, Hirata et al. 1996). Since short-circuit power is delivered by the voltage supply (V_{DD}), the total power can be written as

$$P_{sc} = V_{DD} \int_T I_{sc}(\tau)\mathrm{d}\tau \tag{2.6}$$

where T is the switching period (Acar et al. 2003).

Let us now analyze the short-circuit power component with the aid of a rising ramp input applied to a CMOS inverter as shown in Fig. 2.7. Assuming the input signal begins to rise at origin, the time interval for short-circuit current starts at t_0 when the NMOS device turns ON, and ends at t_1 when the PMOS device turns OFF. During this time interval, the PMOS device moves from linear region of operation

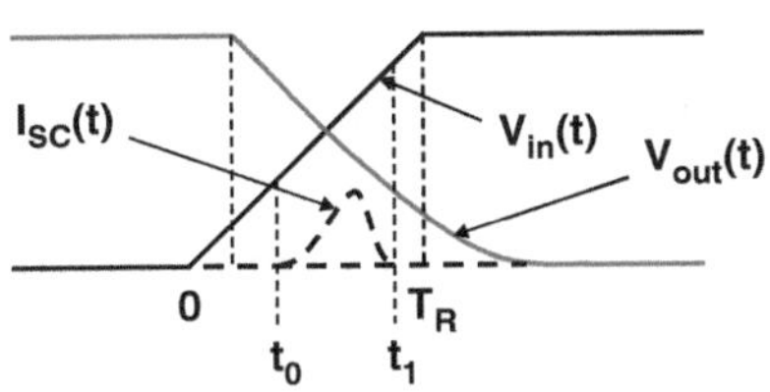

Fig. 2.7 Input and output waveforms for a complimentary metal oxide semiconductor (*CMOS*) inverter when the input switches from low to high and the corresponding short circuit current [figure adopted from Acar et al. (2003)]

to saturation region. On the basis of the ramp input signal with a rise time T_R (as shown in Fig. 2.7), t_0 and t_1 can be expressed as:

$$t_0 = T_R \frac{V_{thn}}{V_{DD}} \tag{2.7a}$$

$$t_1 = T_R \frac{V_{DD} + V_{thp}}{V_{DD}} \tag{2.7b}$$

The average short-circuit power can be specified as the integral of short-circuit current between t_0 and t_1:

$$P_{sc} = V_{DD} \int_{t_0}^{t_1} \frac{I_{sc}(\tau)d\tau}{(t_1 - t_0)} \tag{2.8}$$

2.2.3 Total Power Dissipation

The total power consumption of the CMOS inverter is now expressed as the sum of its three components:

$$P_{total} = P_{stat} + P_d + P_{sc} \tag{2.9}$$

In typical CMOS circuits, the capacitive dissipation was by far the dominant factor. However, with the advent of deep-submicron regime in CMOS technology, the static (or leakage) consumption of power has grown rapidly and account for more than 25% of power consumption in SoCs and 40% of power consumption in high performance logic (ITRS 2007).

2.2.4 Energy Dissipation

Energy is defined as the total power consumed in a CMOS circuit over a period of T. Therefore, mathematically we may express energy dissipated in a CMOS circuit as:

$$E_{total} = \int_T P_{total} d\tau \tag{2.10}$$

Substituting the expression for P_{total} from (2.9), we get:

$$E_{total} = \int_T P_{stat} d\tau + \int_T P_d d\tau + \int_T P_{sc} d\tau \tag{2.11}$$

All the three individual power components are input state dependent. Therefore, the energy dissipated over a period of T will depend on the set of input vectors applied to the circuit during that period as well as the order in which they are applied.

2.3 Manufacturing Test Flow

Testing and diagnosis of VLSI systems can be broadly classified into four types depending on the specific purpose it accomplishes and the current phase of production (from fabrication to shipment) for the circuit under test (Bushnell and Agrawal 2000, Stevens 1986). In the following four subsections, we briefly cover these four types of test methods in the order they are conducted during the design and manufacturing processes.

2.3.1 Characterization Test

Also known as design debug or verification testing, this form of testing is performed on a new design before it is sent to production (Bushnell and Agrawal 2000). The main objective of characterization test is to verify that the design is correct and the device will meet all specifications. Comprehensive AC and DC measurements are made during this test process. The requirement for thoroughness during this testing phase may often lead to probing of internal nodes of a chip, not performed as part of any other test process. Specialized tools such as *scanning electron microscopes* and electron beam testers, and techniques such as *artificial intelligence* and *expert systems* are often used in this form of testing. A characterization test determines the exact limits of device operating values. Generally, the devices are tested for the worst case because it is easier to evaluate than average cases and devices passing this test will work for any other conditions.

2.3.2 Production Test

Every fabricated chip is subjected to production tests, which are less comprehensive than characterization tests yet they must enforce the quality requirements by determining whether the device meets specifications (Bushnell and Agrawal 2000). It may not be possible to cover all possible functions and data patterns, but production tests must have a high coverage of modeled faults. Since every device must be tested before being packaged, test application time is of great importance. Production test should be as brief as possible and is usually different from characterization tests or diagnostic tests.

2.3.3 Burn-in Test

All devices that pass production tests are not identical. When put to actual use, some will fail very quickly whereas others will function for a long time. Burn in screens for long-term reliability of devices by either continuous or periodic testing over a

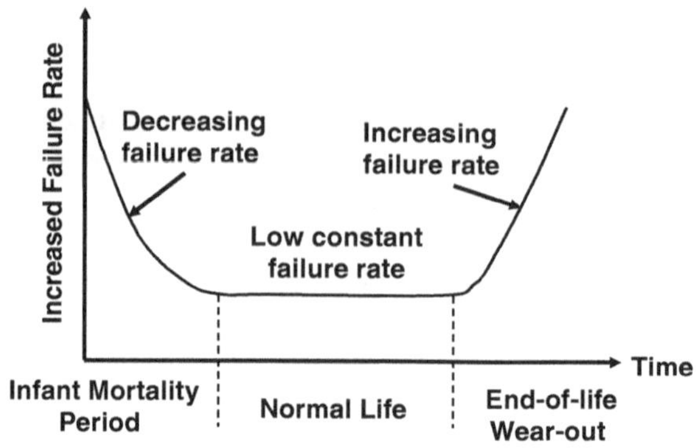

Fig. 2.8 Bathtub curve showing the rate of failure of integrated circuits at different phases of life [figure adopted from Hnatek (1987)]

period, usually under nonrated conditions. Rate of failure of integrated circuits at different phases of life follows a *bathtub* curve (Hnatek 1987) (shown in Fig. 2.8). Correlation studies show that the occurrence of potential failures can be accelerated at elevated temperatures (Jensen and Petersen 1982). Two types of failures are isolated by burn-in: (1) *infant mortality failures*, often caused by a combination of sensitive design and process variation, and may be screened out by a short-term burn-in (10–30 h) in a normal or slightly accelerated working environment, and (2) *freak failures*, that is, devices having the same failure mechanisms as the reliable devices, require long burn-in time (100–1,000 h) in an accelerated environment. In practice, a manufacturer must balance economic considerations against the device reliability. In any case, the elimination of infant mortality failures is considered essential in many applications (Bushnell and Agrawal 2000).

2.3.4 Incoming Inspection

System manufacturers perform incoming inspection (also called *quality assurance*) on the purchased devices before integrating them into the system. Depending upon the context, this testing can be either similar to production testing, or more comprehensive than production testing, or even tuned to the specific systems application. The most important purpose of this testing, performed at the vendor site, is to avoid placing a defective device in a system assembly where the cost of diagnosis may far exceed the cost of incoming inspection.

2.3.5 Typical Test Flow

Actual test selection depends on the manufacturing level (processing, wafer, or package) being tested. Although some testing is done during device fabrication to assess the integrity of the process itself, device testing is predominantly performed after the wafers have been fabricated.

- The first test, known as wafer sort or probe, isolates the potentially good devices from defective ones (Einspruch 1985, Stevens 1986). Historically, the defective dies were used to be inked using a dropper, which has been replaced by digital inking of the defective ones in a die database. After this, the wafer is scribed and cut, and the potentially good devices are packaged. The main objective of wafer sort test is to save on packaging cost by separating the good dies from the defective ones.
- After packaging, burn-in test is performed to accelerate the aging defects on packaged devices. The devices are often shaken mechanically with high *g* forces for a period of time. They are also subjected to high voltage and temperature stresses to accelerate the aging defects. Typically stress conditions are applied one at a time and not together. Usually, device output responses are not measured during burn-in test because the circuit may not be rated to operate under such elevated voltage or temperature conditions.
- After burn-in, the devices go through full specification testing. During class test, comprehensive testing is performed to attain high defect coverage, speed-binning through at-speed testing, and measurement of various DC and AC parameters such as I/O slew rate, standby current, PLL lock range, and lock frequency. Class test is usually quite comprehensive because often it is the last test performed by chip manufacturers before they are shipped to system manufacturers.
- At the system manufacturer end, inspection tests are conducted on the incoming devices. Manufacturers typically apply system level tests (such as high end software applications) on a sample of the incoming lot to perform a statistical study on the quality of the devices received from the fabrication house. Similar tests on a sample of parts may be applied by a chip manufacturer to ensure shipped product quality level.

Figure 2.9 summarizes different types of test methods on the basis of their objective, test metrics, type of patterns applied, and the environment variables involved as part of the testing process.

2.4 Power Delivery Issues During Test

To understand power delivery issues during various testing phases, we have to understand how the power is connected to the device under test. In Sect. 2.4.1 and 2.4.2, we thoroughly examine the power pad and packaging-related issues and power grid-related issues as manifested during testing. In Sect. 2.4.3, we discuss different sources of power supply noise (also known as *droop*) in the context of testing.

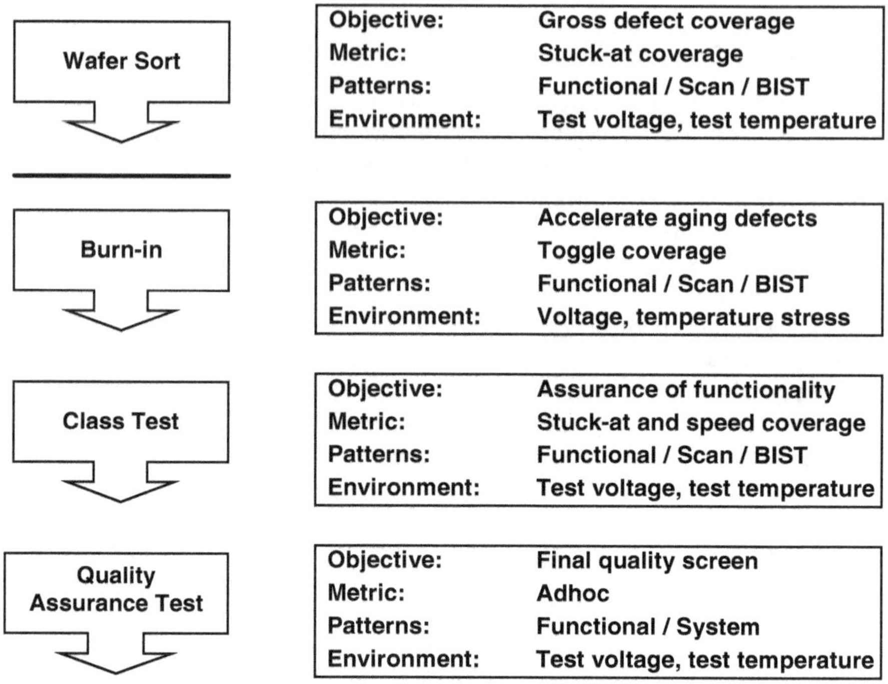

Fig. 2.9 Typical test flow

2.4.1 Packaging

Let us first investigate the role of power supply contacts between the tester and the device-under-test (DUT). To this end, we distinguish between wafer probe test and package test. To facilitate the discussion of package test, a brief description of package types is given below:

- **Wire-bonded packages:** In this technology, the pins of a bare die are situated along the perimeter and wire bonded to the package (Fig. 2.10a). Wire bonding was the only form of packaging before flip-chip technology came along.
- **Flip-chip packaging:** In this technology, the pins of a bare die are arranged as an array. The packaging substrate has a similar pin map. The die and the package are bonded together after the bare die is placed with its pin side facing down to make contact with the package substrate. This is also known as *controlled collapse chip connect* (C4) technology (Fig. 2.10b). Flip-chip technology is the dominant mode of packaging today.

DUT may not be adequately supplied with power during wafer sort test. Central problem here is that a typical C4 power contact may only be good for an average of 50 mA of current delivery to the chip, so a large array of C4 bumps is needed to supply the necessary current needed for the chip to operate at its rated power level.

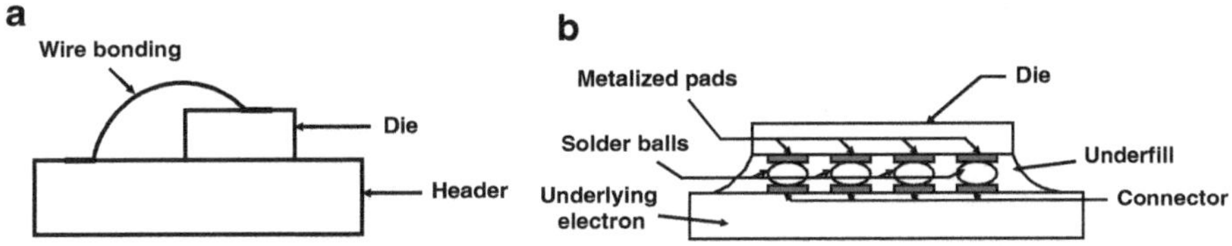

Fig. 2.10 Side-view schematic of different die mounting technology: (**a**) wire bonding and (**b**) flip-chip through C4 solder bump

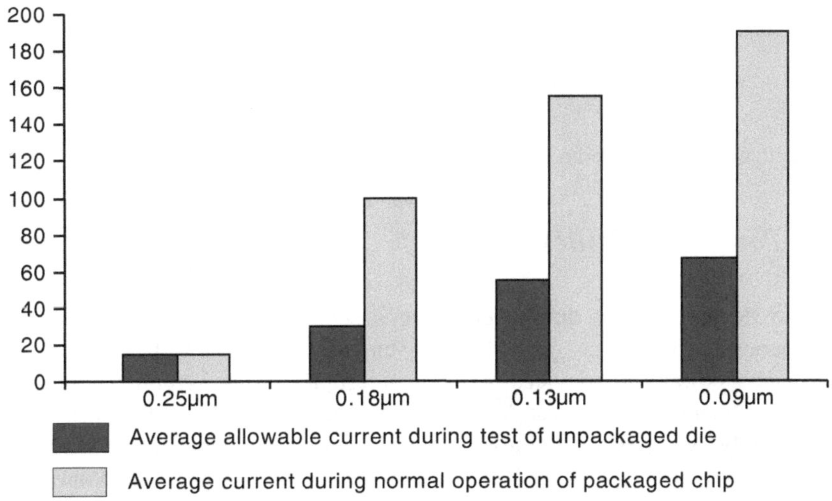

Fig. 2.11 Power availability (shown as Amps in *Y*-axis) during wafer testing [figure adopted from Kundu et al. (2004)]

During wafer sort test, the number of C4 pads that can be contacted by the probe pins is limited by mechanical strength of the wafer.

Since wafer thickness typically ranges from 300 to 500 μm and each probe pin applies a force of 5–10 g on the die, the number of probe contacts per unit area of the die is limited. Consequently, during wafer test all power pads may not be contacted. Power delivery constraint arising out of this limitation is shown in Fig. 2.11 (Kundu et al. 2004).

A second problem that afflicts wafer testing is the inductance of the power delivery path from the TPS to the DUT. This includes pin and C4 pad inductances, as well as inductance of wiring on the probe card and the inductance of the connectors to the tester. A large inductance on power delivery path impedes sudden changes in power consumption pattern by collapsing the supply voltage that in turn may produce false errors at the tester. The same problem can be seen in package test as well. However, package testing is usually done with chip socketing and the socket typically has large local decoupling capacitor to mitigate such problems. Similar capacitors on probe card tend to be farther away from the DUT.

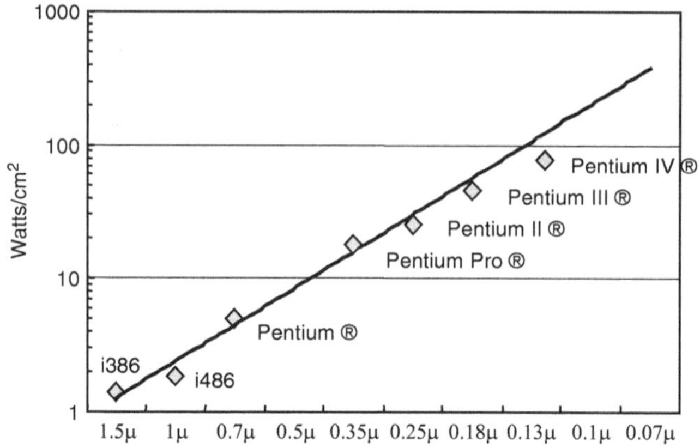

Fig. 2.12 Power density by technology [figure adopted from Tirumurti et al. (2004)]

2.4.2 *Power Grid Issues*

Increased device density due to continuous scaling of device dimensions and simultaneous performance gain has driven up the power density of high-performance computing devices such as microprocessors, graphics chips, and FPGAs. For example, in the last decade, microprocessor power density has risen by approximately 80% per technology generation, whereas power supply voltage has been scaling down by a factor of 0.8. This has lead to 225% increase in current per unit area in successive generation of technologies (Fig. 2.12).

The increased current density demands greater availability of metal for power distribution. However, this demand conflicts with device density requirements. If device density increases, the device connection density will also increase, requiring more metal tracks for signal routing. Consequently, compromises are made for power delivery and power grid becomes a performance limiter. Nonuniform pattern of power consumption across a power distribution grid causes a nonuniform voltage drop. Instantaneous switching of nodes may cause localized drop in power supply voltage, which we call as *droop*. This instantaneous drop in power supply at the point of switching causes excessive delay and a path-delay problem (Tirumurti et al. 2004). There are multiple factors that contribute to power supply droop on a chip including inductance of off-chip power supply lines, inductance of package interconnects, and resistive power distribution network (PDN) on chip. The first two factors can cause large droop and must be addressed in design phase whereas the last factor has no acceptable design solution and must be addressed in test.

2.4.3 *Power Supply Noise*

In this subsection, we discuss the physics behind various types of power supply noise (also known as droop) in more detail. The various droop mechanisms can be classified as low-frequency power droop, mid-frequency power droop, and high-frequency power droop. Next we describe each droop mechanism in detail.

2.4.3.1 Low-Frequency Power Droop

The current generation of microprocessors consumes 50–105 W of power (Intel White Paper 2006). At a supply voltage of 0.9–1.1 V, this translates to 45–95 amps of current. The voltage attenuation on this power line should be as small as possible. If the resistance of the power delivery line is kept in the order of mΩ, the resulting IR drop will be of the order of m$\Omega \times 10^2$ A $\approx$ 100 mV or 10% of the power supply voltage. Such large drop is unacceptable. Therefore, the power delivery line needs to have even smaller resistance. Unfortunately, this tends to increase self-inductance of the power delivery line.

Let, the parasitic inductance of the interconnect be denoted by L. This inductance is associated with the power supply connector *external* to the chip as the inductance of the package pins and solders bumps. We call a sudden increase in current i demanded per unit time t (which is equivalent to a sudden increase in power consumption), a $\mathrm{d}i/\mathrm{d}t$ event. After a $\mathrm{d}i/\mathrm{d}t$ event, the DUT will see its power supply voltage V_{DD} reduced by $L\mathrm{d}i/\mathrm{d}t$. For a current transient of 100 amp, taking place within 10^{-9} s or three cycles of a 3.3 GHz machine, this value is deleterious even for inductances L far below 1 nano-Henry (nH), whereas typical value of this inductance is 1–10 nH.

In reality, the impact of this inductance is mitigated by adding a capacitance C as shown in Fig. 2.13 to meet the short-term demand of current of the DUT during a $\mathrm{d}i/\mathrm{d}t$ event. The voltage droop per unit time induced by the load current is calculated as $\mathrm{d}V/\mathrm{d}t = i/C$ (Bakoglu 1990). The capacitor C needs to be sufficiently large to survive the $L\mathrm{d}i/\mathrm{d}t$ effect. Typically, these transients last 50–100 ns. Even though the worst case magnitude of this drop can be severe, it is not transistor or gate specific. Usually, there is ample time to detect beginning of these events by *on-die droop detector* and respond to these events by modulating clock frequencies or flushing the pipeline and restarting computation. Thus, while these droops are severe in magnitude, they are often handled well at the design level for the functional patterns.

In test mode, self-adaptation is usually turned off because it leads to nondeterminism of the output. For example, if the power supply voltage fluctuates as test patterns are being applied, due to self-governing mechanisms of a chip, accurate speed-binning cannot be performed because performance changes with supply voltage.

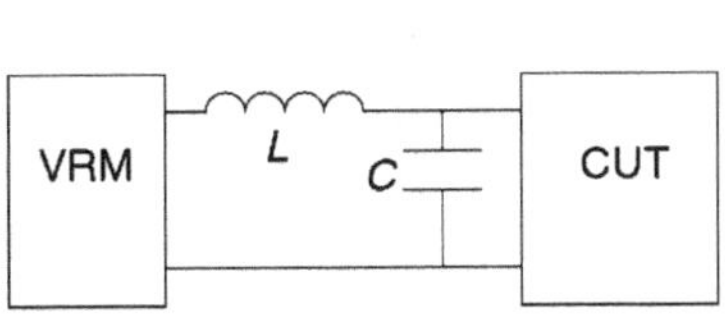

Fig. 2.13 Circuit-under-test (CUT) connected to voltage regulator module (VRM), including capacitor C and parasitic inductance of interconnect L

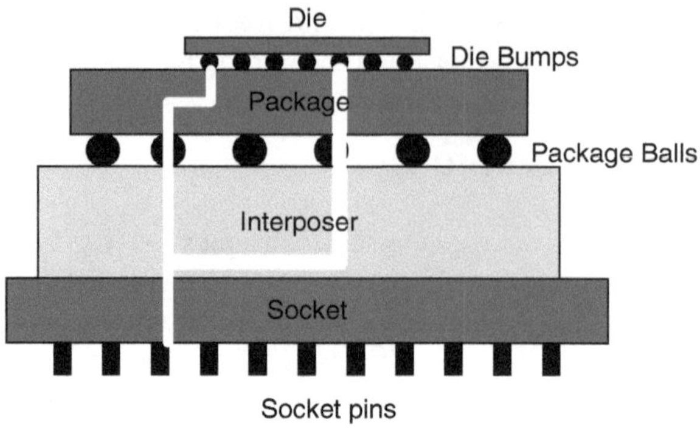

Fig. 2.14 Package showing current paths from socket pins to die bumps (figure courtesy: Intel Corporation)

On the contrary, if voltage levels are never changed, the logic associated with controlling such changes may not be tested. Thus, the onus of managing such power-level changes falls on pattern generation and test ordering mechanisms. Often such test ordering is done in an ad hoc fashion.

2.4.3.2 Mid-Frequency Power Droop

Mid-frequency voltage droop is associated with inductance at the package level. In Fig. 2.14, we show a typical package. From the socket pins to the die bumps, there are low resistance conduction paths that have reasonably high inductance (0.1–0.5 nH). During execution of instructions, if power demand shifts from one area of the die to a different area as shown with solid white line in the figure, one area of the die will experience a drop in voltage while the area where the power demand went down will experience an increase in the voltage. The package also integrates decoupling capacitance. However, owing to the scale of these interconnects, the values of both L and C are significantly smaller and the effect of voltage droop lasts 5–10 ns. For lack of a better term, droops associated with package is often called mid-frequency droop. Typically, these droops affect an entire region (integer execution unit, floating point unit, bus unit etc.) and can be addressed at the functional level by introducing multiple sensors (Clabes et al. 2004).

However, during test, if such droop is not managed well, it will lead to yield loss defined as the loss of good parts due to measurement errors during test.

2.4.3.3 High-Frequency Power Droop

High-frequency droop is associated with the PDN on the die. The PDN is usually a grid structure (Fig. 2.15). The cell library is designed with a fixed height, so that they can connect to power grid at regular points, thereby vastly simplifying the physical design process.

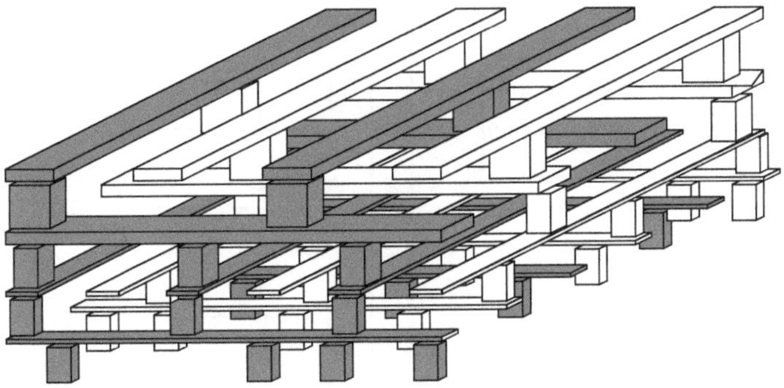

Fig. 2.15 Power distribution grid on a chip [figure adapted from Polian et al. (2006)]

The topmost metal layers (M5–M8) are often reserved for power rails and clock distribution network while lower layers are shared with logic signal lines (M2–M4). In general, the power delivery capacity of a power rail is given by its width and pitch. In microprocessor design, the width is tapered for the interconnect layers where the upper metal layers are wider. This is driven by interconnect density requirement at the lower layers and the power delivery requirement at the upper layers. There is pressure to increase the pitch of power rails as the area consumed by them is not available to logic signal lines. The vias connecting power rails of different layers transfer supply voltage from one metal layer to the next.

High-frequency power droop occurs when multiple cells drawing current from the same power grid segment suddenly increase their current demand. If the current cannot be provided quickly enough from other parts of the chip, power starvation results in a voltage drop. In contrast to low-frequency or mid-frequency power droop, this is a highly transient phenomenon lasting several hundred picoseconds.

On-die droop detector cannot be used for responding to high-frequency droops because droop detection time is usually longer than duration of the droop. Fortunately, high-frequency power droop is much smaller in magnitude. Such droops are handled in functional mode by adding a *frequency guard band.*

A similar guard band is necessary during test mode. Therefore, typically scan tests are not performed at the rated clock frequency (Xiong et al. 2008). If scan test is attempted at rated clock frequency, voltage droop due to excess switching or impedance of power supply path during test may in fact reduce the performance of a good chip below its rated level and manufacturing yield loss may occur.

2.4.3.4 Voltage Drop During At-Speed Scan

Abnormally high levels of state transitions and voltage drop during scan or BIST mode can also lead to degradation of clock frequency. It has been reported that while performing at-speed transition delay testing, fully functional devices are often discounted as "bad" causing manufacturing yield loss (Shi and Kapur 2004). During scan shift, circuit activity increases causing higher power consumption.

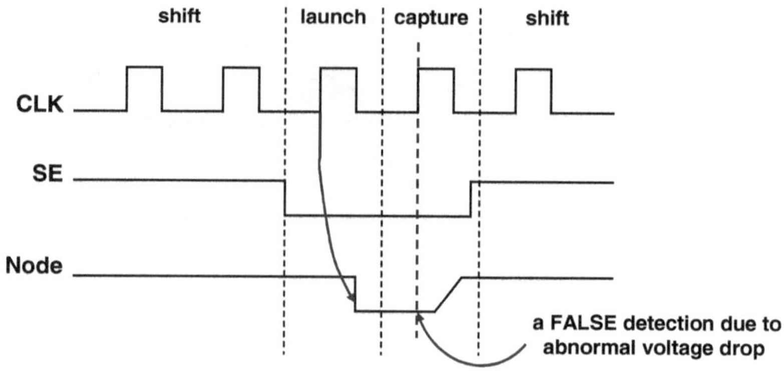

Fig. 2.16 The impact of voltage drop on shippable yield during at-speed testing [figure adopted from Shi and Kapur (2004)]

This in turn may lead to drop of power supply voltage due to IR drop where higher current or I associated with larger power dissipation causes greater voltage drop in PDN (Fig. 2.16). Such drop in voltage increases path delay requiring clock period to be stretched accordingly. If the clock period is not stretched to accommodate this increase in delay, yield loss may occur (Rearick and Rodgers 2005). IR drop not only increases path delay but also increases clock distribution latency. During structural test, a circuit toggles between system mode and scan mode. Performance of such toggle between system clock mode and scan mode may also be impacted due to reduced voltage. Thus, a chip may fail either due to excessive logic path delay or altered clock latencies or both. By contrast, in functional mode, a temporary voltage drop that increases path delay also increases clock latency that may offset increase in path delay (Wong et al. 2006). The interaction between path delay and clock latency is complex as it depends on the magnitude of each parameter as well as rise and fall times of the clock signals. However, it is safe to assume that voltage drop will increase the time it takes to toggle between scan mode and functional mode and will introduce uncertainty if the clock period itself is subject to modulation as *slow-fast-slow* as in launch off-capture or *fast-fast-slow* as in launch off-shift or any other at-speed test mechanism where the goal is to apply functional clock at speed, whereas the scan may proceed at slower speed. Such delay or uncertainty calls for capture clock to be somewhat delayed or stretched to avoid yield loss. An increase of $\sim$15% in cycle time has been reported (Rearick and Rodgers 2005).

2.5 Thermal Issues During Test

The correlation between consecutive test vectors applied to a CUT is often significantly lower than that between two consecutive functional input vectors applied during its normal operation. It directly relates to higher switching activity, and therefore higher power dissipation, during test compared to normal operation mode. The elevated levels of power dissipation during test inherently lead to higher die temperatures compared to the normal operation.

To mitigate these problems, the tests are typically applied at rates much lower than a circuit's normal clock rate in the past, since only the stuck-at fault coverage was deemed to be important. There are two recent developments in the domain of testing integrated circuits that make the power and heat dissipation during testing an extremely important issue. First, aggressive timing to improve performance of the ICs has made it essential for the tests to identify slow chips via delay testing that requires circuits to be tested at higher clock rates – if possible, at the circuit's normal clock rate (called *at-speed* testing). Second, with the advent of systems-on-chips, it is often required to test multiple cores simultaneously to reduce test application time to meet the market demand. High power and heat dissipation in neighbor cores cause undesirable thermal stress and formation of thermal hotspots. In the following subsection, we present a thorough and extensive study of various thermal hot spot-induced issues evolved during test.

Silicon die hot spots result from localized overheating, which occurs much faster than chip-wide overheating due to the nonuniform spatial on-die power distribution (Rosinger et al. 2006). Recent research supported by industrial observations suggests that spatial temperature gradients exceeding 30°C are possible even under typical operating conditions (Skadron et al. 2003), which suggest that there exist large variations in power density across the die. These gradients, especially between active and inactive blocks, are likely to increase during package testing since test power dissipation can be significantly higher compared with functional power (Pouya and Crouch 2000, Shi and Kapur 2004).

In metal oxide semiconductor (MOS) devices, there are two parameters that are predominantly sensitive to temperature: (1) the carrier's mobility μ and (2) the device threshold voltage V_t. The mobility of carriers in the channel is affected by temperature and a good approximation to model this effect is given by (Tsividis 1989):

$$\mu(T) = \mu(T_0)\left(\frac{T}{T_0}\right)^{-k_1} \tag{2.12}$$

where T is the absolute temperature of the device, T_0 is a reference absolute temperature (usually room temperature), and k_1 is a constant with values between 1.5 and 2 (Klaasen 1995). The device threshold voltage V_t exhibits a linear behavior with temperature (Klaasen and Hes 1986):

$$V_t(T) = V_t(T_0) - k_2(T - T_0) \tag{2.13}$$

where the factor k_2 is between 0.5 and 4 mV/K. The range becomes large with more heavily doped substrates and thicker oxides.

Applying these considerations to the behavior of a MOS transistor, we can predict that a temperature increment causes an increment of the drain current due to the decrease in V_t and a decrease of the drain current due to decrease in mobility. Among these two conflicting effects, the effect of mobility dominates for circuits with large overdrive voltage (which is typically the case with ultra deep submicron devices)

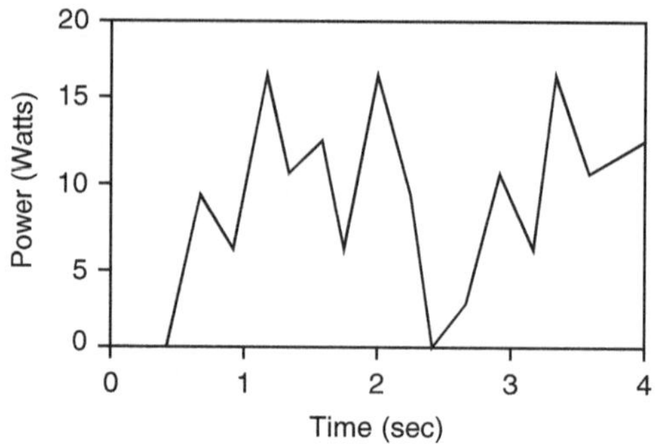

Fig. 2.17 Test patterns arranged in such a way that power cycles through high and low during the entire test period (source: Intel Technology Journal)

resulting in slowing the devices in the thermal hot spot-affected region of the chip. This will manifest as delay failures in the circuit under test causing some "good" chips to be rejected lowering the shippable yield.

In summary, local hot spots lead to (1) increased delays in gates that may register incorrectly at the tester as a defect or (2) excessive leakage that reduces electrical capacity of the local power grid that may indirectly contribute to further increased delay. Therefore, thermal hot spots during test need additional attention.

The thermal hot spot issue during test is often resolved by arranging test patterns in such a way that power is cycled high and low through the entire testing period (Fig. 2.17) so that it does not cross the temperature limits at any given time. However, applying test patterns in this way significantly increases the rate of change in supply current (*di/dt*). This leads to problems described in detail in Sect. 2.4.

2.6 Test Throughput Problem

Test throughput is defined as number of devices tested per test equipment over a given period. Higher the test throughput, higher is the profitability from chip manufacturing business point of view. Power consumption during testing plays a pivotal role in enhancing test throughput. In the following four subsections we discuss few of the test power related issues that directly influence test throughput.

2.6.1 Limited Power Availability During Wafer Sort Test

During wafer sort test, only a fraction of all the power pins could be used (Fig. 2.11). The fine contact pitch and the force required to create an ohmic contact with a C4 bump limits the availability of power from a mechanical point of view. Reduced power supply forces the tests to be performed at a lower frequency implying reduction in test throughput.

2.6.2 *Reduction in Test Frequency During Package Test*

As mentioned in Sect. 2.1, switching activity is often several times higher during testing than in normal operation mode. One ad-hoc way to reduce the dynamic power consumed during test is to lower the operating frequency, but this solution adversely affects the test throughput.

Moreover, with reduced test frequency, it takes longer to complete the entire test process. Therefore, the total energy consumed during test remains unchanged. Also, modern test requires at-speed testing to isolate slow chips, which makes testing at a lower frequency not a viable option.

2.6.3 *Constraint on Simultaneous Testing of Multiple Cores*

If multiple cores placed in a SoC are tested in parallel, it will reduce the overall test application time and therefore, enhance the test throughput. However, testing multiple cores in parallel may result in excessive energy dissipation and may develop thermal hot spots across the chip, which may eventually cause permanent damage to the chip. In order to control the heat dissipation during test, parallel testing of multiple sites is highly restricted contributing to a reduction in test throughput.

2.6.4 *Noisy Power Supply During Wafer Sort Test*

During wafer sort test, the probe card pins establish contact with the wafer metal pads, whereas the tester gets connected to the probe card connection points (Fig. 2.18). The long interconnects from the tester to the wafer metal pads offer high inductance (L). The rate of change of current ($\mathrm{d}i/\mathrm{d}t$) is also high due to thermal constraint as discussed in Sect. 2.5.1; it causes *low-frequency power droop*, quantified as $L\mathrm{d}i/\mathrm{d}t$ (as discussed in detail in Sect. 2.4.3.1). To reduce this voltage noise effect, test frequency is dropped accordingly causing a drop in test throughput.

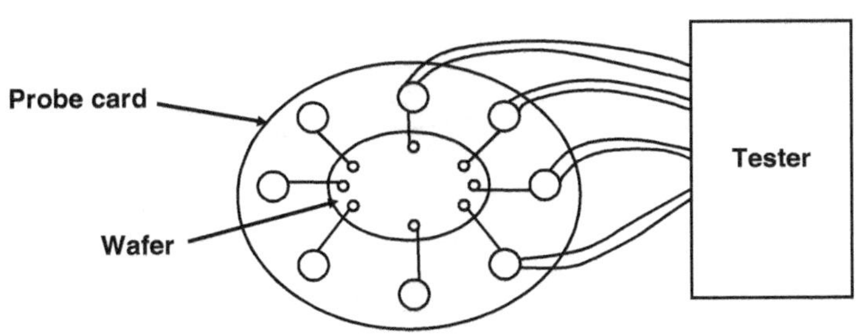

Fig. 2.18 Schematic showing connection between a wafer and the tester through a probe card

2.7 Manufacturing Yield Loss

The profitability of integrated circuits manufacturing depends heavily on the fabrication *yield*, defined as the proportion of operational circuits to the total number of fabricated circuits (Koren and Koren 1998). When a "good" circuit is falsely considered as a faulty circuit, it leads to *manufacturing yield loss*. There are several reasons behind yield loss during test, which we describe in the following five subsections.

2.7.1 ATE Timing Inaccuracy

Overall tester timing accuracy is determined by skews and parasitics between the tester and the DUT. If the test frequency is increased to the overall tester accuracy limits, it may cause significant yield loss. Unless test system timing accuracy improves in tandem with device speed, alternative test methods are necessary. For example, in PC processors the front side bus frequency has increased to 1366 MHz in recent years. When a tester is connected to front side bus, IO signals at different IOs may arrive at different times due to skew in test environment.

If the skew is greater than the rated IO period of $1/1366\,\mu s$, the chip yield will go to zero. In such a scenario, IOs cannot be tested at full frequency. On the contrary, if IOs are not tested at the rated frequency, test is incomplete and alternative tests must be devised. In PC platform chips, IO wrap test or IO loopback test is often used, where signal from one IO of a chip is returned to a different IO on the same chip through a short local path allowing the chip to test its own IO speed (Fig. 2.19) (Kundu et al. 2004).

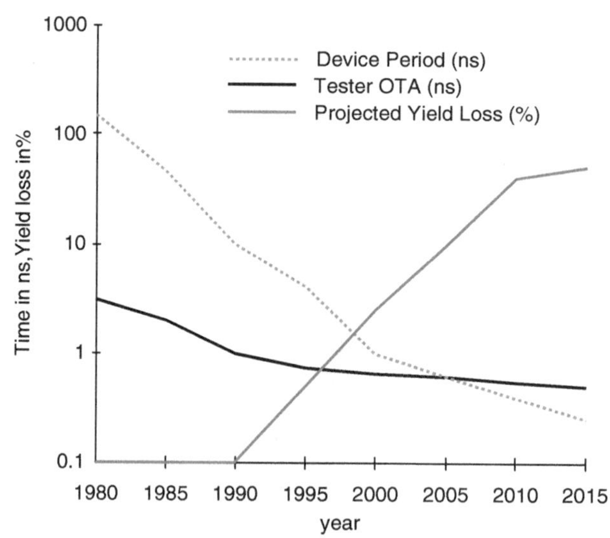

Fig. 2.19 Yield loss projection due to overall tester timing accuracy [figure adopted from Kundu et al. (2004)]

This problem is somewhat mitigated by the move to DFT-enabled IO testing. For structural testing, the only accurate timing needed is that of the system clock. Most processors or large SoCs use the tester supplied clock to generate an internal core clock, which can be many multiples of the tester clock. *On product clock generation* (OPCG) poses problems with controlling launch and capture clocks and requires additional DFT to enable such features. Circuit level verification of DFT for OPCG is also a critical issue for test. Without effective and precise launch and capture, products cannot be tested.

2.7.2 Application of Illegal Test Vectors

During structural testing, pseudorandom or deterministic patterns are applied through scan chain(s). Many of these patterns are not functional patterns and sometimes application of such a nonfunctional pattern to the DUT may perform some illegal operation from circuit perspective resulting in faulty behavior or even permanent damage of the DUT (Ogihara et al. 1983, Van der Linden et al. 1994).

The following example illustrates one such situation. Figure 2.20a shows the schematic for a 4-to-1 multiplexer (MUX), where under normal operation mode, only one input among A, B, C, or D is selected by applying appropriate selection signals (viz. S_1, S_2, S_3, and S_4). If more than one selection signal becomes active, it may possibly cause a *bus contention* by driving a 0 and a 1 on the bus output at the same time (Wohl et al. 1996). Figure 2.20b shows a gate-level schematic of the 4-to-1 MUX. To make the MUX operation faster, the combinational logic for the MUX is partitioned into two parts by inserting flip-flops to store the selection signals (viz. S_1, S_2, S_3, and S_4) generated by the MUX control signals (C_0 and C_1). In the second partition, these selection signals are used to select one of the four inputs as the output of the bus. We show the transistor level schematic of the second logic partition in Fig. 2.20(c). The flip-flops inserted in between the two logic partitions are part of normal scan chain during structural testing. Let us consider a bus contention caused by activating both selection signals S_1 and S_2 during scan shift operation. It will select both the inputs A and B. Now if $A = 0$ while $B = 1$, a DC path is established between V_{DD} and ground (indicated by the arrow in Fig. 2.20c). Such contending buses draw excess current that may result in a voltage drop. While a single bus contention problem may not cause a large drop in supply voltage, such contentions on a datapath consisting of wide buses may be significant. If the power supply voltage drops significantly, delays or intermediate voltage levels may cause faulty behavior in "good" chips causing manufacturing yield loss.

In summary, power supply voltage affects circuit delays as well as output voltage levels. A drop in power supply voltage increases circuit delay while its output voltage level may not saturate at expected strength levels. Alone or together, these factors contribute to manufacturing yield loss. Power delivery problems as described earlier may be artifacts of power delivery path during test, or DFT problems such as bus contention or OPCG issues or pattern related such as abrupt power level changes or contention problems.

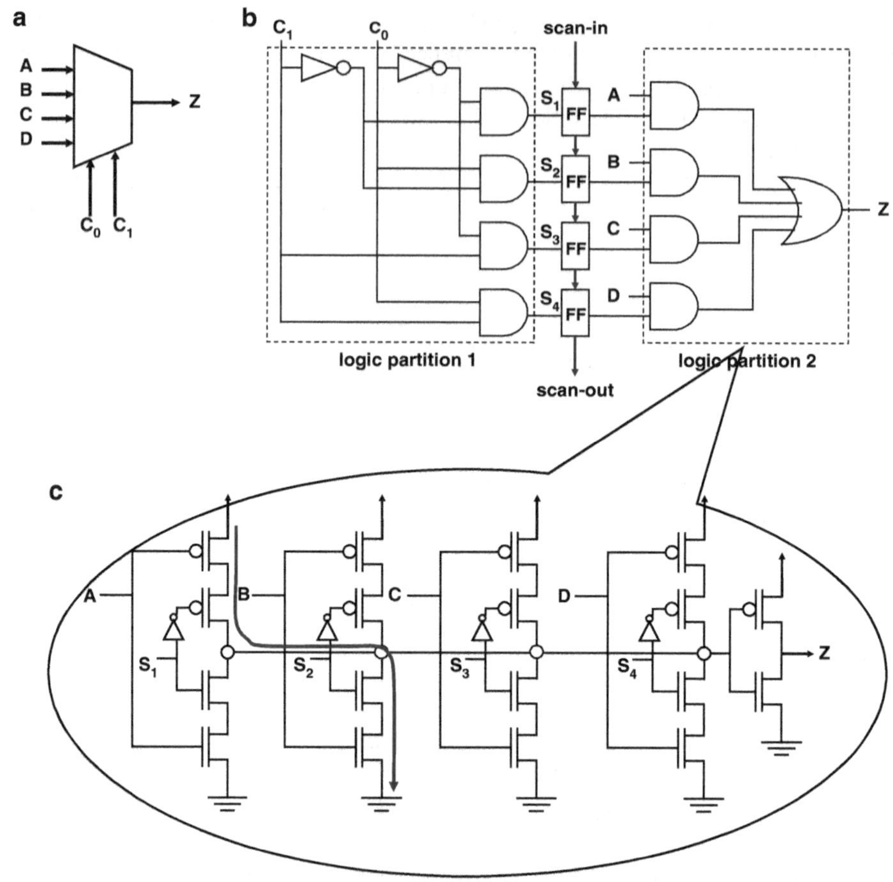

Fig. 2.20 An example illustrating a yield loss scenario due to application of an illegal test vector: (**a**) schematic of a 4-to-1 multiplexer, (**b**) gate-level schematic of the multiplexer with a scan chain partitioning the logic into two partitions, and (**c**) a transistor level schematic of the logic partition 2 of (**b**)

2.8 Test Power Metrics and Estimation

Power consumption is now considered an important constraint during test. Power estimation is required to measure the saving in power and evaluate the effectiveness of a given test power reduction technique. As both the SoC designs and the ultra deep submicron geometry becomes prevalent, larger designs, tighter timing constraints, higher operating frequencies, and lower applied voltages all affect the power consumption of silicon devices. Accurate analysis of power consumption during normal operation as well as test is necessary. Therefore, it is important to define test power metrics and their estimation.

2.8.1 Power Metrics

Following are the four major power and energy metrics that should be quantified accurately to analyze the power dissipation effects during test (Pouya and Crouch 2000):

- **Energy:** Energy is estimated as the total switching activity generated during test application. It affects the battery lifetime during power up or periodic self-test of battery-operated devices.
- **Average power:** Average power is the average distribution of power over the entire test period. Elevated average power increases the thermal load that must be vented away from the DUT to prevent structural damage (hot spots) to the silicon, bonding wires, or package.
- **Instantaneous power:** Instantaneous power is the value of power consumed at any given instant. Usually, it is defined as the power consumed right after the application of a synchronizing clock signal. Elevated instantaneous power might overload the power distribution systems of the silicon or package, causing brown-out.
- **Peak power:** The highest power value at any given instant, peak power determines the component's thermal and electrical limits and system packaging requirements. If peak power exceeds a certain limit, designers can no longer guarantee that the entire circuit will function correctly. In fact, the time window for defining peak power is related to the chip's thermal capacity, and forcing this window to one clock period is sometimes just a simplifying assumption.
- **Rate of change of power:** The highest rate of change of power affects the Ldi/dt drop and highlights deficiencies in decoupling capacitor placement or sizing. As described earlier in Sect. 2.5.3, they may cause manufacturing yield loss. Consequently, this is an important metric in characterizing power consumption during test.

2.8.2 Modeling of Power and Energy Metrics

From (2.5), we know that the average energy consumed at node i per rising transition is $C_i V_{\mathrm{DD}}^2$, where C_i is the equivalent output capacitance and V_{DD} is the power supply voltage (Cirit 1987). Therefore, a good approximation of the energy consumed in a period is $C_i s_i V_{\mathrm{DD}}^2$ where s_i is the number of rising transitions during the period. Nodes connected to more than one gate experience higher parasitic capacitance. On the basis of this fact, as a first approximation we assume capacitance C_i to be proportional to the fan-out count F_i of node I (Wang and Roy 1995). Therefore, an estimation of energy E_i consumed at node i during one clock period is

$$E_i = s_i F_i c_0 V_{\mathrm{DD}}^2 \tag{2.14}$$

where c_0 is the circuit's minimum parasitic capacitance.

According to this expression, estimating energy consumption at the logic level requires the calculation of fan-out F_i and the number of rising transitions of node i, s_i over a period. Circuit topology defines the fan-out of the nodes, and a logic simulator can estimate the switchings (Girard 2002). Product $s_i F_i$ is called the weighted rising transition activity of node i and represents the only variable part in the energy consumed at node i during test application.

According to the previous formulation, the energy consumed in the circuit after application of successive input vectors $\langle V_{k-1}, V_k \rangle$ is

$$E_{vk} = c_0 V_{\mathrm{DD}}^2 \sum\nolimits_i s(i,k) F_i \tag{2.15}$$

where i ranges all the circuit's nodes and $s(i,k)$ is the number of rising transitions caused by V_k at node i.

Let us now consider a pseudorandom test sequence of length m, required to achieve the targeted fault coverage. The total energy consumed in the circuit during application of the complete test sequence is:

$$E_{\mathrm{total}} = c_0 V_{\mathrm{DD}}^2 \sum_{k=1}^{m-1} \sum\nolimits_i s(i,k) F_i \tag{2.16}$$

By definition, the instantaneous power is the power consumed during a small instant of time t_{small} such as the portion of a clock cycle immediately following the system clock rising or falling edge. Therefore, we can express the instantaneous power consumed in the circuit after application of vectors V_k as

$$P_{\mathrm{inst}}(V_k) = \frac{E_{vk}}{t_{\mathrm{small}}} \tag{2.17}$$

The peak power consumption corresponds to the maximum instantaneous power consumed during the test session. It, therefore, corresponds to the highest energy consumed during the same small instant of time, t_{small}. More formally we can express it as

$$P_{\mathrm{peak}} = \max_k [P_{\mathrm{inst}}(V_k)] = \frac{\max_k (z)}{t_{\mathrm{small}}} \tag{2.18}$$

Finally, the average power consumed during the test session is the total energy divided by the test time:

$$P_{\mathrm{ave}} = \frac{E_{\mathrm{total}}}{mT} \tag{2.19}$$

where m is the number of test vectors applied during the test session.

This model for power and energy computation during test is definitely crude and simplified, but it suffices quite well for power analysis during test.

According to these expressions of power and energy consumption, and assuming a given CMOS technology and supply voltage for the circuit design, number of rising transitions s_i of a node i in the circuit is the only parameter that affects the energy, peak power, and average power consumption. Similarly, the clock frequency used during testing affects computation of the average power. Finally, test length affects only the total energy consumption. Consequently, when deriving a solution for power and energy minimization during test, a designer or a test engineer has to keep these relationships in mind (Girard 2002).

Static power dissipation is defined as the power dissipation that occurs after all signal transitions have settled in a circuit. Therefore, static power dissipation depends only on (1) the pattern and (2) the temperature. Temperature dependence arises from transistor subthreshold leakage that increases exponentially with temperature. Consequently, static power dissipation is not a concern except in (1) I_{DDQ} testing and (2) burn-in test. In I_{DDQ} test, test application is slow and the current consumption is pattern dependent, whereas in burn-in test, static power dissipation is large due to elevated temperature. For a large circuit in nano-CMOS technology, the total leakage current does not vary greatly from pattern to pattern. Consequently, leakage current for such circuits is also defined as I_{SB}, or standby current. The main difference between I_{DDQ} and I_{SB} is that the former is pattern specific whereas the latter is not.

2.8.3 *Test Power Estimation*

During conventional design, power consumption in functional mode is estimated in one of the following three levels of abstraction (Najm 1994): (1) architecture-level, (2) RT-level, and/or (3) gate-level. Each one of these estimation strategies represents different tradeoffs between accuracy and estimation time (*see* Fig. 2.21 below).

Estimation of power consumption during test is not only required for sign-off to avoid destructive testing but also to facilitate power-aware test space exploration (during DFT or ATPG) early in the design cycle. A very inaccurate though early and fast way to estimate test power is to use architecture-level power calculators that compute switching activity factor based on architectural pattern simulation and use gate count, and various library parameters to estimate a power value (Ravi et al. 2008). However, in today's design, testing is mostly based on structural patterns

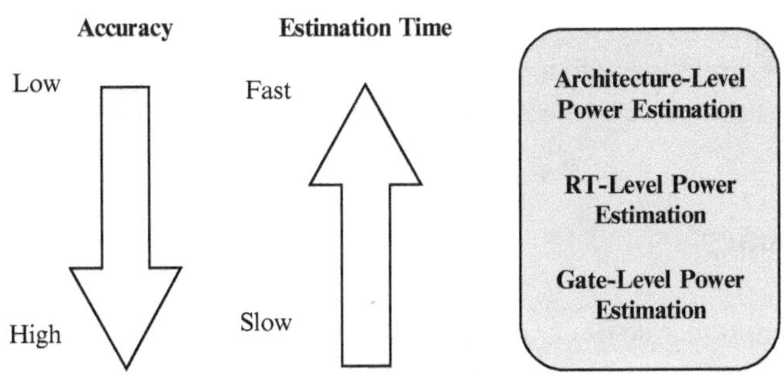

Fig. 2.21 Accuracy versus time in power estimation

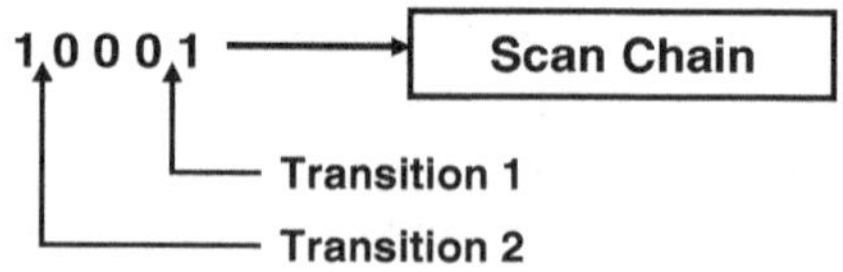

Fig. 2.22 Transitions in scan vector [figure adopted from Sankaralingam et al. (2000)]

applied through a scan chain. The architectural or RT-level designs usually do not contain any scan information that is added later in the design flow, and therefore appear only at the gate-level abstraction. Hence, gate-level test power estimator is needed. A limitation of gate-level estimation is that it is time consuming and therefore, cannot be invoked frequently early during the design cycle. Moreover, gate-level simulators are expensive in terms of memory and run time for multimillion gate SoCs. Such simulators are more suited for final analysis rather than during design iteration. RT-level test power estimators can only be used if DFT insertion and test generation can be done at the RT level (Midulla and Aktouf 2008).

Quick and approximate models of test power have also been suggested in the literature. The weighted transition metric proposed by Sankaralingam et al. (2000) is a simple and widely used model for scan testing, wherein transitions are weighted by their position in a scan pattern to provide a rough estimate of test power.

This is illustrated with an example adopted from the authors. Consider a scan vector in Fig. 2.22 consisting of two transitions. When this vector is scanned into the CUT, Transition 1 passes through the entire scan chain and toggles every flip-flop in the scan chain. On the other hand, Transition 2 toggles only the content of the first flip-flop in the scan chain, and therefore, dissipates relatively less power compared with Transition 1. In this example with five scan flip-flops, a transition in position 1 (in case of Transition 1) is considered to weigh four times more than a transition in position 4 (in case of Transition 2). The weight assigned to a transition is the difference between the size of the scan chain and the position of the transition in the scan-in vector. The total number of weighted transitions for a given scan vector can be computed as follows (Sankaralingam et al. 2000):

$$\text{Weighted transitions} = \sum(\text{Scan chain length} - \text{Transition position in vector}) \tag{2.20}$$

Although the correlation with the overall circuit test power is quite good, a drawback of this metric is that it does not provide an absolute value of test power dissipation.

2.9 Summary

Power consumption, rate of change of power consumption, and overall energy consumption are important factors during test. Availability of power during test may be limited. Abrupt changes in power consumption introduce unwanted changes to the voltage levels in a chip. Excessive power consumption may change the operating temperatures within a chip. Such unwanted changes may invalidate tests and cause yield loss. To mitigate the impact of such changes, an array of approaches is needed. They range from modeling power delivery to heat flux analysis; test strategies, and DFT to support high-throughput power friendly tests as well as electrical verification of final test patterns to ensure that no problems are expected during testing. In this chapter, we outlined the broader set of issues and their interconnectedness. Subsequent chapters will deal with specifics.

References

M. Abramovici, M. A. Breuer, and A. D. Friedman. "Digital Systems Testing and Testable Design". *IEEE Press*, New York City, NY, 1990

E. Acar, R. Arunachalam, and S. R. Nassif. "Predicting Short Circuit Power from Timing Models," In *Proc IEEE Asia-South Pacific Design Automation Conference*, 277–282, 2003

H. Bakoglu. "Circuits, Interconnections, and Packaging for VLSI". *Addison-Wesley*, Reading, MA, 1990

M. L. Bushnell, and V. D. Agrawal. "Essentials of Electronic Testing for Digital, Memory and Mixed-Signal VLSI Circuits," *Kluwer Academic Publishers*, Boston, MA, 2000

M. Cirit. "Estimating Dynamic Power Consumption of CMOS Circuits," In *Proc International Conference on Computer Aided Design (ICCAD)*, 534–537, 1987

J Clabes et al. "Design and Implementation of the POWER5 Microprocessor," In *Proc Design Automation Conference* (DAC), 670–672, 2004

V. Dabholkar, S. Chakravarty, I. Pomeranz et al. "Techniques for Minimizing Power Dissipation in Scan and Combinational Circuits during Test Application," *IEEE Transactions on Computer-Aided Design of Integrated Circuits and Systems*, Vol. 17, No. 12, pp.1325–1333

M. Drazdziulis, and P. Larsson-Edefors "A Gate Leakage Reduction Strategy for Future CMOS Circuits," In *Proc European Solid-State Circuits Conference*, pp. 317–320, 2003

N. G. Einspruch. "VLSI Handbook," *Academic Press*, Orlando, FL, 1985

P. Girard "Survey of Low-Power Testing of VLSI Circuits," *IEEE Design & Test of Computers*, Vol. 19, No. 3, pp. 82–92, 2002

E. R. Hnatek. "Integrated Circuit Quality and Reliability," Mercel Dekker, New York City, NY, 1987

C. Hu et al. "BSIM4 Gate Leakage Model Including Source-Drain Partition," In *Proc International Electron Device Meeting*, pp. 815–818, 2000

A. Hirata, H. Onodera, and K. Tamaru. "Estimation of Short-Circuit Power Dissipation for Static CMOS Gates," *IEICE Transactions on Fundamentals of Electronics, Communication and Computer Sciences*, Vol. E79, No. A, pp. 304–311, 1996

Intel White Paper (online resource): Intel® Multi-Core Processors: Making the Move to Quad-Core and Beyond. http://www.intel.com/technology/architecture/downloads/quad-core-06.pdf

International Roadmap for Semiconductors – System Drivers (online resource): http://www.itrs.net/links/2007ITRS/2007_Chapters/2007_SystemDrivers.pdf

F. Jensen, and N. E. Petersen. "Burn-In," John Wiley & Sons, Chichester, UK, 1982

F. M. Klaasen, and W. Hes. "On the Temperature Co-efficient of MOSFET Threshold Voltage," *Solid State Electronics*, Vol. 29, no. 8, pp. 787–789, 1986

F. M. Klaasen. "MOS Devices Modelling. In: Design of VLSI Circuits for Communications," *Prentice Hall*, Upper Saddle River, NJ, 1995

Z. Kohavi "Switching and Finite Automata Theory," McGraw-Hill, New York City, NY, 1978

I. Koren, and Z. Koren. "Defect Tolerance in VLSI Circuits: Techniques and Yield Analysis," *Proceedings of the IEEE*, vol. 86, No. 9, pp. 1819–1836, 1998

S. Kundu, T. M. Mak, and R. Galivanche. "Trends in Manufacturing Test Methods and Their Implications," In *Proc IEEE International Test Conference*, pp. 679–687, 2004

I. Midulla, and C. Aktouf. "Test Power Analysis at Register Transfer Level," *ASP Journal of Low Power Electronics*, Vol. 4, No. 3, pp. 402–409, 2008

S. Mukhopadhyay, A. Raychowdhury, and K. Roy. "Accurate Estimation of Total Leakage Current in Scaled CMOS Logic Circuits Based on Compact Current Modeling," In *Proc IEEE/ACM Design Automation Conference*, pp. 169–174, 2003

F. Najm. "A Survey of Power Estimation Techniques in VLSI Circuits," *IEEE Transactions on Very Large Scale Integrated Systems*, Vol. 2, No. 4, pp. 446–455, 1994

N. Nicolici, and X. Wen. "Embedded Tutorial on Low Power Test," In *Proc IEEE European Test Symposium*, pp. 202–210, 2007

T. Ogihara, S. Murai, and Y. Takamatsu et al. "Test Generation for Scan Design Circuits with Tri-state Modules and Bidirectional Terminals," In *Proc. IEEE/ACM Design Automation Conference*. pp. 71–78, 1983

R. Pierret. "Semiconductor Device Fundamentals," Ch. 6, pp. 235–300. *Addison-Wesley*, Reading, MA, 1996

I. Polian, A. Czutro, and S. Kundu et al. "Power Droop Testing," In Proc. *IEEE International Conference on Computer Design*, pp. 135–138, 2006

B. Pouya, A. Crouch. "Optimization Trade-offs for Vector Volume and Test Power," In *Proc. IEEE International Test Conference*, pp. 873–881, 2000

J. M. Rabaey, A. Chandrakasan, and B. Nikolic. "Digital Integrated Circuits: A Design Perspective," *Prentice Hall*, Upper Saddle River, NJ, 1996

S. Ravi, S. Parekhji, and J. Saxena. "Low Power Test for Nanometer System-on-Chips (SoCs)," *ASP Journal of Low Power Electronics*, Vol. 4, No. 1, pp. 81–100, 2008

J. Rearcick, and R. Rodgers. "Calibrating Clock Stretch During AC Scan Testing," In *Proc. International test Conference*, 2005

P. Rosinger, B. M. Al-Hashimi, and K. Chakrabarty. "Thermal-Safe Test Scheduling for Core-Based System-on-Chip Integrated Circuits," *IEEE Transactions on Computer-Aided Design of Integrated Circuits and Systems*, Vol. 25, No. 11, pp. 2502–2512, 2006

K. Roy, S. Mukhopadhyay, and H. Mahmoodi-Meimand. "Leakage Current Mechanisms and Leakage Reduction Techniques in Deep-Submicrometer CMOS Circuits," *Proceedings of the IEEE*, Vol. 91, No. 2, pp. 305–327, 2003

R. Sankaralingam, R. Oruganti, and N. A. Toub. "Static Compaction Techniques to Control Scan Vector Power Dissipation," In *Proc. IEEE VLSI Test Symposium*, pp. 35–42, 2000

C. Shi, and R. Kapur. "How power aware test improves reliability and yield," EETimes. http://www.eetimes.com/news/design/features/showArticle.jhtml?articleId = 47208594&kc = 4235. Accessed 21 November 2008

K. Skadron, M. Stan, and W. Huang et al. "Temperature-Aware Microarchitecture. In Proc. International Symposium on Computer Architecture, pp. 2–13, 2003

A. K. Stevens. "Introduction to Component Testing," *Addison-Wesley*, Reading, MA, 1986

Y. Taur, and T. H. Ning. "Fundamentals of Modern VLSI Devices," *Cambridge University Press*, New York City, NY, 1998

C. Tirumurti, S. Kundu, S. Sur-Kolay et al. "A Modeling Approach for Addressing Power Supply Switching Noise Related Failures of Integrated Circuits," In *Proc. IEEE Design, Automation, and Test in Europe Conference*, pp. 1078–1083, 2004

Y. P. Tsividis. "Operation and modeling of the MOS Transistor," *McGraw-Hill*, New York City, NY, 1989

J. T. H. Van der Linden, M. H. Konijnenburg, and A. J. Van de Goor. "Test Generation and Three-State Elements, Busses and Bidirectionals," In *Proc IEEE VLSI Test Symposium*, pp. 114–121, 1994

H. J. M. Veendrick. "Short-Circuit Dissipation of Static CMOS Circuitry and Its Impact on the Design of Buffer Circuits," *IEEE Journal of Solid State Circuits*, Vol. 19, No. 4, pp. 468–473, 1984

S. Vemuri, and N. Scheinberg. "Short-Circuit Power Dissipation Estimation for CMOS Logic Gates," *IEEE Transactions on Circuits and Systems-I*, Vol. 41, No. 11, pp. 762–765, 1994

C. Y. Wang, and K. Roy. "Maximum Power Estimation for CMOS Circuits Using Deterministic and Statistical Approaches," In *Proc. IEEE VLSI Conference*, pp. 364–369, 1995

S. Wang, and S. K. Gupta. "ATPG for Heat Dissipation Minimization During Test Application," *IEEE Transactions on Computers*, Vol. 47, No. 2, pp. 256–262, 1998

N. H. E. Weste, and K. Eshraghian. "Principles of CMOS VLSI Design: A Systems Perspective," *Addison-Wesley*, Reading, MA, 1988

P. Wohl, J. Waicukauski, and M. Graf. "Testing "Untestable" Faults in Three-State Circuits," In *Proc. VLSI Test Symposium*, pp. 324–331, 1996

K. L. Wong, T. R.-Arabi, and M. Ma et al. "Enhancing Microprocessor Immunity to Power Supply Noise with Clock-Data Compensation," *IEEE Journal of Solid State Circuits*, Vol. 41, No. 4, pp. 749–758, 2006

K. N. Yang, H. T. Huang, and M. J. Chen et al. "Characterization and Modeling of Edge Direct Tunneling (EDT) Leakage in Ultra Thin Gate Oxide MOSFETs," *IEEE Transactions on Electron Devices*, Vol. 48, No. 6, pp. 1159–1164, 2001

J. Xiong, V. Zolotov, and C. Visweswariah et al. "Optimal Margin Computation for At-Speed Test," In *Proc. IEEE Design, Automation and Test in Europe Conference*, pp. 622–627, 2008

Chapter 3
Low-Power Test Pattern Generation

Xiaoqing Wen and Seongmoon Wang

Abstract Test pattern generation is an important part of the VLSI testing flow that offers many possibilities that can be explored for reducing test power dissipation. The issue of test power reduction can be addressed at various stages of test generation for logic circuits, by employing low-power automatic test pattern generation, low-power test compaction, low-power *X*-filling, and low-power test vector ordering. In addition, power dissipation in memory testing can be reduced through low-power memory test generation. The most significant advantage of reducing test power through low-power test generation is that this approach causes neither circuit overhead nor performance degradation. However, low-power test generation is a complex technical field, in which many important factors in addition to the effect of test power reduction such as test vector count inflation, potential fault coverage loss, test generation time increase, compatibility with compressed scan testing, and test generation flow modification should be taken into careful consideration. Therefore, the objective of this chapter is to provide a comprehensive overview of the basic principals and fundamental approaches to low-power test generation, along with detailed descriptions of typical methods, so as to help researchers devise more innovative solutions and practitioners build better flows in order to achieve the goal of optimally reducing test power through low-power test generation.

3.1 Introduction

Test generation is the process of creating test data, including test stimuli and test responses, for a *circuit-under-test* (CUT). Ideal test generation should create test stimuli that cause neither *under-test* nor *over-test* (Bushnell and Agrawal 2000). Under-test occurs when the test stimuli generated pass defective chips, and over-test occurs when they fail good chips. Since under-test may compromise quality

---

X. Wen (✉)
Kyushu Institute of Technology, Iizuka, Japan
e-mail: wen@cse.kyutech.ac.jp

S. Wang
NEC Labs, Princeton, NJ, USA

P. Girard et al. (eds.), *Power-Aware Testing and Test Strategies for Low Power Devices*, DOI 10.1007/978-1-4419-0928-2_3,

and over-test may cause die/chip/package damage, reliability degradation, and yield loss, realistic test generation should avoid both of them to the greatest extent possible.

Under-test is caused by inadequate fault coverage, unrealistic fault models, insufficient coverage of small-delay defects, circuit model difference between test mode and functional mode, etc. Over the years, this problem has been tackled by improving *automatic test pattern generation* (ATPG) algorithms to increase fault coverage and using more realistic defect representations, such as the transition delay, path delay, and bridging fault models (Abramovici et al. 1994; Jha and Gupta 2003; Wang et al. 2006a), in test generation. Recently, new metrics and ATPG algorithms that directly target small-delay defects have also emerged (Sato et al. 2005; Yilmaz et al. 2008).

Over-test can be classified as one of two types: *fault-induced* and *power-induced.* Fault-induced over-test occurs when some faults detected in test mode are actually benign in functional mode. This type of over-test has been addressed by utilizing functional constraints in scan test generation (Pomeranz 2004). In contrast, power-induced over-test occurs when fully-functional chips fail during testing because of excessive heat and/or power supply noise caused by test vectors with excessive switching activity. It has been shown that excessive heat may cause die/chip/package damage and reliability degradation (Zorian 1993), and excessive power supply noise may cause a significant circuit delay increase that results in timing failures (Saxena et al. 2003; Yoshida and Watari 2003). Power-induced over-test is rapidly becoming one of the most serious problems in scan testing, especially for low-power, high-speed, and deep-submicron integrated circuits.

This chapter focuses on how to reduce or (more preferably) avoid power-induced over-test through low-power test pattern generation (Girard 2002; Nicolici and Al-Hashimi 2003; Girard et al. 2007; Ravi 2007; Nicolici and Wen 2007; Ravikumar et al. 2008). The purpose is to provide comprehensive information on various approaches and typical methods of low-power test generation, which is cost-effective in solving the problem of power-induced over-test without causing circuit overhead or performance degradation.

The organization of this chapter is depicted in Fig. 3.1. Low-power test generation for logic circuits is covered in Sects. 3.2 through 3.5 based on the following scenario: partially-specified test cubes are first created by deterministically assigning logic values to some inputs for fault detection and test power reduction (Sect. 3.2: *Low-Power ATPG*); the test cubes are then merged by taking test power

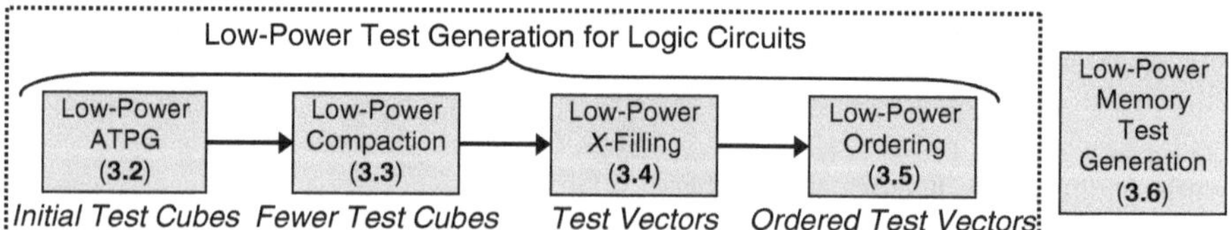

Fig. 3.1 Organization of Chap. 3

into consideration (Sect. 3.3: *Low-Power Test Compaction*); after that, logic values are assigned to the remaining unspecified bits so as to create fully-specified test vectors with reduced test power (Sect. 3.4: *Low-Power X-Filling*); and finally, the test vectors are ordered to further reduce test power (Sect. 3.5: *Low-Power Test Ordering*). Low-power memory test generation is described in Sect. 3.6. A summary of this chapter is provided in Sect. 3.7, with comments on the future direction of research in the field of low-power test pattern generation.

3.2 Low-Power ATPG

The primary function of ATPG is to determine the necessary logic values for some inputs of a CUT so as to detect one or more targeted faults (Bushnell and Agrawal 2000; Jha and Gupta 2003; Wang et al. 2006a). The direct result of ATPG is a test cube, composed of specified bits as well as unspecified or don't-care bits (referred to as *X-bits*). Low-power ATPG is an advanced form of ATPG that targets test power reduction in addition to fault detection during test cube generation (Girard et al. 2007). In this section, low-power ATPG methods for general logic circuits are introduced, followed by a description of low-power ATPG methods for full-scan circuits.

3.2.1 General Low-Power Test Generation

General test generation targets combinational and sequential circuits, including full-scan, enhanced-scan, partial-scan, and nonscan logic circuits. The goal of general low-power test generation is to create a sequence of test vectors that cause a minimal number of transitions at inputs between any two consecutive cycles.

A typical method for combinational low-power test generation is based on the PODEM algorithm (Goel 1981). It replaces cost functions that measure difficulty of fulfilling objectives during test generation, such as those utilized in SCOAP (Goldstein and Thigpen 1980), with cost functions that measure the number of transitions (Wang and Gupta 1994, 1998). Three new cost functions are defined in Wang and Gupta (1994, 1998). For each line l in a circuit and each logic value s, *transition controllability* $TC_s(l)$ and *transition observability* $TO_s(l)$ are calculated to reflect the minimum numbers of weighted transitions required to set l to s and propagate a fault effect (the corresponding fault-free value of which is s) from l to a primary output, respectively. Test vectors with fewer transitions can be generated by using transition controllability to guide the backtrace procedure and transition observability to select D-frontiers for fault propagation. In addition, the *transition test generation cost* for detecting the stuck-at-$\bar{s}$ fault at line l is defined as $TG_s(l) = TC_s(l) + TO_s(l)$, and the stuck-at-$\bar{s}$ fault with the smallest $TG_s(l)$ is selected as the next target fault so that transitions can be further reduced during testing. It is obvious that this approach of using power-oriented cost functions to guide ATPG operations can be readily applied to other fault models, such as the transition delay fault model.

General low-power ATPG methods have also been proposed for sequential circuits. For example, redundancy can be introduced into initial test vectors by detecting a fault multiple times before dropping it, and a final test sequence with low switching activity can be selected from the initial test vectors (Corno et al. 1998).

3.2.2 Low-Shift-Power Scan Test Generation

Most complex logic designs now support full-scan that uses shift and capture operations for testing (Wang et al. 2006a). In shift mode, scan chains are configured into shift registers to shift-out the test response for the previous test vector and shift-in the next test vector. Scan shift causes scan-FF transitions in the scan chains; these transitions then propagate into the combinational logic, resulting in more switching activity. The accumulated impact of the overall shift switching activity due to the repeated application of shift clock pulses may result in excessive heat dissipation. This makes it necessary to reduce average shift power.

Shift power can be reduced by utilizing two types of don't-care bits (Wang and Gupta 1997a). For example, a test vector for the combinational logic shown in Fig. 3.2 has three *primary input* (PI) bits and four *pseudo primary input* (PPI) bits that correspond to the outputs of the scan FFs SFF_1 through SFF_4. The shift-in of the previous test vector is completed at T_{i-1}, and scan capture is conducted at T_i. After that, four shift clock pulses are applied in order to shift-out the captured test response and simultaneously shift-in the PPI bits of the next test vector. Note that the PI bits need to be set at only T_{i+4}. In other words, the don't-care bits in ① are independent and can be assigned any logic values from T_i through T_{i+3}. On the other hand, the don't-care bits in ② move through different numbers of scan FFs, and the total number of transitions at scan FFs depends on the contents of the test response

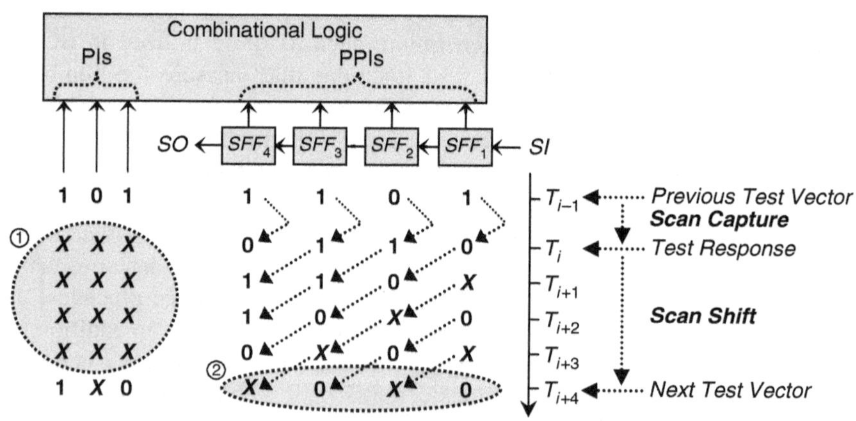

Fig. 3.2 Shift power reduction by blocking and reducing scan-FF transitions

and the next test vector. These two types of don't-care bits can be utilized to reduce shift power by employing the following two techniques:

- *Using the don't-care bits in ① to block the impact of scan-FF transitions*: This technique attempts to prevent transitions at the scan FFs from propagating into the combinational logic by assigning proper logic values to PIs. One logic value assignment that can block transitions at the maximum number of scan FFs is computed and applied for the duration of the entire shift operation for all test vectors in order to reduce test data volume. Note that multiple primary logic assignments may be used for a better blocking effect, at the cost of larger test data volume.
- *Using the don't-care bits in ② to reduce scan-FF transitions*: First, PODEM (Goel 1981) is modified using new cost functions to guide backtrace and D-frontier selection so that more PPI bits within a given test cube are left unspecified. Next, the K-L algorithm (Kernighan and Lin 1970) is used to find a logic assignment for the unspecified bits so that transitions are minimized for the entire shift operation. If the run time is too long, one may use simple heuristics, such as minimizing only the transitions for PPI bits of the next test vector at T_{i+4}.

Blocking-based low-shift-power test vectors can also be generated via *input control* (Huang and Lee 1999). This technique uses a modified D-algorithm (Roth 1966) to generate a fixed logic assignment for PIs, called *control pattern*, for preventing the transitions at scan FFs from propagating into the combinational logic. Note that one control pattern is applied to all PIs, all but the last of the shift clock pulses for each scan shift operation, and for all test vectors. This helps reduce shift power without significantly increasing test data volume.

Another technique for low-shift-power test generation is based on the fact that in scan testing, the PIs can be frozen at the values of the previous test vector until any time between the first and the second-to-last shift clock pulses. For example, PI values in Fig. 3.2 can be frozen, that is, need not change to the PI values of the next test vector until any time between T_{i+1} and T_{i+3}. An algorithm, called *best primary input change* (BPIC) (Nicolici et al. 2000), can be used to compute the best primary input change time for each test vector, so that the number of transitions in the combinational logic is minimized during the scan shift operation.

All of the techniques above require modification to ATPG algorithms. To avoid this inconvenience the constraint of the total shift transition count can be encapsulated in a *power constraint circuit* (PCC) (Ravi et al. 2007). The PCC is combined with the original circuit to form a complete circuit model, on which normal ATPG is conducted. The advantage of using a PCC is that low-shift-power test vectors can be readily generated by using a normal ATPG program without any modification.

3.2.3 Low-Capture-Power Scan Test Generation

Most complex logic designs now support full-scan that uses shift and capture operations for testing (Wang et al. 2006a). In capture mode, all scan FFs operate

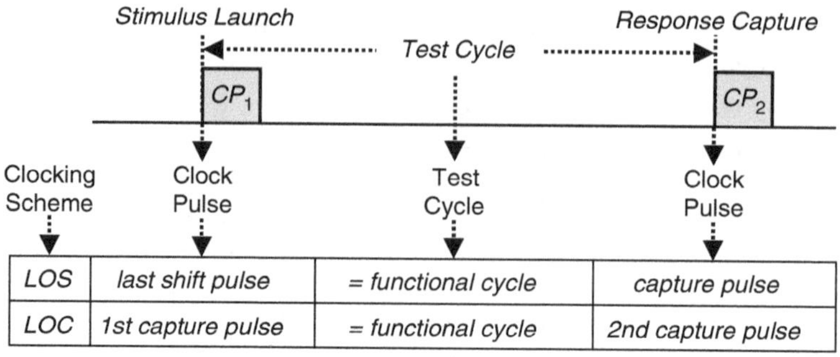

Fig. 3.3 General capture operation in at-speed scan testing

individually as functional FFs, updating their outputs with data from the outputs of the combinational logic of the scan circuit with one (*single-capture*) or two (*double-capture*) capture clock pulses (Wang et al. 2006a). The former is utilized in slow-speed scan testing for structural faults (e.g., stuck-at, bridging, etc.), as well as in *launch-on-shift* (LOS)-based at-speed scan testing for timing-related faults (e.g., transition delay); the latter is utilized by *launch-on-capture* (LOC)-based at-speed scan testing for timing-related faults (Savir and Patil 1994). While benign in slow-speed scan testing, test power dissipated in capture mode, often called *capture power*, is a big issue in at-speed scan testing.

Figure 3.3 illustrates the general capture operation in at-speed scan testing. In order to conduct delay testing for timing-related faults, the *stimulus launch pulse* CP_1 (applied in shift or capture mode) launches a transition at the start-point scan FF of a path, and the *response capture pulse* CP_2 (applied only in capture mode) captures the test response at the end-point scan FF of the path. Two potential problems with capture power may occur in at-speed scan testing, as described below.

The first problem is caused by the (stimulus) *launch switching activity* (LSA) at CP_1. LSA-induced IR-drop increases gate delay (Wang et al. 2006b), the accumulation of which may result in timing failures at CP_2 (Saxena et al. 2003), leading to undue yield loss. This is an especially severe problem for high-speed designs, whose short test cycles make them highly susceptible to unexpected delay increase. In addition, excessive LSA makes it risky to use long sensitization/propagation paths for improving the delectability of small-delay defects since IR-drop may reduce slack (Kajihara et al. 2006; Lin et al. 2006). Therefore, peak LSA should be reduced so that capture-safety can be guaranteed in at-speed scan testing.

The second problem is caused by the (response) *capture switching activity* (CSA) at CP_2. If excessive CSA causes severe IR-drop, scan FFs may consequently malfunction, resulting in capturing of erroneous test responses (Wen et al. 2008a). However, since CSA is usually much lower than LSA, and CSA does not affect timing, it is generally more important to reduce LSA.

LSA in LOS-based at-speed scan testing (*LOS-type LSA*) for a test vector, whose PPI portion is $<v:\text{PPI}> = <b_1 b_2 \ldots b_{n-1} b_n>$, is dominated by $\Sigma(b_{i-1} \oplus b_i)$ for $i = 2, \ldots, n$, as shown in Fig. 3.4a. This means that LOS-type LSA can be readily

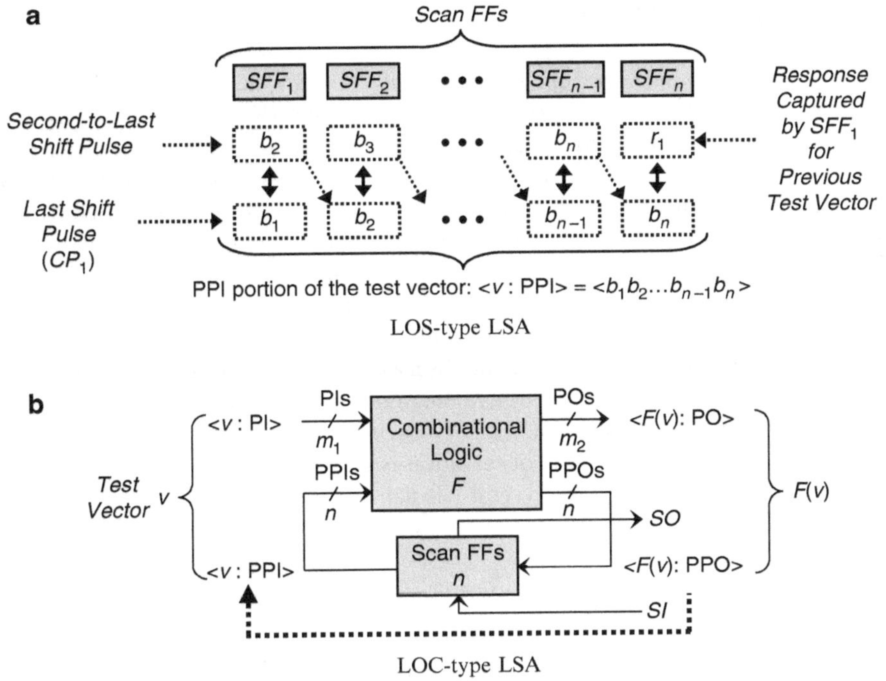

Fig. 3.4 LSA reduction in at-speed scan testing

reduced by minimizing the difference between neighboring bits in $<v\!:\!\text{PPI}>$ through simple *X*-filling techniques, such as *0-fill* and *1-fill* (to be described in Sect. 3.4.2.1). In contrast, LSA in LOC-based at-speed scan testing (*LOC-type LSA*) is dominated by $\Sigma(<v\!:\!\text{PPI}> \oplus <F(v)\!:\!\text{PPO}>)$, where $<F(v)\!:\!\text{PPO}>$ is the PPO portion of the test response to the combinational logic F for test vector v, as shown in Fig. 3.4b. This means that test responses must be taken into consideration in order to reduce LOC-type LSA, making it more difficult to reduce than LOS-type LSA. In the rest of this chapter, low-capture-power test generation will be discussed in terms of reducing LSA in LOC-based at-speed scan testing.

Since low-capture-power test generation for reducing LSA in LOC-based at-speed scan testing is time-consuming, a two-pass flow is usually needed. First, normal ATPG is conducted to generate initial vectors, and *capture safety checking* is conducted to identify capture-unsafe vectors due to excessive LSA. Low-capture-power test generation is then conducted for the faults detected only by the capture-unsafe vectors. This way, final capture-safe test vectors can be obtained more efficiently. In the following three subsections, various metrics for capture safety checking are first discussed, followed by descriptions of two typical low-capture-power ATPG techniques, namely *reversible backtracking* and *clock manipulation*.

3.2.3.1 Capture-Safety Checking

A test vector v is said to be *capture-safe* if its LSA does not cause the delay of any path activated by v to exceed the test cycle (Kokrady and Ravikumar 2004; Wen et al. 2007a). Ideal capture-safety checking is time-consuming and memory-intensive since it requires timing-based logic simulation and IR-drop/delay calculation. Simplification techniques, such as those to be described below, are therefore needed for practical capture-safety checking.

A non-simulation-based technique for capture-safety checking infers LSA from fault detection information (Lee and Tehranipoor 2008). If a transition delay fault is *detected* (DT) by a test vector, transitions must occur along a path passing through the fault from a controllable input to an observable output. Additionally, if a transition delay fault is undetected due to its being *not observable* (NO) by a test vector, some transitions related to the fault likely occur but fail to reach any observable output. The metric based on this observation is called *fault list inferred switching* (FLIS). The FLIS value for a test vector v is defined as

$$FLIS(v) = N_{DT}(v) + \sum_{i=0}^{N_{NO}(v)} P_{TR_i}$$

where $N_{DT}(v)$ is the number of DT faults of v, $N_{NO}(v)$ is the number of NO faults of v, and P_{TR_i} is the probability of a transition occurring at the ith NO fault site of v. Experimental results on ISCAS'89 benchmark circuits indicate a close correlation between FLIS and the amount of node switching activity in a circuit.

The simplest and most widely used simulation-based metrics for capture-safety checking are *toggle count* (TC) and *weighted switching activity* (WSA) for FFs or nodes (FFs and gates) in a circuit. TC and WSA are calculated as the number of transitions and the number of weighted transitions, respectively. A transition at a node with large load capacitance has a larger impact on IR drop than that at a node with small load capacitance. Ideally, the weight of a node is best set to reflect its output capacitance. Due to a lack of load capacitance data, the number of fanouts plus 1 was used as the weight of each node in experiments in (Wang and Gupta 1994). Note that real load capacitances need to be used if higher accuracy is required.

Region-based capture-safety checking takes power-grid information into consideration (Devanathan et al. 2007a). As shown in Fig. 3.5, a circuit is divided into regions bounded by power straps, and a test vector is judged to be capture-safe if (1) *global toggle constraint* (for limiting transitions in all regions throughout the test cycle); (2) *global instantaneous toggle constraint* (for limiting transitions at any time instant in all regions); and (3) *regional instantaneous toggle constraint* (RITC) (for limiting transitions at any time instant in a specific region), are satisfied. These limits are obtained through static IR-drop analysis, and Fig. 3.5b shows sample *RITC* limits. This capture-safety checking technique is more accurate since it takes both spatial and temporal aspects of LSA into consideration. Similar techniques are used in (Devanathan et al. 2007b; Wen et al. 2008b).

a

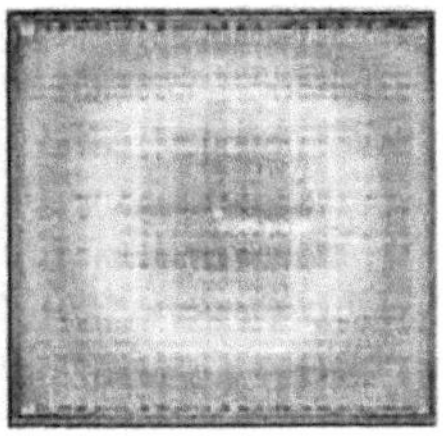

Static IR-drop plot

b

50	50	50	50
50	45	45	50
50	45	45	50
50	50	50	50

Sample RITC limits

Fig. 3.5 Region-based capture-safety checking

Another technique also partitions a circuit layout into regions by using straps/rings as midpoints (Lee et al. 2008). The switching activity of a test vector v in a region $R(i,j)$ is estimated by extended WSA for all gates in the region as follows:

$$WSA_{ij}(v) = \sum_{k \in R(i,j)} (t_k \times (g_{w_k} + f_k \times f_{w_k}))$$

where g_{w_k} is the weight of gate k in $R(i,j)$, f_k is the number of its fanouts, f_{w_k} is its fanout load weight, and t_k is 1 (0) if a transition occurs (does not occur) at gate k. The maximum WSA for each region $R(i,j)$ is defined as

$$WSAmax_{ij} = \sum_{k \in R(i,j)} (g_{w_k} + f_k \times f_{w_k})$$

The switching activity limit for $R(i,j)$ can be set as a percentage of $WSAmax_{ij}$, and used as a threshold to determine whether a test vector is capture-safe or not.

In addition to the number of transitions, the time period during which they occur is also important for a more accurate estimation of their impact on IR-drop. In order to address this issue, *switching cycle average power* (SCAP) for a test vector v is calculated as follows:

$$SCAP(v) = \frac{\sum C_i \times VDD^2}{STW(v)}$$

where C_i is the output capacitance of gate i with a transition caused by v, VDD is the power supply voltage, and $STW(v)$ is the *switching time window* (STW) of v (Ahmed et al. 2007a). $STW(v)$, the time-frame during which the transitions caused by v occur, can be obtained approximately as follows:

$$STW(v) = t_i + T/2 \text{ for } t_d(v) \le T/2$$
$$STW(v) = t_i + t_d(v) \text{ for } t_d(v) > T/2$$

where t_i is the clock insertion delay, T is the test cycle, and $t_d(v)$ is the maximum path delay of v (Ahmed et al. 2007b). Experimental results on ISCAS'89 benchmark circuits indicate a good correlation between SCAP and IR-drop.

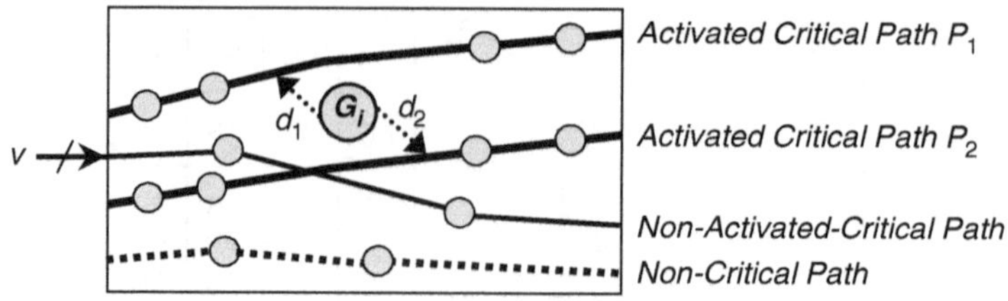

Fig. 3.6 Critical path and critical weight

From the definition of capture-safety, it is obvious that activated critical or long paths are the most susceptible to the impact of IR-drop (i.e., delay increase). As a capture-safety checking metric that takes this fact into consideration, *critical capture transition* (CCT) for a test vector v is calculated as follows:

$$CCT(v) = \sum (CW_i(v) \times w_i \times t_i)$$

where $CW_i(v)$ is the *critical weight* of node i for v, w_i is the weight of node i as in WSA, and t_i is 1 (0) if a transition occurs (does not occur) at node i (Wen et al. 2007a). Here, $CW_i(v) = \Sigma(1/d_j)$, where d_j is the distance of node i from activated critical path j. Note that node i is a critical node (i.e., its distance from any critical path activated by v is within a given radius). In the example shown in Fig. 3.6, $CW_i(v) = 1/d_1 + 1/d_2$ reflects the impact of a transition at G_i on the two activated critical paths, P_1 and P_2. Simulation results for the impact of LSA on path delay indicate that CCT is more accurate than WSA.

Generally, capture-safety checking is important not only for low-power test generation but also for low-power test design. It should be accurate enough so as to avoid any underestimation and too much overestimation. It should also be efficient enough so that it can be used iteratively in test generation and test design to maximize test power reduction while minimizing pattern and/or circuit overhead.

3.2.3.2 LCP ATPG Technique 1: Reversible Backtracking

In PODEM (Goel 1981), a backtracking occurs when *X-path-checking* finds a *detection conflict* (D-conflict), which means that no path with unspecified values exists between D-frontiers and any primary or pseudo primary output. A D-conflict means that the current input value assignment cannot detect the target fault, and backtracking is triggered to undo the current input assignment that leads to the conflicts.

Low-capture-power test vectors can be generated by modifying PODEM so that a backtracking can be triggered by constraints not only for fault detection but also for capture power reduction. One technique uses the concept of *capture conflict* (C-conflict) that occurs when the input and the output of a scan FF have opposite logic values for the current input value assignment (Wen et al. 2006). As shown in Fig. 3.7, a C-conflict at a scan FF means that a transition will occur when the stimulus launch

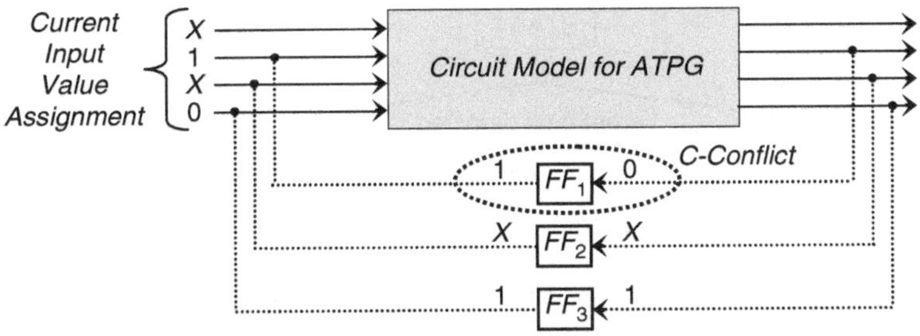

Fig. 3.7 C-conflict in test generation for launch-on-capture (*LOC*)-based at-speed scan testing

pulse (CP_1 in Fig. 3.3) is applied in LOC-based at-speed scan testing. This indicates that reducing C-conflicts leads to low LSA.

A C-conflict, like a D-conflict, may be avoided through backtracking. However, a C-conflict-triggered backtracking may prevent a fault from being detected. This problem can be solved by utilizing *reversible backtracking* (Wen et al. 2006). With this technique, if test generation fails due to a C-conflict, the backtracking conducted for the C-conflict is reversed, and the C-conflict will not be checked again. This technique prevents fault coverage loss, while generating a test cube with relatively fewer C-conflicts (Wen et al. 2006; Devanathan et al. 2007a).

Figure 3.8 shows a modified PODEM procedure with reversible backtracking to generate a test cube for a target fault with as few C-conflicts as possible (Wen et al. 2006). When a D-conflict or a C-conflict is found (①, ②), the procedure conducts the backtracking (⑤), which is managed by two types of stacks: a *primary implication stack* for managing the search space and a *restoration implication stack* as a copy of the primary implication stack. One restoration implication stack is created for each C-conflict; all of the stacks are then placed in a list, called a *restoration implication stack list* (④). When the primary implication stack is exhausted (③), it is checked whether the restoration implication stack list is empty or not (⑥). If not, it means that at least one C-conflict occurred, and that the backtracking conducted for the C-conflict may have blocked the detection of the target fault. Therefore, the top stack S in the restoration implication stack list is removed from the list and restored as the primary implication stack (⑦), and the C-conflict corresponding to the stack S is suppressed from incurring any future backtracking (⑧). Then, test generation is resumed from the restored primary implication stack. This way, a test cube that detects the target fault with a reduced number of C-conflicts is generated, at the cost of a possibly larger final test set.

3.2.3.3 LCP ATPG Technique 2: Clock Manipulation

LSA originates from scan-FF transitions. A transition occurs at a scan FF because of (1) *input-output difference* (i.e., the input and the output of the scan FF must have opposite values) and (2) *clock activation* (i.e., the clock for the scan FF must

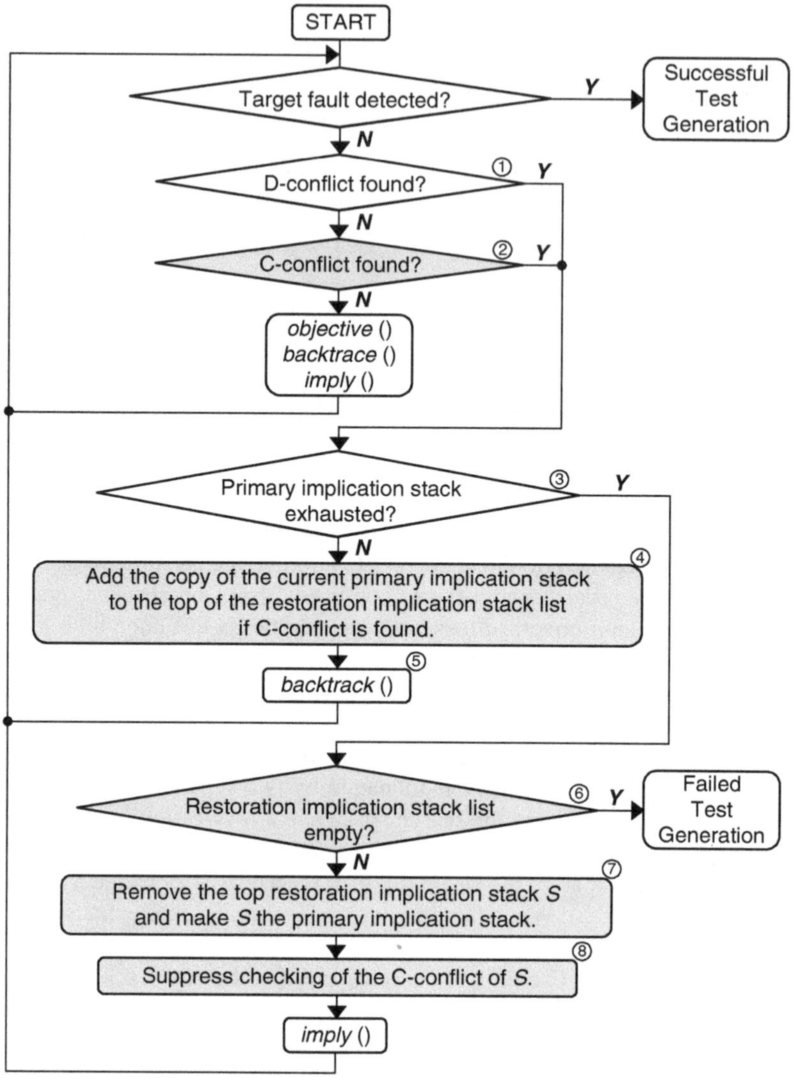

Fig. 3.8 Low-capture-power test cube generation with reversible backtracking

be applied). Clearly, transitions can be reduced not only by equalizing input-output values as in reversible backtracking but also by manipulating test clocks. Note that instead of manipulating test clocks, the scan enable signal can be selectively controlled to allow only a portion of scan FFs to capture (Wang and Wei 2005).

Test clocks can be manipulated *externally* by *automatic test equipment* (ATE) to reduce capture power for a circuit with multiple clock domains. Using one clock for the capture operation of all clock domains simplifies ATPG and reduces test data volume, but it may result in excessive capture power dissipation since all clock domains capture simultaneously. To solve this problem, two low-capture-power clocking schemes, *one-hot* and *multi-capture*, can be used (Wang et al. 2006a). During the capture cycle, the former allows only one clock domain to capture, while the latter staggers capture clocks for different clock domains. No ATPG change is needed for the one-hot clocking scheme, but test data volume often increases due to the serialization of capture operations for different clock domains. On the other hand, the multi-capture clocking scheme leads to fewer test vectors, but memory-consuming multi-time-frame circuit expansion is needed, and fault coverage loss may occur.

Test clocks can also be manipulated *internally* by using the clock-gating mechanism to disable clocks for some scan FFs in capture mode. Given the fact that typically over 80% of FFs in low-power devices have gated clocks and that capture mode is actually functional mode, it is highly desirable and feasible to explore gated clocks for capture power reduction.

Figure 3.9 shows a typical clock-gating mechanism with enhancement for scan testing. In shift mode ($SE = 1$), the test clock *CK* constantly drives the FF clock *GCK* so that the shift operation is properly conducted. In capture mode ($SE = 0$), whether or not *CK* drives *GCK* depends on the clock control signal *EN*, which is driven by the functional clock enable logic. Clearly, test generation can be used to set proper logic values in a test vector in order to disable capture clocks for capture power reduction (Keller et al. 2007; Czysz et al. 2008; Furukawa et al. 2008).

A typical clock-disabling technique is to obtain *clock control cubes* (CCCs) that disable each clock control signal, and order all the CCCs of a clock by the decreasing number of FFs that are disabled by each CCC (Czysz et al. 2008). In test generation, after a test cube is generated for detecting one or more target faults, compatible CCCs are merged with the test cube in the aforementioned order so as to maximize the gating-off of FFs that do not need to be clocked for fault detection. Capture power can be effectively reduced, especially when a clock control signal for a large number of FFs are disabled, at the cost of a larger final test set.

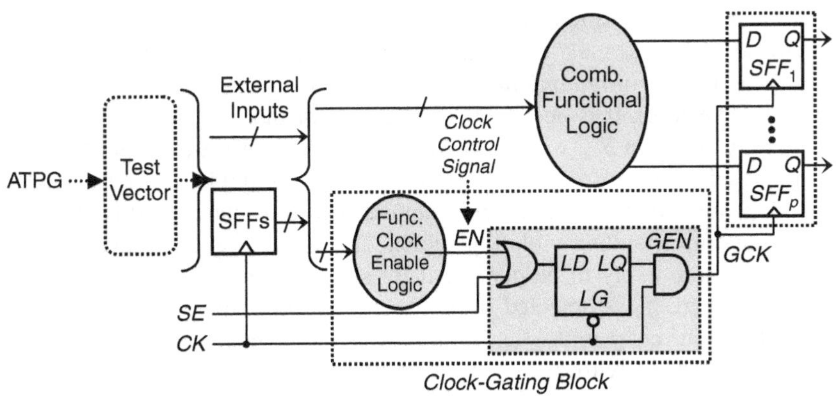

Fig. 3.9 Typical clock-gating mechanism

3.3 Low-Power Test Compaction

The purpose of test compaction, both *dynamic* and *static*, is to reduce the number of final test vectors (Wang et al. 2006a). Test compaction exploits X-bits that are present in most test vectors. Such a test vector containing unspecified bits, or X-bits, is called a *test cube*. Dynamic compaction specifies the X-bits in a test cube in order to detect more faults than the initially targeted fault, while static compaction uses the X-bits in compatible test cubes to merge them into one test cube.

In addition to test data reduction, test power reduction can also be targeted in test compaction. Typical techniques for low-power dynamic compaction and low-power static compaction are described in Sects. 3.3.1 and 3.3.2, respectively.

3.3.1 Low-Power Dynamic Compaction

In ATPG, the test cube generated for the first or primary target fault usually contains many X-bits. Hence, after a test cube is generated for a primary target fault, a new or secondary target fault is selected and the X-bits are specified to detect it. This process, called *dynamic compaction*, makes one test vector to target more than one fault so that a smaller final test set can be obtain (Abramovici et al. 1994; Bushnell and Agrawal 2000). Dynamic compaction can also be conducted to reduce test power by taking test power into consideration. Such a process is called *low-power dynamic compaction*.

One low-power dynamic compaction technique is based on the following flow: A normal ATPG generates an initial test set, capture-safety checking identifies the violating test set (S_{vio}), and low-capture-power test generation creates a new test set (S_{new}) such that all faults detected only by S_{vio} are still detected by S_{new} (Wen et al. 2006; Devanathan et al. 2007a). To avoid a significant increase in test vector count, each new vector is generated as a replacement for its corresponding violating vector. This requires careful selection of faults to be targeted by each new vector. All faults detected by S_{vio} are classified as *vector-essential faults* (i.e., faults detected by one test vector alone in S_{vio}), *set-essential faults* (i.e., faults detected by multiple test vectors alone in S_{vio}), or *redundant faults* (i.e., faults detected by violating and non-violating test vectors). It is necessary to target all of the vector-essential faults of each violating vector in S_{vio} with one new vector in S_{new}. A set-essential fault, however, is selected as the target fault of a replacement vector if it has the least reachable area overlap (towards both inputs and outputs) with those target faults already selected by the corresponding violating vector. This is because a low-capture-power test vector is more likely to be generated for such a target fault. Table 3.1 shows an example, where f_9 is detected by both v_1 (currently with f_1 and f_6 as target faults) and v_4 (currently with f_4, f_7, and f_8 as target faults). If f_9 has the least reachable area overlap with f_1 and f_6, it will be selected as the target fault for the new vector r_1.

Table 3.1 Replacement-based target fault selection

Generated → S_{vio}	Fault Detection Information of S_{vio}	Target Fault List for S_{new}	S_{new} ← To be generated
v_1	○f_1 ○f_6 □f_9 □f_{10}	f_1 f_6 f_9	r_1
v_4	○f_4 ○f_7 ○f_8 □f_9 ◇f_{12}	f_4 f_7 f_8	r_2
v_5	○f_5 □f_{10} ◇f_{11}	f_5 f_{10}	r_3

○: vector-essential fault □: set-essential fault ◇: redundant fault

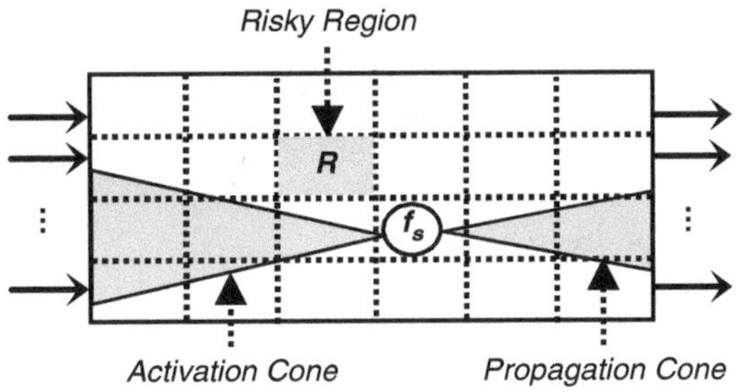

Fig. 3.10 Region-based target fault selection

Another low-power dynamic compaction technique selects a target fault in such a manner that the risk of violating capture-power limits is minimized (Wen et al. 2008b). For this purpose, region-based capture-safety checking (Devanathan et al. 2007a, b) is conducted to identify risky regions, each with at least one capture-power metric value being close to its limit. Then, a target fault sharing the least reachable area (activation cone and propagation cone) overlap with the risky regions is selected. Figure 3.10 shows an example, where f_s is selected as the target fault since its reachable area has no overlap with the risky region R. This way, chances of generating a capture-safe test vector are increased.

3.3.2 *Low-Power Static Compaction*

Static compaction is the process of merging multiple test cubes into one if they are compatible (Wang et al. 2006a). Two test cubes, c_1 and c_2, are *compatible* if no two corresponding bits in c_1 and c_2 have opposite logic values. Conventionally, static compaction is conducted to reduce the final test vector count; it can also reduce test power if power is considered when compatible test cubes are merged.

3.3.2.1 Low-Shift-Power Static Compaction

Conventional static compaction tries to merge as many compatible test cubes as possible without considering test power. By carefully selecting the order in which

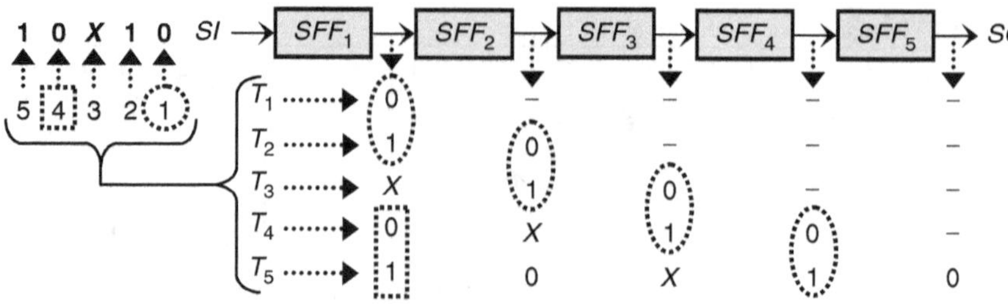

Fig. 3.11 Weighted transitions for estimating shift-in power

compatible test cubes are merged, average shift-in power can be effectively reduced (Sankaralingam et al. 2000). A merging order is selected by comparing the cost of merging two test cubes in terms of shift-in power, and the concept of *weighted transition* can be used for this purpose.

Figure 3.11 shows a scan chain of five scan FFs. Test cube <01X01> has two transitions at bit positions of 1 and 4. Since the test cube is shifted into the scan chain, the transition at bit position 1 causes four shift-in transitions in the scan chain, whereas the transition at bit position 4 causes one shift-in transition in the scan chain. This shows that transitions at different bit positions in a test cube have different degrees of impact on shift-in power. This fact is reflected in the number of *weighted transitions*, defined as follows:

$$Weighted\_Transitions = \sum(Scan\_Chain\_Length - Transition\_Position)$$

In Fig. 3.11, the number of weighted transitions for <01X01> is 5, which can be used as an estimate of shift-in power caused by the test cube.

The cost of merging two compatible test cubes can be set as the number of weighted transitions for the resulting new test cube (Sankaralingam et al. 2000). The complete cost information for a set of test cubes can be expressed with a cost graph, in which each node corresponds to a test cube and each edge between two nodes indicates that the two corresponding test cubes are compatible. The weight on an edge is the cost of merging the two corresponding test cubes. Static compaction can be conducted by selecting the edge with the smallest weight, merging the two corresponding test cubes, updating the cost graph, and repeating the process until no edges (indicating two compatible test cubes) remain.

3.3.2.2 Low-Capture-Power Static Compaction

Most low-capture-power static compaction techniques aim to reduce LSA in LOC-based at-speed scan testing. This can be achieved by having static compaction check not only for compatibility between two test cubes but also the impact of merging them in terms of capture power.

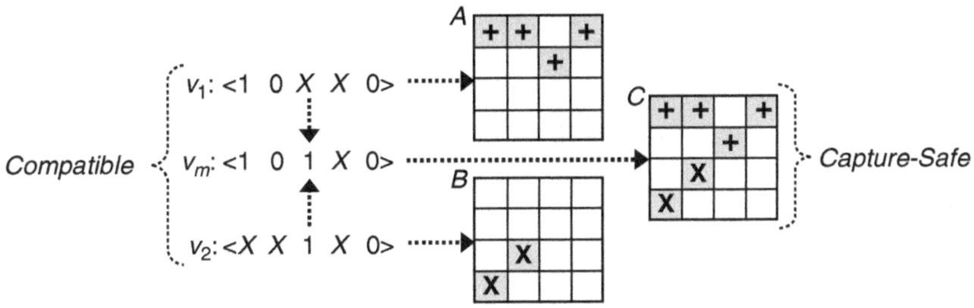

Fig. 3.12 Regional-switching-activity-based low-capture-power static compaction

One low-capture-power static compaction technique takes layout and LSA distribution into consideration (Lee et al. 2008). Its goal is to evenly distribute the LSA of a test vector across the entire chip rather than allowing high LSA to occur in a small area, which could cause excessive regional IR-drop. It uses the extended WSA metric described in Sect. 3.2.3.1, and calculates the LSA profile for each region. Figure 3.12 is an example of compatible test cubes v_1 and v_2 and their LSA profiles (A and B), where "+" and "X" mean high-LSA regions for v_1 and v_2, respectively. Merging v_1 and v_2 will create a new LSA profile (C). Since no overlap exists for the high-LSA regions of v_1 and v_2, the resulting test cube v_m is capture-safe.

Another low-capture-power static compaction technique further estimates the impact of capture power on path delay in static compaction (Wang et al. 2005a, b). It conducts a vector-dependent power supply noise analysis, in which IR-drop is estimated by using the layout of the chip and the switching activity of the test vector. The impact of the IR-drop on gate delay is assessed, and timing violations are checked against targeted (usually critical) paths. This power supply noise analysis is added to normal static compaction to guarantee that no merged test vector causes excessive capture power. Obviously, the usefulness of this technique depends on the accuracy of the process used for power supply noise analysis. In experiments on benchmark circuits, voltage error was found to range from -1.5% to 1.7% with an average error of 1%, while path delay error was found to range from –3% to 6% with an average error of 1.9% (Wang et al. 2005b).

3.4 Low-Power *X*-Filling

A large number of *X*-bits usually remain in test cubes even after test compaction is conducted (Wohl et al. 2003; Hiraide et al. 2003). Since final test stimuli can only contain specified bits, *X*-filling needs to be conducted on *test cubes* (which are partially-specified) to create *test vectors* (which are fully-specified).

Traditionally, *random fill* has been conducted on test cubes in order to reduce final test vector count by increasing the chances of fortuitous detection. Test cubes can also be filled for test power reduction. Because abandoning *random fill* increases final test vector count (Remersaro et al. 2006), another approach is to first

Table 3.2 Typical low-power *X*-filling methods

Low-shift-power *X*-filling methods		
Shift-in power reduction	Shift-out power reduction	Total shift power reduction
0-fill	*output-justification-based X-filling*	*MTR-fill*
1-fill		
MT-fill		
adjacent fill / repeat fill		
Low-capture-power *X*-filling methods		
FF-oriented	Node-oriented	Critical-area-oriented
PMF-fill	*PWT-fill*	*CCT-fill*
LCP-fill	*state-sensitive X-filling*	
preferred fill		
JP-fill		
CTX-fill		
Low-shift-and-capture-power *X*-filling methods		
impact-oriented X-filling	*hybrid X-filling*	*bounded adjacent fill*
Low-power *X*-filling methods for compressed scan testing		
0-fill	*PHS-fill*	*CJP-fill*

conduct *random fill* to generate a compact initial test set and then identify as many *X*-bits from the test set as possible without reducing its fault coverage (Miyase and Kajihara 2004). The identified *X*-bits are then filled for test power reduction. The advantage of this approach is that it does not increase final test vector count.

Test cube preparation is first discussed in Sect. 3.4.1, followed by descriptions of *X*-filling methods for reducing shift power, capture power, and both in normal or noncompressed scan testing in Sects. 3.4.2, 3.4.3, and 3.4.4, respectively. After that, low-power *X*-filling for compressed scan testing is discussed in Sect. 3.4.5. Table 3.2 lists the typical *X*-filling methods that will be described in these sections.

3.4.1 Test Cube Preparation

There are two approaches, namely *direct generation* and *test relaxation*, to preparing test cubes for *X*-filling. Direct generation is to explicitly leave some bits unfilled in ATPG by suppressing *random fill*, while test relaxation is to turn some bits in a set of fully-specified test vectors into *X*-bits without reducing its fault coverage. While both approaches are capable of obtaining test cubes with 50–90% of their bits as *X*-bits, each has its own advantages and disadvantages.

3.4.1.1 Direct Generation

In order to reduce test vector count, ATPG usually attempts to detect as many faults as possible with one test vector. This can be achieved through dynamic compaction and *random fill* (Abramovici et al. 1994; Bushnell and Agrawal 2000; Wang et al. 2006a). Dynamic compaction makes use of the remaining *X*-bits in a test cube to detect additional faults. When it becomes difficult or too time-consuming to detect more faults with dynamic compaction, *random fill* is conducted to assign random logic values to the remaining *X*-bits in a test cube. This random logic value assignment increases the chances of accidental or fortuitous detection.

Obviously, test cubes can be obtained by simply suppressing *random fill* (Butler et al. 2004). This *direct generation* method may result in test cubes with over 95% of their total bits as *X*-bits. *X*-bits in the test cubes can be filled by taking test power into consideration so that the resulting test vectors have lower test power.

However, suppressing *random fill* cannot control the numbers of *X*-bits in an individual test cube meaning that some test cubes requiring many *X*-bits for test power reduction could potentially not have enough *X*-bits. A more sophisticated direct generation method manipulates the target fault list for ATPG by excluding all or part of the faults in some high-test-power blocks (Ahmed et al. 2007a). This results in test cubes with more *X*-bits corresponding to the high-test-power blocks, thus increasing the chances of sufficiently reducing their test power.

Note that direct generation of test cubes generally increases final test vector count. This is because leaving *X*-bits alone without conducting *random fill* significantly reduces fortuitous detections. It has been reported that, for seven industrial circuits ranging in size from 229K to 2.01M gates, test vector count increased by 144.8% on average when *random fill* was not conducted (Remersaro et al. 2006). This problem can be alleviated to some extent by conducing partial *random fill*. For example, a 10% *random fill* limited test vector count increase to 37.5% on average in experiments on the seven industrial circuits (Remersaro et al. 2006).

3.4.1.2 Test Relaxation

X-bits can also be obtained from a set of fully-specified test vectors while preserving one or more of its properties. This approach is called *test relaxation* or *X-identification* (XID). The primary property to be preserved is fault coverage (Sankaralingam and Touba 2002; El-Maleh and Al-Suwaiyan 2002; Miyase and Kajihara 2004; El-Maleh and Al-Utaibi 2004). Other properties such as sensitized paths for delay fault detection can also be preserved (Wen et al. 2007b).

The concept of test relaxation is illustrated in Fig. 3.13. First, dynamic compaction and *random fill* are fully utilized to generate a compact initial test set, with high fault coverage (and usually high test power as well). Next, test relaxation is conducted to turn some bits into *X*-bits. Then, the *X*-bits are filled for test power reduction. This causes the resulting final test set to have lower test power but the same fault coverage and the same test vector count as the initial test set.

Compared with direct generation of test cubes, test relaxation does not increase final test vector count. In addition, test relaxation has stronger control over *X*-bit distribution among test cubes (Remersaro et al. 2007; Miyase et al. 2008). This means that test cubes generated by test relaxation are usually of higher quality since

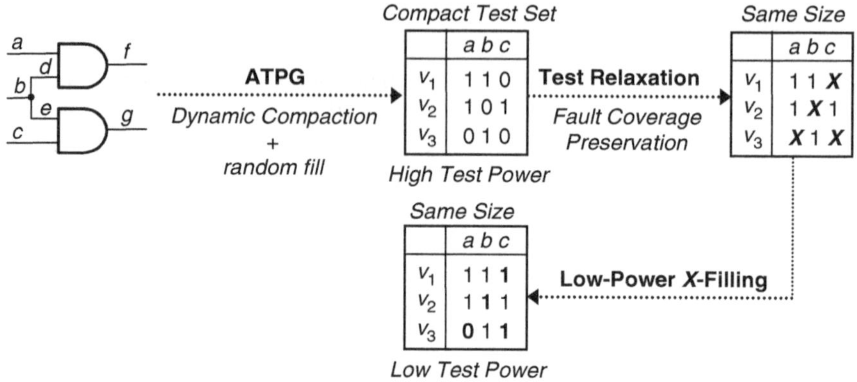

Fig. 3.13 Concept of test relaxation

test power can be more efficiently reduced by relaxing proper bits to *X*-bits and filling them with proper logic values. On the other hand, test relaxation is algorithm-based, and thus more time-consuming than direct test cube generation.

In the remaining part of this subsection, basic techniques for test relaxation are first described, an advanced technique for preserving sensitized paths is then introduced, and finally techniques for controlling *X*-bit distribution are discussed.

Basic Techniques

The simplest test relaxation technique is called *bit-stripping* (Sankaralingam and Touba 2002). It changes one bit in a test vector into an *X*-bit and conducts 3-value fault simulation to determine whether all the faults that are only detected by the test vector remain detected. If so, the bit is kept as an *X*-bit; otherwise it is restored to its original value. This process is repeated for all the bits in the test vector. Obviously, bit-stripping can turn a number of bits into *X*-bits without reducing the overall fault coverage of the initial test set.

Bit-stripping can be time-consuming since it processes one bit at a time. To solve this problem, a test relaxation technique attempts to identify multiple *X*-bits at once (El-Maleh and Al-Suwaiyan 2002; El-Maleh and Al-Utaibi 2004). The basic idea of this technique is to identify faults newly detected by a test vector, and mark all the lines whose values are required for the faults to be detected. All bits for the unmarked input lines are turned into *X*-bits since they do not affect fault detection.

X-identification (XID) is another test relaxation technique capable of identifying multiple *X*-bits simultaneously (Miyase and Kajihara 2004). Its basic procedure is shown in Fig. 3.14. At Step-1, the *essential faults* of an initial test vector v_i (i.e., the faults that are detected only by v_i) are identified via 2-value fault simulation. At Step-2, implication and justification techniques are applied to identify the necessary input bits from internal values for activating and propagating the essential faults. An

```
Procedure XID (C, T)
C: circuit model;
T: initial test vector set;
{
  /* Pass-1 */
  for each test vector v_i in T {
    EF = find_essential_fault (v_i);                 /* Step-1 */
    initial_c_i = create_test_cube (EF);             /* Step-2 */
    3_valued_fault_simulation (initial_c_i);         /* Step-3 */
  }
  /* Pass-2 */
  for each test vector t_i in T {
    UF = find_undetected_fault (v_i, c_i);           /* Step-4 */
    final_c_i = adjust_test_cube (UF);               /* Step-5 */
    3_valued_fault_simulation (final_c_i);           /* Step-6 */
  }
}
```

Fig. 3.14 Basic procedure of *X*-identification (*XID*)

initial test cube, $initial\_c_i$, is obtained by turning all but the necessary input bits into *X*-bits. At Step-3, 3-value fault simulation is conducted to identify all faults detected by $initial\_c_i$. Note that only essential faults are guaranteed to be detected in Pass-1, so Pass-2 is needed to guarantee that all nonessential faults are also detected. At Step-4, all nonessential faults that are detected by v_i but not by $initial\_c_i$ are identified. At Step-5, implication and justification techniques are applied to identify necessary input bits from internal values for activating and propagating these nonessential faults. If an identified input bit is an *X*-bit, its original value is restored. This adjustment results in a final test cube $final\_c_i$. At Step-6, 3-value fault simulation is conducted to identify all faults detected by $final\_c_i$.

Even for a highly compact initial test set, the XID procedure can usually identify over 70% of its total bits as *X*-bits. For example, for a 2M-gate circuit with 6,748 transition delay test vectors, XID identified 96% of all bits as *X*-bits.

Sensitization-Path-Keeping Test Relaxation

It has been noted that the longest sensitization path (the combination of activation and propagation paths) for a transition delay fault detected by a test vector, called a *characteristic path*, is important since its length determines the test vector's capability of detecting small-delay defects (Sato et al. 2005). Therefore, in test relaxation for a transition delay test set, it is preferable to keep the characteristic path of each detected fault, in addition to avoiding any fault coverage loss.

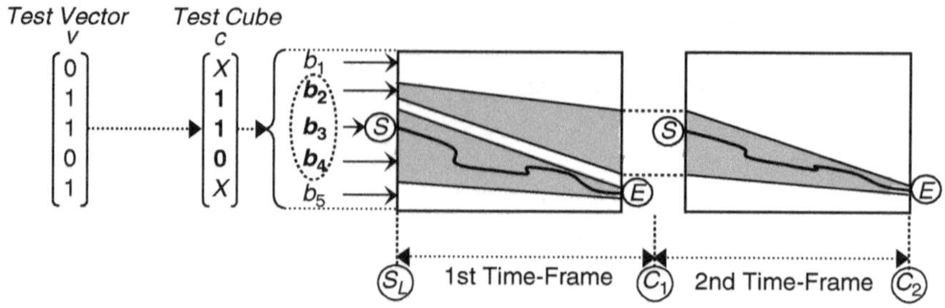

Fig. 3.15 Sensitization-path-keeping X-identification (XID)

Sensitization-path-keeping test relaxation can be accomplished with a simple but powerful cone-analysis-based technique (Wen et al. 2007b). The example shown in Fig. 3.15 has one characteristic path P, which starts from S and ends at E in the two-time-frame circuit model for LOC-based at-speed scan testing. Cone analysis is conducted from the end-point (E) of P in both time-frames in order to identify the inputs that could potentially affect the sensitization of P. In this case, only b_2, b_3, and b_4 are capable of affecting P in terms of sensitization. Therefore, the first and fifth bits in the test vector v can be turned into X-bits to create a test cube c, without affecting the sensitization state of the characteristic path P.

X-Distribution-Controlled Test Relaxation

The effect of X-filling on a test cube depends on the number of its X-bits. However, XID (Miyase and Kajihara 2004) usually results in an X-bit distribution, in which the first few test cubes have a smaller number of X-bits than last ones. This is because at Step-4 in Fig. 3.14, some X-bits in the initial test cube $initial\_c_i$ have to be restored so as to detect all nonessential faults that are detected by v_i but not by $initial\_c_i$. As shown in Table 3.3b, the first few test cubes tend to target more nonessential faults. This results in unbalanced X-bit distribution across final test cubes. To solve this problem, a technique, called *distribution-controlling X-identification* (DC-XID), distributes nonessential faults in a specific manner (Miyase et al. 2008). Table 3.3c shows an example, in which nonessential faults are evenly distributed across test cubes. This helps create an even X-bit distribution. Generally, DC-XID can be used to create any desired distribution of X-bits across resulting test cubes.

Figure 3.16 shows the comparison results of applying XID and DC-XID to 652 transition delay test vectors of an industrial circuit of 600K gates (Miyase et al. 2008). XID identified less than 60% of all bits as X-bits in the first few test vectors, resulting in an uneven X-bit distribution. DC-XID, on the other hand, resulted in an almost totally even X-bit distribution, with over 90% of all bits identified as X-bits on average. Note that other methods of distributing nonessential faults, such as matching the LSA distribution across initial test vectors, can also be used.

Table 3.3 Control of nonessential fault distribution in XID

Test cubes	Essential faults	Nonessential faults				
(a) Detectable faults						
c_1	f_1	f_5	f_6	f_7	f_8	f_{10}
c_2	f_2	f_4	f_5	f_7	f_9	f_{10}
c_3	f_3	f_4	f_5	f_6	f_8	f_9
(b) Faults targeted by XID						
c_1	f_1	f_5	f_6	f_7	f_8	f_{10}
c_2	f_2	f_4	f_9			
c_3	f_3					
(c) Faults targeted by DC-XID						
c_1	f_1	f_6	f_{10}			
c_2	f_2	f_4	f_7	f_9		
c_3	f_3	f_5	f_8			

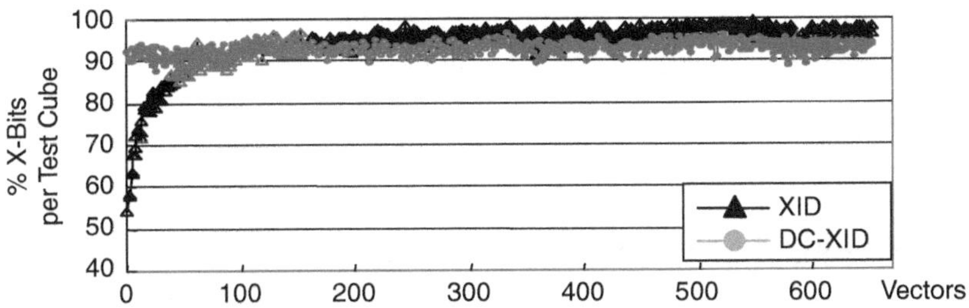

Fig. 3.16 Effect of controlling *X*-bit distribution in test relaxation

3.4.2 Low-Shift-Power X-Filling

Shift transitions in scan chains, together with the resulting switching activity in the combinational logic, occur over a series of shift clock pulses. The resulting IR-drop may cause timing failures in scan chains, especially in high-speed scan shift (Yoshida and Watari 2003). More significantly, the accumulation of excessive shift switching activity may cause excessive heat dissipation (Zorian 1993).

Since shift transitions in scan chains have a good correlation with the switching activity throughout the entire circuit (Sankaralingam et al. 2000), low-shift-power *X*-filling basically attempts to reduce shift transitions. As shown in Fig. 3.17, there are two types of shift transitions: *shift-in transitions* (due to the shifting-in of the next test vector) and *shift-out transitions* (due to the shifting-out of the test response for the previous test vector). Shift-in transitions can be readily reduced by properly filling *X*-bits in test cubes, but it is relatively more difficult to reduce shift-out transitions as they also depend on the combinational logic. Typical *X*-filling methods for reducing shift-in, shift-out, and the total shift power are described below.

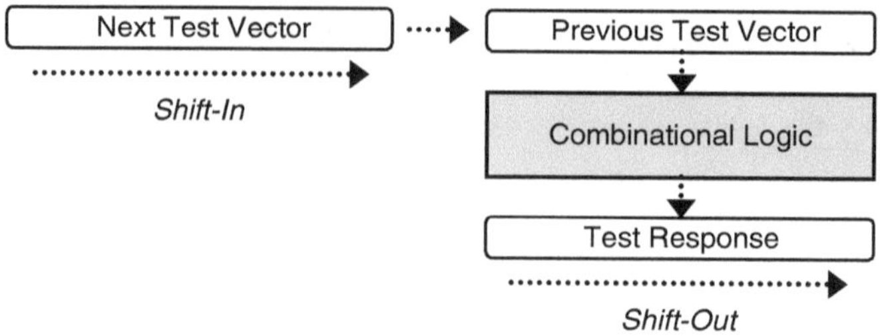

Fig. 3.17 Shift-in transitions and shift-out transitions

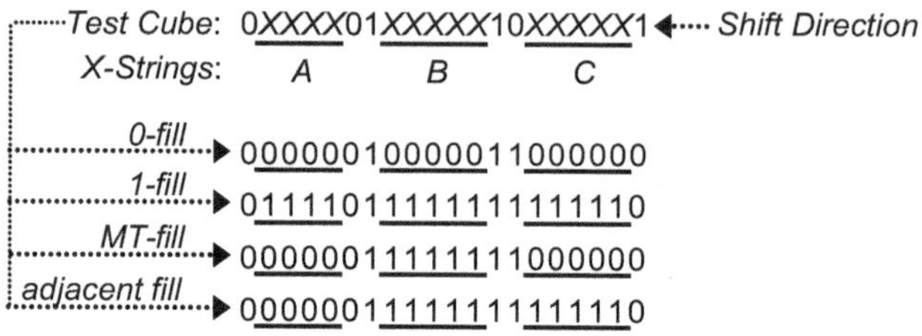

Fig. 3.18 *X*-Filling for shift-in power reduction

3.4.2.1 Shift-In Power Reduction

A test cube can be viewed as having several groups of consecutive *X*-bits, called *X-strings*, separated by specified bits. For example, Fig. 3.18 shows a test cube with three *X*-strings, *A, B,* and *C*. Clearly, it is preferable to fill all of the *X*-bits in an *X*-string with the same logic value in order to reduce shift-in transitions. Various *X*-filling methods are available for determining the filling logic value.

Generally, if the *X*-strings in a test cube are long, *0-fill* and *1-fill*, which simply fill all *X*-bits in all *X*-strings with 0 and 1, respectively, are effective in reducing shift-in transitions (Butler et al. 2004). Examples of *0-fill* and *1-fill* are shown in Fig. 3.18. However, shorter and fragmented *X*-strings require more sophisticated *X*-filling methods for shift-in power reduction. These methods determine logic values for *X*-bits based on some specified bit values in the test cube (Wang and Gupta 1997a).

One such method is *minimum transition fill* (MT-fill) (Sankaralingam and Touba 2002). It fills each *X*-string with a logic value determined as follows: If the specified bits on both sides of an *X*-string have the same logic value, the *X*-string is filled with that logic value; if the specified bits on the two sides of an *X*-string have opposite logic values, the *X*-string is filled with an arbitrary logic value. Fig. 3.18 shows an example where 0 is selected to fill the *X*-string *C*.

Another method, which is similar to *MT-fill*, always fills an *X*-string with the logic value of the last specified bit in the shift direction. This method is called *adjacent* fill (Butler et al. 2004). Figure 3.18 shows an example, where 1 is used to fill the *X*-string *C*. Note that *adjacent fill* is also referred to as *repeat fill*.

Some low-shift-in-power X-filling methods may also help reduce shift-out and even capture power. For example, *0-fill* tends to produce many 0s in an AND-dominated circuit, and consequently many 0s in the test response. This reduces switching activity in shift-out and capture operations (Butler et al. 2004). Modified *adjacent fill* also helps reduce shift-out and capture power (Chandra and Kapur 2008).

3.4.2.2 Shift-Out Power Reduction

X-filling can also be conducted explicitly to reduce shift-out transitions (Sankaralingam and Touba 2002), as illustrated in Fig. 3.19. First, test vector v is turned into test cube c via bit-stripping. Then, controllability values are calculated from inputs to outputs under the initial conditions shown in Fig. 3.19. After that, the candidate output that eliminates shift-out transitions if its value (logic 0 in this example) is flipped and is easiest to control (i.e., has the smallest controllability value) is identified. Finally, line justification is conducted to set the candidate output to the opposite value (logic 1 in this example). If this operation succeeds, the number of shift-out transitions is reduced.

3.4.2.3 Total Shift Power Reduction

X-filling can also be conducted explicitly to reduce both shift-in and shift-out transitions simultaneously. One such method is *minimum transition random X-filling* (MTR-fill) (Song et al. 2008). This method is a simulated annealing process guided by the *total weighted transition metric* (TWTM), which is the sum of the weighted shift-in and shift-out transitions of a test vector and a test response (Sankaralingam et al. 2000). The TWTM for a test vector t_i and a test response r_i is as follows:

$$TWTM(t_i, r_i) = \sum_{j=1}^{L-1} (t_{i,j} \otimes t_{i,j+1}) \times j + \sum_{j=1}^{L-1} (r_{i,j} \otimes r_{i,j+1}) \times j$$

where L is the scan chain length and $t_{i,j}\,(r_{i,j})$ is the jth bit of $t_i\,(r_i)$.

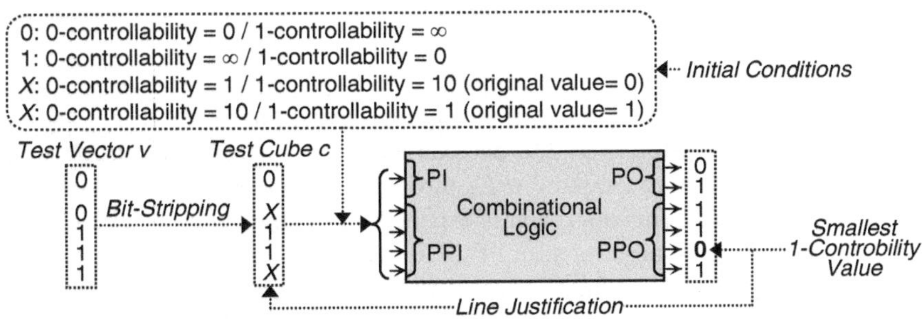

Fig. 3.19 X-filling for shift-out power reduction

The basic flow of *MTR-fill* is as follows: Suppose that c_i is a test cube with n *X*-bits. First, c_i is filled with *MT-fill* to create an initial solution test vector t_i. Suppose that the test response for the previous test vector is r_i. $TWTM(t_i, r_i)$ is calculated as the current cost value. Then, a new test vector t_s is created by flipping the kth bit in the n *X*-bit positions of c_i. After that, $TWTM(t_s, r_i)$ is calculated, and if $TWTM(t_s, r_i) < TWTM(t_i, r_i)$, t_s replaces t_i to become the new solution test vector. This process is repeated until there is no cost reduction in *TWTM*.

3.4.3 Low-Capture-Power X-Filling

In capture mode, the contents of all scan FFs are updated by the outputs of the combinational logic, resulting in capture transitions at the outputs of some scan FFs. Capture transitions, together with subsequent gate transitions in the combinational logic, cause LSA in LOC-based at-speed scan testing, and excessive LSA may lead to test-induced yield loss (Saxena et al. 2003).

Low-capture-power *X*-filling reduces LSA by assigning proper logic values to *X*-bits in test cubes. This can be realized by utilizing one of three types of methods: *FF-oriented*, *node-oriented*, and *critical-area-oriented*, as described below.

3.4.3.1 FF-Oriented *X*-Filling

FF-oriented low-capture-power *X*-filling attempts to reduce capture transitions at scan FFs. This is because capture transition count has a strong correlation with the total switching activity in the circuit (Sankaralingam et al. 2000). Note that a capture transition occurs at the output of a scan FF if (1) the functional input value of the scan FF is different from its current output value; and (2) the clock for the scan FF is applied. This means that capture transitions can be reduced using two approaches: *FF-silencing* and *clock-disabling* (Furukawa et al. 2008).

FF-silencing is to equalize the functional data input value and the current output value of a scan FF. Since this operation causes the value to be loaded into the scan FF to be equivalent to the current value currently stored in it, no capture transition occurs when a capture clock pulse is applied. That is, this approach reduces the number of capture transitions at active scan FFs individually (i.e., one by one) instead of collectively.

Clock-disabling is to disable the capture clock for a group of scan FFs, usually through gated clocks. For example, if a test cube causes an *X* value at the clock gater signal *EN* as shown in Fig. 3.9, one can attempt to justify 0 at the *EN* signal so as to disable the capture clock of SFF_1 through SFF_p, thereby reducing capture transitions collectively. This approach is especially effective when one clock-gater signal controls a large number of scan FFs. However, fewer capturing scan FFs due to clock-disabling may cause fault coverage loss and/or test vector count inflation.

FF-silencing basics, typical FF-silencing methods, and a hybrid method that combines clock-disabling and FF-silencing are described below.

FF-Silencing Basics

For a test cube c and its circuit response $F(c)$, $<c\colon \text{PPI}>$ and $<F(c)\colon \text{PPO}>$ denote the pseudo primary input (PPI) portion of c and the pseudo primary output (PPO) portion of $F(c)$, respectively. The goal of FF-silencing is to minimize the Hamming distance between $<c\colon \text{PPI}>$ and $<F(c)\colon \text{PPO}>$. Here, it is assumed that all scan FFs are active. However, it is easy to extend the following discussions to a case where some scan FFs are inactive due to clock-disabling.

A simple FF-silencing method starts from a fully-specified test vector v and its fault-free test response $F(v)$ (Sankaralingam and Touba 2002). First, bit-stripping is conducted on v to create a test cube c with X-bits. After that, one X-bit whose original value in $<v\colon \text{PPI}>$ is different from the value of the corresponding bit in $<F(v)\colon \text{PPO}>$ is selected. Then, the X-bit is assigned the logic value opposite to its original value, and fault-free simulation is conducted to determine whether the number of capture transitions has decreased. If so, the change is kept; otherwise, it is undone and the original value is restored to the X-bit. This process is repeated for all X-bits in $<c\colon \text{PPI}>$ until a new fully-specified test vector is obtained.

Other FF-silencing methods rely on more systematic handling of X-bits in a test cube and its corresponding circuit response. Given a test cube c and its circuit response $F(c)$, a bit p in $<c\colon \text{PPI}>$ and its corresponding bit q in $<F(c)\colon \text{PPO}>$ correspond to the output and the functional data input of the same scan FF, respectively. $<p, q>$ is called a *bit-pair*. Obviously, a capture transition occurs if $p \neq q$. As illustrated in Table 3.4, bit-pairs can be classified into four types, depending on the possible values (0, 1, X) of p and q (Wen et al. 2005).

Clearly, there is no need to consider Type-A bit-pairs for FF-silencing. As for Type-B bit-pairs, most FF-silencing methods use the *assignment* approach that, for a bit-pair $<p, q>$ where p is an X-bit and q is a logic value, p is assigned the value of q. As for Type-C and Type-D bit-pairs, different FF-silencing methods use different approaches, namely *random*, *justification-based*, *probability-based*, and *justification-probability-based*, to determine logic values for X-bits.

Random FF-Silencing

The typical method based on this approach is *progressive match filling* (PMF-fill) (Li et al. 2005). First, for each Type-B bit-pair in the form of $<X_{\text{PPI}}, \textit{logic\_value}>$, *logic_value* is assigned to X_{PPI}. After that, logic simulation is conducted for the new test cube, which may turn some Type-D bit-pairs into Type-B bit-pairs. Such

Table 3.4 Types of bit-pairs in FF-silencing

		Bit q in $<F(c)\colon \text{PPO}>$	
		0 or 1	X
Bit p in $<c\colon \text{PPI}>$	0 or 1	Type-A	Type-C
	X	Type-B	Type-D

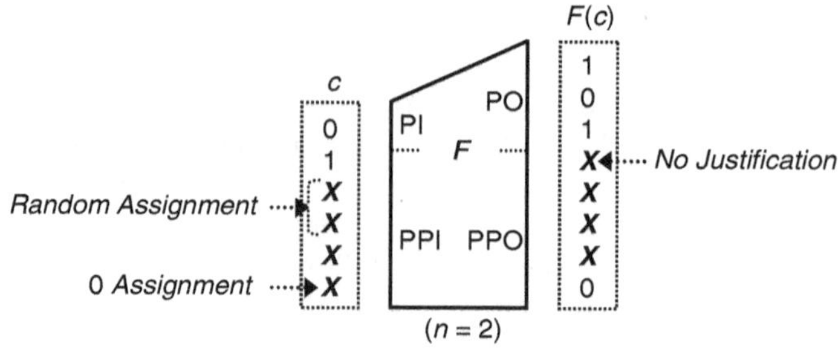

Fig. 3.20 Example of progressive match filling (*PMF-fill*)

X-filling and logic simulation are repeated until only Type-C and Type-D bit-pairs remain. *PMF-fill* does not process Type-C bit-pairs; instead, it randomly selects *n* Type-D bit-pairs in the form of $<X_{PPI}, X_{PPO}>$, and randomly assigns logic values to the PPI *X*-bits in the selected bit-pairs. After this logic value assignment, logic simulation is conducted to check whether there are any new Type-B bit-pairs. This process is repeated until there are no more *X*-bits. An example is shown in Fig. 3.20. The parameter *n* in *PMF-fill* is user-specified. Generally, a smaller *n* leads to a more effective reduction, but at the cost of a longer execution time.

Justification-Based FF-Silencing

The typical method based on this approach is *low-capture-power X-filling* (LCP-fill) (Wen et al. 2005). As in *PMF-fill*, all Type-B bit-pairs are processed by assignment, followed by logic simulation, until only Type-C and Type-D bit-pairs remain. Then, a Type-C bit-pair in the form of $<logic\_value, X_{PPO}>$ is selected, and a justification procedure, like the one in PODEM (Goel 1981), is used in an attempt to set *logic_value* to X_{PPO}. One or more Type-C bit-pairs can be selected in the process based on the difficulty of justifying a logic value on a PPO line and/or the impact or the weight of a capture transition at the corresponding scan FF. After that, logic simulation is conducted for the new test cube, which may turn some Type-D bit-pairs into Type-B or Type-C bit-pairs. This process is repeated until only Type-D bit-pairs, in the form of $<X_{PPI}, X_{PPO}>$, remain. At this time, assignment and justification are first conducted in an attempt to set the same logic value to X_{PPI} and X_{PPO}; if unsuccessful, different logic values are set to the *X*-bits. An example is shown in Fig. 3.21. *LCP-fill* effectively reduces capture reductions via highly deterministic justification, at the cost of a longer execution time.

Probability-Based FF-Silencing

The typical method based on this approach is *preferred fill* (Remersaro et al. 2006). It is a one-pass process, in which all *X*-bits in a test cube are filled at once. First, assignment-based *X*-filling is conducted on all Type-B bit-pairs. Then, signal

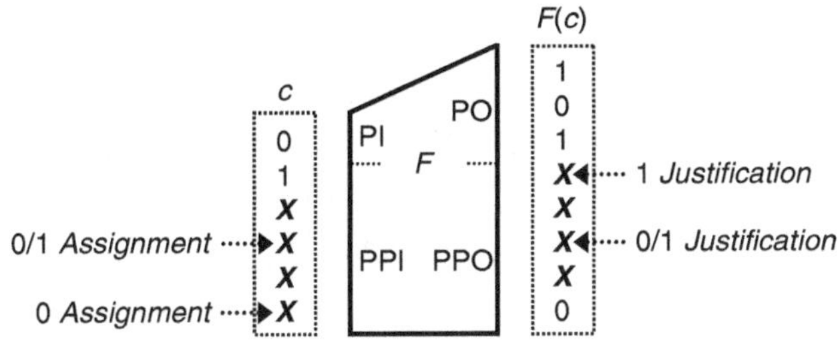

Fig. 3.21 Example of low-capture-power (*LCP*)-fill

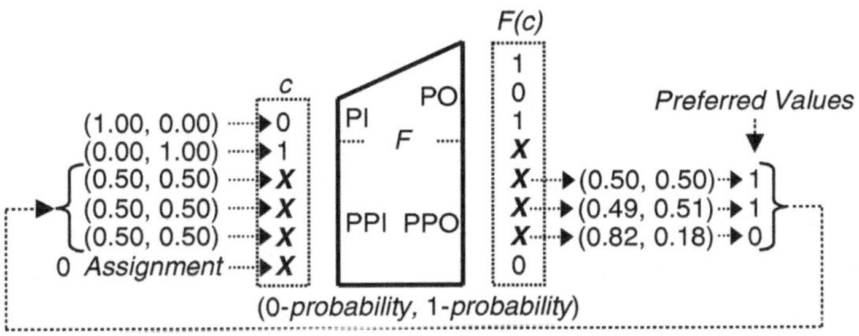

Fig. 3.22 Example of preferred fill

probability calculation is conducted to obtain the 0-probability and 1-probability of all PPO X-bits. For this purpose, 1.0 (0.0) 0-probability and 0.0 (1.0) 1-probability are assumed for each circuit input with logic value 0 (1), 0.5 0-probability and 0.5 1-probability are assumed for each circuit input with X, and probability propagation is conducted (Parker and McCluskey 1975; Papoulis 1991). Based on signal probabilities, the *preferred value pv* of each PPO X-bit is determined in the following manner: *pv* is 0 (1) if the 0-probability of the PPO X-bit is greater (less) than its 1-probability; otherwise, a random logic value is selected as *pv*. After that, each Type-D bit-pair in the form of $<X_{PPI}, X_{PPO}>$ is processed by filling X_{PPI} with the preferred value of X_{PPO}. An example is shown in Fig. 3.22. *preferred fill* is highly scalable due to its one-pass nature.

Justification-Probability-Based FF-Silencing

The typical method based on this approach is *justification-probability-based X-filling* (JP-fill) (Wen et al. 2007b), which attempts to achieve a balance between scalability and effectiveness in low-capture-power X-filling. Type-B and Type-C bit-pairs are processed using assignment and justification, as in *LCP-fill* (Wen et al. 2005). When only Type-D bit-pairs remain, probability-based logic value determination is conducted. However, unlike the one-pass method of *preferred fill* (Remersaro et al. 2006), *JP-fill* uses a multipass procedure. Figure 3.23 shows an example that has three Type-D bit-pairs, $<X_1, X_a>$, $<X_2, X_b>$, and $<X_3, X_c>$.

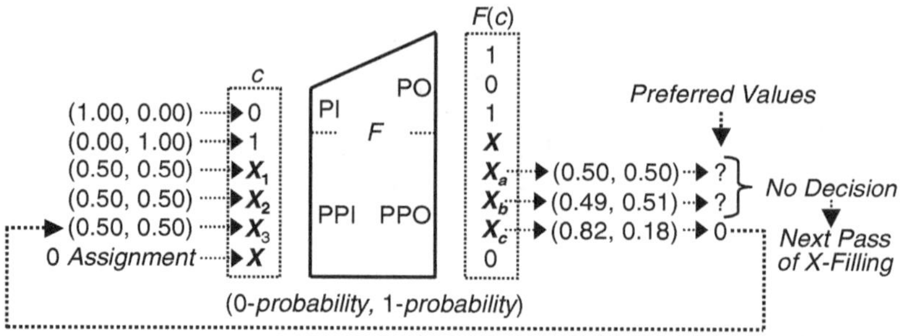

Fig. 3.23 Example of justification-probability-based *X*-filling (*JP-fill*)

As in *preferred fill*, the preferred value of X_c is set to 0 since its 0-probability and 1-probability are significantly different. However, unlike *preferred fill*, *JP-fill* does not determine preferred values for X_a and X_b. This is because the difference between their 0-probability and 1-probability is insignificant, resulting in low confidence in setting preferred values. In the current pass, only X_3 is assigned the preferred value of X_c; logic simulation is then conducted, followed by the next pass of processing. In essence, *JP-fill* uses justification and multiple passes to improve its effectiveness, and probability-based multi-bit logic value determination to improve its scalability.

Combination of Clock-Disabling and FF-Silencing

Clock-disabling is a powerful capture-power-reduction approach since it can reduce capture transitions effectively in a collective manner. However, it has two problems. First, fault overage loss and test vector count inflation may occur, especially when clock-disabling is conducted directly using ATPG in test cube generation (Keller et al. 2007; Czysz et al. 2008). Second, clock-disabling cannot reduce capture transitions for scan FFs whose capture clock must be active for the purpose of fault detection. The first problem can be alleviated by first generating a compact initial test set without conducting clock-disabling during ATPG, and then conducting test relaxation to create test cubes with *X*-bits that allow some clocks to be disabled via *X*-filling. The second problem can be alleviated by conducting FF-silencing for the scan FFs driven by active capture clocks. Therefore, a hybrid approach of combining clock-disabling and FF-silencing is needed for *X*-filling.

The typical method based on the hybrid approach is *clock-gating-based test relaxation and X-filling* (CTX-fill) (Furukawa et al. 2008). *CTX-fill* consists of two stages, as shown in Fig. 3.24. The first stage is based on clock-disabling, in which test relaxation is conducted to convert as many active clock control signals (*EN*=1, as shown in Fig. 3.9) as possible into neutral ones (*EN*=*X*) without fault coverage loss. Justification is then conducted to set as many neutral clock control signals into inactive ones (*EN*=0) as possible. Capture transitions are reduced

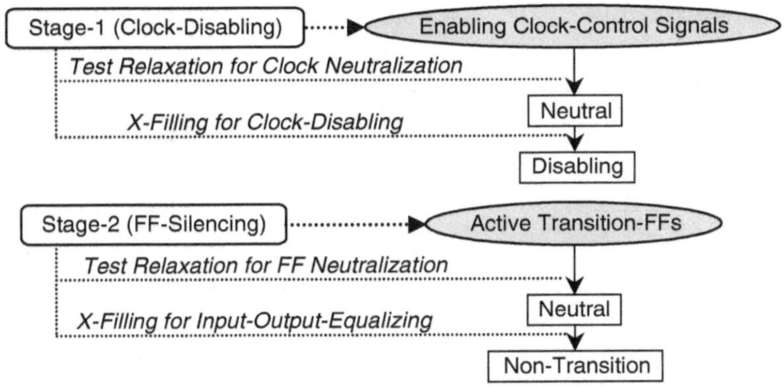

Fig. 3.24 General flow of clock-gating-based test relaxation and *X*-filling (*CTX-fill*)

collectively in the first stage. The second stage is based on FF-silencing, in which constrained test relaxation is conducted to create test cubes with neither fault coverage loss nor any value change at inactivated clock control signals. *JP-fill* is then conducted for the test cubes. Capture transitions are reduced one by one in the second stage.

Combining clock-disabling and FF-silencing in *X*-filling enables greater capture transition reduction than applying either of the methods individually. This hybrid approach is especially useful when the number of *X*-bits available for capture power reduction is limited (such as in compressed scan testing, where *X*-bits are also required for test data compression) (Li et al. 2006; Touba 2006).

3.4.3.2 Node-Oriented *X*-Filling

FF-oriented low-capture-power *X*-filling is indirect for capture power reduction, in the sense that it reduces capture transitions at scan FFs instead of transitions at nodes (including scan FFs and gates in the combinational logic). As described below, this issue can be addressed via node-oriented *X*-filling that generally has a greater success in reducing the switching activity of the entire circuit.

One node-based *X*-filling method uses *X-score* to select a target *X*-bit, and *probabilistic weighted capture transition count* (PWT) to determine a proper logic value for the selected *X*-bit, so as to reduce the switching activity throughout the entire circuit (Wen et al. 2006).

The *X*-score of an *X*-bit is a value reflecting its impact on transitions at nodes. The *X*-score can be calculated as simply the number of nodes structurally reachable from the *X*-bit. More accurate (though also more time-consuming) *X*-score calculation takes into consideration the logic values of specified bits in a test cube and simple logic functions (such as inversion) in the combinational logic. Figure 3.25 shows an example that is based on *set-simulation* (Wen et al. 2006). The *X*-bit with the highest *X*-score is selected as the target *X*-bit in each *X*-filling run.

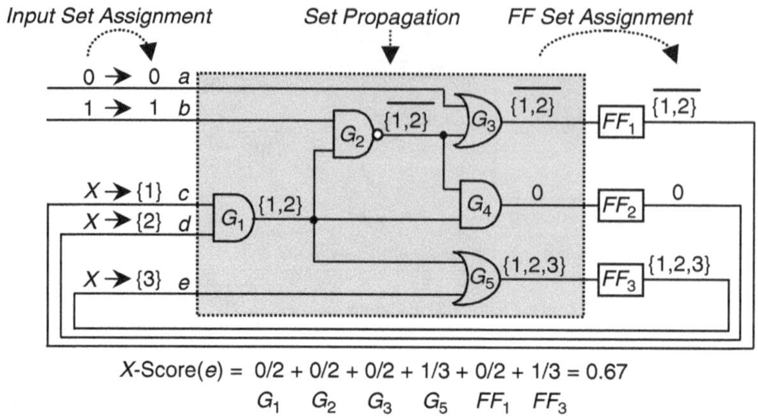

Fig. 3.25 Set-simulation

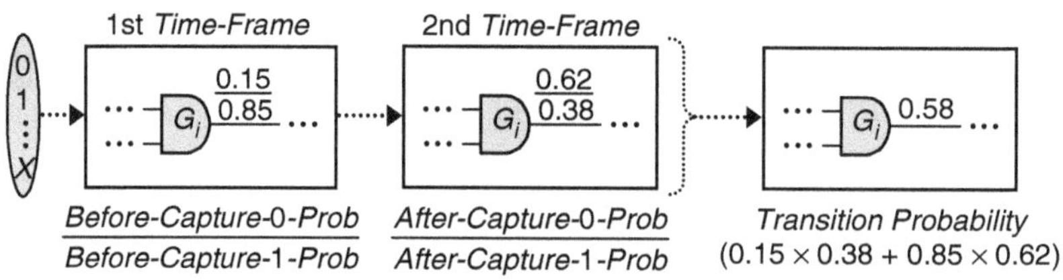

Fig. 3.26 Node probability calculation

The logic value for the target X-bit is determined by comparing the PWT values of the two test cubes obtained by filling the target X-bit with 0 and 1. The PWT of test cube c, denoted by $PWT(c)$, is defined as follows:

$$PWT(c) = \sum_{i=1}^{n} (w_i \times p_i)$$

where n is the number of all nodes in the circuit, w_i is the weight of node i, and p_i is the transition probability at the output of node i. The weight represents the load capacitance of node i. The transition probability of node i can be computed in a manner similar to the one used in *preferred fill* (Remersaro et al. 2006), but it should be computed for two time-frames. An example is shown in Fig. 3.26.

Another node-based X-filling method attempts to minimize the number of node transitions by taking the spatial relationship among state lines (i.e., PPO bits in a test cube) into consideration (Yang and Xu 2008). This method first obtains the potential vector sets and then determines logic values for X-bits that minimize the number of node transitions. In addition, all the PPO X-bits are filled in parallel, resulting in a shorter execution time.

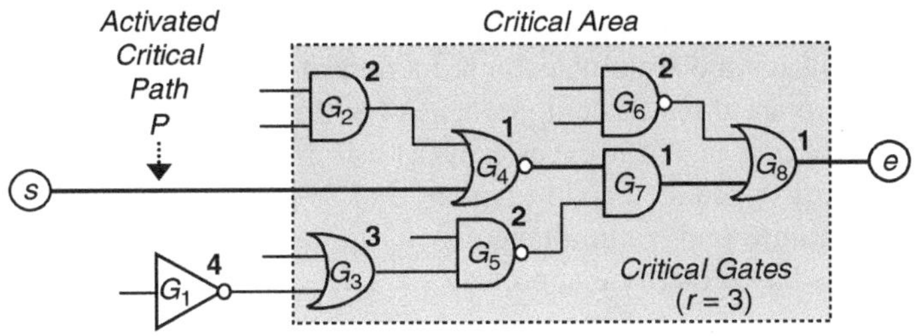

Fig. 3.27 Critical area

3.4.3.3 Critical-Area-Oriented *X*-Filling

The ultimate goal of low-capture-power test generation for LOC-based at-speed scan testing is to guarantee the capture-safety of each test vector *v*, meaning that the LSA caused by *v* does not increase the delay of any path activated by *v* so much that it exceeds the test cycle (Kokrady and Ravikumar 2004; Wen et al. 2007a). Generally, activated critical paths are the most susceptible to the impact of IR-drop caused by LSA. As discussed in Sect. 3.2.3.1, the capture-safety of a test vector is better assessed with the CCT metric (Wen et al. 2007a), which is the weighted transition count using two types of weights: *capacitance weight* (ideally calculated from layout information but often simply set as the number of fanout branches of a node plus 1 in practice) and *distance weight* (calculated using the distance from activated critical paths). CCT provides a good assessment of the impact of LSA on the critical area that is composed of critical nodes whose distance from any activated critical path is within a given radius. A sample critical area is shown in Fig. 3.27, where the distance of a gate from a path is defined as $d+1$ if its output is directly connected to a gate of distance d, where $d \geq 1$ and the distance of any on-path gate is 1.

Obviously, targeting CCT reduction in *X*-filling directly contributes to the improvement of capture-safety. For this purpose, one can first select a target *X*-bit based on its impact on the LSA in the critical area. After that, the CCT values for assigning 0 and 1 to the target *X*-bit are calculated, and 0 (1) is selected to fill the target *X*-bit if the CCT value for 0 (1) is smaller. Note that CCT calculation in *X*-filling is time-consuming, since signal transition probabilities are needed due to the *X*-bits in a test cube (Wen et al. 2007a). As an alternative, a genetic algorithm-based method can be used to find a CCT-minimizing logic assignment for all *X*-bits in a test cube. In this method, no transition probability is needed since only fully-specified test vectors are simulated (Yamato et al. 2008).

3.4.4 Low-Shift-and-Capture-Power X-Filling

Since a scan circuit operates in two modes (*shift* and *capture*), test power includes both shift and capture power, with shift power further including shift-in and shift-out power. Clearly, both shift and capture power need to be reduced to meet safety limits. This motivates the simultaneous reduction of both shift and capture power in X-filling. There are three basic approaches for this purpose: (1) using X-bits to reduce the type of power that is excessive; (2) using some of X-bits to reduce shift power and the rest to reduce capture power; and (3) filling X-bits so as to reduce both shift and capture power simultaneously. Typical low-shift-and-capture-power X-filling methods based on these approaches are described below.

3.4.4.1 Impact-Oriented *X*-Filling

Generally, not all X-bits in a test cube are needed to reduce capture power under a safe limit. In addition, some test vectors resulting from low-shift-power X-filling may not cause excessive capture power. These observations lead to an iterative two-phase X-filling method, called *iFill* (Li et al. 2008a). In the first phase, X-filling is conducted on a test cube to reduce shift power. If the resulting test vector violates the capture power limit, the second phase is executed, in which the result of the previous X-filling is discarded and new X-filling is conducted to reduce capture power. X-filling in both phases repeats two operations, *target X-bit selection* and *logic value determination*, until no X-bits remain in the test cube. Both operations are based on the impact an X-bit has on the type of power to be reduced by X-filling. Note that *iFill* targets both shift-in and shift-out power in shift power reduction.

Target X-bit selection in the first phase (low-shift-power X-filling) of *iFill* is based on *S-impact*. The impact of an X-bit X_i on shift-in power, denoted by S_{in}, can be estimated using its distance to the input of the scan chain (*Scan-In*). This is because the closer an X-bit is to *Scan-In*, the fewer shift-in transitions it can cause. On the other hand, the impact of an X-bit X_i on shift-out power, denoted by S_{out}, can be estimated using the sum of the distances to the output of the scan chain (*Scan-Out*) from the FFs affected by X_i in the test response. For example, the distance to *Scan-In* from X_3 in the test cube shown in Fig. 3.28 is 3. In addition, the FFs affected by X_3 are SFF_{12}, SFF_{13}, and SFF_{15}, and their distances to *Scan-Out* are 5, 4, and 2, respectively. Once S_{in} and S_{out} are obtained, *S-impact* is calculated as $S_{in} + S_{out}$.

In low-shift-power X-filling, the X-bit with the highest *S-impact* is selected as the target X-bit X_i. The logic value for X_i is determined by comparing the *shift transition probability* (STP) values of the test cubes obtained by filling X_i with 0 and 1, denoted by $STP(X_i{=}0)$ and $STP(X_i{=}1)$, respectively. Here, $STP(X_i{=}v) = SITP(X_i{=}v) + SOTP(X_i{=}v)$, where $SITP(X_i{=}v)$ and $SOTP(X_i{=}v)$ are the shift-in transition probability and shift-out transition probability of the test cube obtained by filling X_i with logic value v, respectively. $SITP(X_i{=}v) = p_{in} \times (d_i - 1) + p_{out} \times d_i$, where d_i is the distance-to-scan-input of X_i, while p_{in} and p_{out} are the probabilities that the bits neighboring X_i on the near-scan-input side and the near-scan-output side have logic values different than that of X_i, respectively. For example, if the target X-bit in Fig. 3.28 is X_3, $SITP(X_i{=}1) = 0 \times 2 + 0.5 \times 3 = 1.5$. On the other

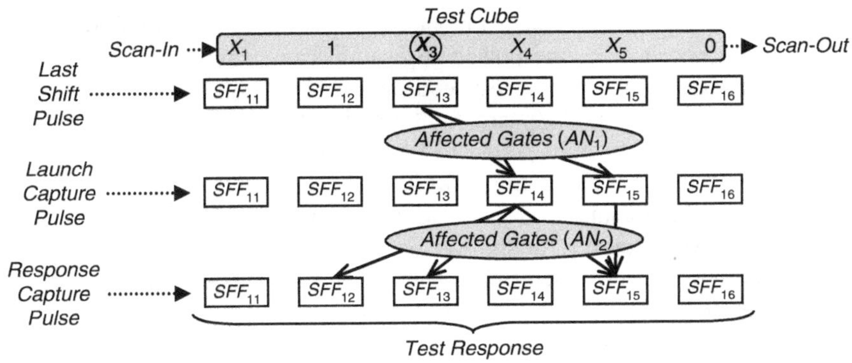

Fig. 3.28 Concept of *iFill*

hand, $SOTP(X_i{=}v) = \sum_{X_j \in A} (p_{in} \times (d_j - 1) + p_{out} \times d_j)$, where A is the set of FFs in the test response affected by X_i, d_j is the distance-to-scan-output of X_j, and p_{in} and p_{out} are the probabilities that the bits neighboring X_j on the near-scan-input side and the near-scan-output side have logic values different than that of X_j, respectively.

Target X-bit selection in the second phase (low-capture-power X-filling) of *iFill* is based on a metric called *C-impact*. The *C-impact* of an X-bit is the total number of FFs and gates that are reachable from the X-bit and have undetermined logic values in the test cycle of LOC-based at-speed testing. As shown in Fig. 3.28, the nodes reachable from X_3 in the test cube are SFF_{14}, SFF_{15}, and all gates in AN_2.

In low-capture-power X-filling, the X-bit with the highest *C-impact* is selected as the target X-bit X_i. The logic value for X_i is determined by comparing the *capture transition probability* (CTP) values of the test cubes obtained by filling X_i with 0 and 1, denoted by $CTP(X_i{=}0)$ and $CTP(X_i{=}1)$, respectively. $CTP(X_i{=}v)$ is the sum of transition probabilities at the nodes reachable from X_i for the test cube obtained by filling X_i with v in the test cycle of LOC-based at-speed testing.

3.4.4.2 *X*-Distribution-Controlled Test Relaxation and Hybrid *X*-Filling

Hybrid X-filling is a straightforward approach to reducing both shift and capture power by using some X-bits to reduce shift power and the rest to reduce capture power. However, the effect of reducing each type of scan test power may not be sufficient if the number of X-bits available for each purpose is too small. To address this issue, a low-shift-and-capture-power X-filling method combines hybrid X-filling with X-distribution-controlled test relaxation (Remersaro et al. 2007).

The basic idea is to match the percentage of X-bits in a test cube with the capture power profile of the test cube. This method converts an initial test set $T_{initial}$ into a final test set T_{final} with reduced shift and capture power by utilizing a procedure comprised of the following three steps:

Step 1: All test vectors in $T_{initial}$ are placed in decreasing order of WSA for capture power (e.g., the power dissipation caused by LSA in LOC-based scan testing). The new test set is denoted by T_{temp}.

Step 2: Test vectors in T_{temp} are fault-simulated in reverse order with fault dropping. All faults found to be detected by a test vector in the fault simulation are called *target faults* of the test vector.

Step 3: Steps 3a, 3b, and 3c are repeated for each test cube in T_{temp}.

3a: The test vector v_t at the top of T_{temp} is removed and relaxed into a partially-specified test cube c by turning some bits in v_t into X-bits while guaranteeing that all target faults of v_t are still detected by c.

3b: Some of the PPI X-bits in c are randomly selected and filled with preferred values as in *preferred fill* (Remersaro et al. 2006), and the remaining X-bits are filled with *adjacent fill* (Butler et al. 2004). The resulting fully-specified test vector v_f is placed into a new test set T_{final} if it has lower WSA than the original vector v_t; otherwise, v_t is placed into T_{final}.

3c: v_f is fault-simulated if it is placed into T_{final}, and all faults detected by v_f are dropped from the set of the target faults of each vector in T_{temp}. Vectors without any corresponding target fault are deleted from T_{temp}, leading to more compact T_{final}.

The WSA-based vector ordering in Step 1 and reverse fault simulation in Step 2 result in the test set T_{temp}, in which a test vector with higher WSA has a smaller number of target faults. Since fewer target faults for a vector lead to more X-bits in the resulting test cube, a vector with higher WSA will be relaxed into a test cube with more X-bits. Because more X-bits are available, WSA is more likely to be sufficiently reduced. This is illustrated in Fig. 3.29.

In Step 3b, it is possible to fill different proportions of X-bits with *preferred fill* and *adjacent fill* to reduce capture and shift power, respectively. However, from experiments on ISCAS'89 benchmark circuits, it has been found that filling 50% of X-bits with each of the aforementioned X-filling techniques seems to be the most effective way to simultaneously reduce shift and capture power (Remersaro et al. 2007).

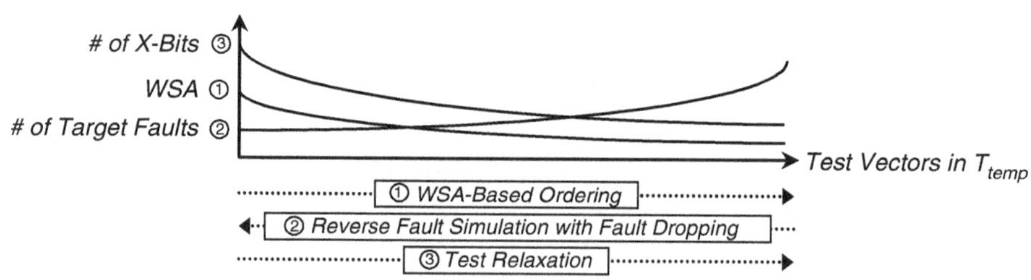

Fig. 3.29 WSA-based ordering and reverse simulation for X-distribution control

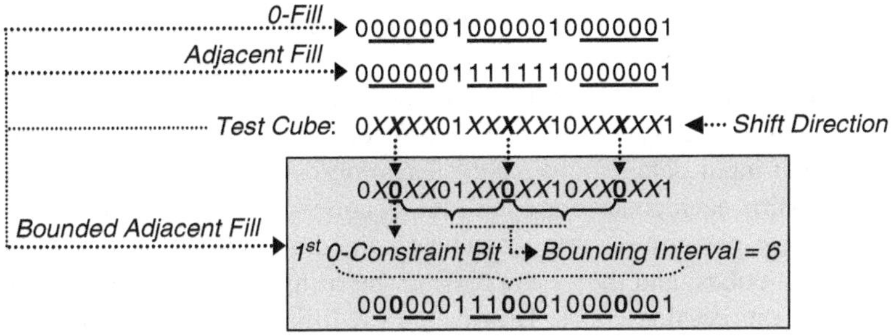

Fig. 3.30 Example of bounded adjacent fill (*BA-fill*)

3.4.4.3 Bounded Adjacent Fill

Adjacent fill, 0-fill, and *1-fill* are the major *X*-filling methods for reducing shift-in power. Although these methods perform similarly in terms of shift-in power reduction, *adjacent fill* is preferable with respect to test data reduction. This is because *0-fill* and *1-fill* greatly reduce the chances of fortuitous fault detection, leading to a larger final test set. However, *0-fill* performs the best with respect to the reduction of shift-out and capture power. This is because *0-fill* tends to result in similar circuit response data, which means reduced shift-out and capture power.

Based on these observations, an *X*-filling method, called *bounded adjacent fill* (BA-fill), attempts to combine the benefits of *adjacent fill* and *0-fill* (Chandra and Kapur 2008). The basic idea is to first constrain or set several *X*-bits in a test cube to 0 and then conduct *adjacent fill*. This operation increases the occurrence of 0 in the resulting fully-specified test vector that helps reduce shift-out and capture power. At the same time, applying *adjacent fill* helps reduce shift-in power. Figure 3.30 shows an example, where the first 0-constraint bit is the third bit in the test cube from the scan input side, and the bounding interval is 6 (i.e., every seventh bit from the third bit in the test cube is set to 0). After that, *adjacent fill* is conducted. The results of *BA-fill, 0-fill*, and *adjacent fill* are also shown in Fig. 3.30.

3.4.5 Low-Power X-Filling for Compressed Scan Testing

Test data volume has been growing due to ever-increasing circuit scales, more fault models to be targeted in ATPG, and the need to improve scan testing's capability to detect small-delay defects. Because of this, *compressed scan testing* is beginning to be adopted for reducing test costs by compressing test data with a *code-based, linear-decompressor-based*, or *broadcast-scan-based* scheme (Touba 2006; Li et al 2006). Typical methods for low-power *X*-filling in a compressed scan testing environment are described below. General power-aware code-based and LFSR-based test compression methods are discussed in Sects. 5.2 and 5.3.

3.4.5.1 *X*-Filling for Code-Based Test Compression

Code-based test compression partitions the original fully-specified test input data into symbols, and each symbol is replaced or encoded with a codeword to form compressed test input data (Touba 2006). Decompression is conducted with a decoder that converts each codeword back into its corresponding symbol.

Generally, low-power test vectors can be obtained by first conducting low-power *X*-filling on test cubes, and then compressing the resulting fully-specified test vectors with data compression codes. However, an *X*-filling technique that is good for test power reduction may be bad for test data compression. Therefore, it is necessary to conduct *X*-filling by taking both test power reduction and test data reduction into consideration. The typical low-shift-power and low-capture-power *X*-filling methods for code-based test compression are described below.

Shift Power Reduction

The *X*-bits in a test cube can be filled with logic values to create a fully-specified test vector, which is then compressed using a data compression code, such as *Golomb code* (Chandra and Chakrabarty 2001a). From the point of view of shift-in power reduction, it is preferable to use *MT-fill* or *adjacent fill* for the test cube to reduce its *weighted transition metric* (WTM) (Sankaralingam et al. 2000). However, these *X*-filling techniques tend to cause difficulty in test data compression, and may even increase final test data volume in some cases. A simple solution to this problem is to use *0-fill*. Using *0-fill* results in long runs of 0s that provide a high test data compression ratio with Golomb code (Chandra and Chakrabarty 2001b). An example is shown in Table 3.5. Another benefit of using *0-fill* is that shift-out transitions are often reduced, especially in AND-type circuits.

Capture Power Reduction

Capture power reduction in code-based test compression can be achieved by *capture-power-aware selective encoding* (Li et al. 2008b), *preferred Huffman symbol-based X-filling* (PHS-fill) (Lin et al. 2008), etc. These methods take test responses into consideration so as to minimize the impact of capture power reduction on test data compression. *Capture-power-aware selective encoding* is based on the

Table 3.5 Impact of *X*-filling on Golomb-code-based test compression

Partially-specified test cube	Fully-specified vector (*adjacent fill*)	Fully-specified vector (*0-fill*)
01XXX10XXX01	011111000001	010001000001
Group size = 4	Golomb code length: 19 *WTM* = 18	Golomb code length: 10 *WTM* = 37

selective encoding scheme (Wang and Chakrabarty 2005), whereas *PHS-fill* is based on Huffman code (Huffman 1952). *PHS-fill* is described below as an example.

PHS-fill attempts to reduce capture transitions in *X*-filling test cubes, and the resulting fully-specified test vectors are encoded with Huffman code. First, three *PHSs* (PHS^1, PHS^2, and PHS^3) are identified for the CUT. This is conducted by obtaining preferred values for all scan FFs (Remersaro et al. 2006), determining a scan FF block size with respect to Huffman coding (e.g., 4) and counting the occurrences of each possible preferred value combination for the scan FF blocks. The top-3 combinations are set as PHS^1, PHS^2, and PHS^3. An example is shown in Fig. 3.31a. *PHS-fill* is applied in two forms: *compatible PHS-fill* and *forced PHS-fill*. Compatible *PHS-fill* is applied in dynamic compaction. Whenever a new test cube is generated, scan FF blocks are compared with PHS^1, PHS^2, and PHS^3 (in that order). If a block is compatible with PHS^i, it is filled with PHS^i. This process simultaneously reduces capture transitions and enhances Huffman coding efficiency. Forced *PHS-fill* is applied after dynamic compaction instead of *random fill*. In this case, the compatibility check is skipped, and each unspecified bit is filled with the value of the corresponding bit in PHS^1. This process focuses on reducing capture transitions. An example is shown in Fig. 3.31b.

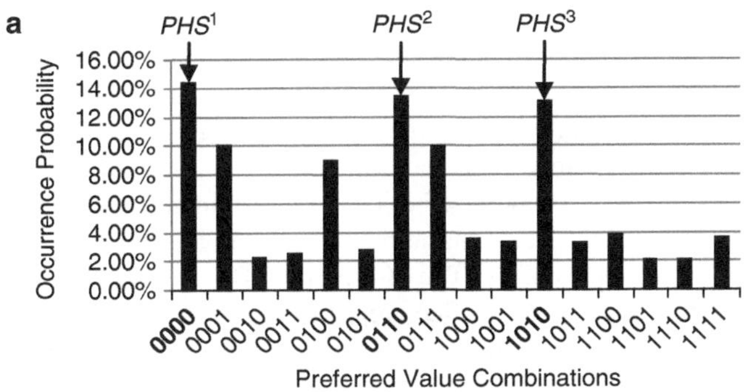

Preferred Huffman symbols

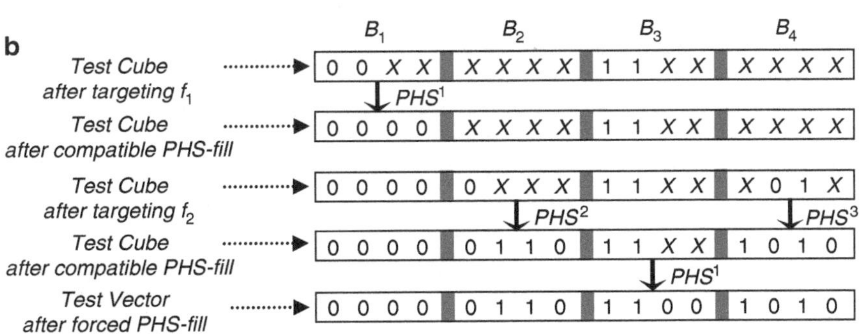

Compatible *PHS-fill* and forced *PHS-fill*

Fig. 3.31 Example of preferred Huffman symbol-based *X*-filling (*PHS-fill*)

3.4.5.2 *X*-Filling for Linear-Decompressor-Based Test Compression

Generally, linear-decompressor-based test compression is capable of achieving a higher compression ratio than other approaches (Touba 2006; Li et al. 2006). As shown in Fig. 3.32, a linear decompressor, which consists of a finite state machine (composed of only XOR gates, wires, and D flip-flops) and a phase shifter, is used to bridge the gap between a small number of external scan input ports and a large number of internal (and shorter) scan chains. A typical example is the *embedded deterministic test* (EDT) scheme (Rajski et al. 2004). Compressed test vectors are generated in two passes. First, an internal test cube is generated for the combinational logic. Then, the compressibility of the test cube is checked by solving a system of linear equations corresponding to the decompressor and the test cube in order to obtain an external compressed test vector for the internal test cube.

Two challenges exist with low-capture-power *X*-filling in linear-decompressor-based test compression. One is *X-bit limitation* (i.e., both test data compression and capture power reduction need *X*-bits), and the other is *compressibility assurance* (i.e., low-capture-power *X*-filling may negate compressibility). *X*-limitation can be alleviated by improving the effectiveness of low-capture-power *X*-filling and utilizing gated clocks (Czysz et al. 2008; Furukawa et al. 2008). On the other hand, compressibility assurance can be addressed by utilizing two techniques from *compressible JP-fill* (CJP-fill) (Wu et al. 2008), namely *X-classification* and *compatible free bit set* (CFBS) identification. *X*-classification is to separate *implied X-bits* (which must actually be assigned certain logic values in order to maintain compressibility) from *free X-bits* (which may have any logic values and do not affect compressibility, provided that they are filled one at a time). Furthermore, in order to improve the efficiency of filling the free *X*-bits, CFBS identification is conducted to identify a set of free *X*-bits that can be filled with any logic values simultaneously without affecting compressibility. The *X*-bits in CFBS are filled using *JP-fill* (Wen et al. 2007b). This way, *CJP-fill* effectively reduce capture power without significantly increasing test vector count.

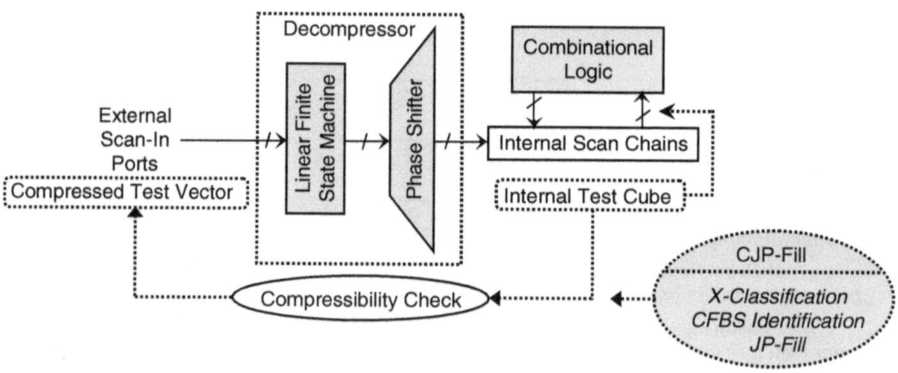

Fig. 3.32 Test generation flow in linear-decompressor-based test compression

3.4.5.3 *X*-Filling in Broadcast-Based Test Compression

In broadcast-based test compression, a broadcaster is placed between external scan-input ports and the inputs of internal scan chains. A broadcaster can be as simple as a set of direct connections (as in *Broadcast Scan* (Lee et al. 1998) and *Illinois Scan* (Hamzaoglu and Patel 1999)) or a piece of combinational circuitry (as in *VirtualScan* (Wang et al. 2004) and *Adaptive Scan* (Sitchinava et al. 2004)). Broadcast-based test compression uses a one-pass ATPG flow. In other words, the constraints posed by the broadcaster are expressed as part of the circuit model, and normal ATPG is used to generate compressed test vectors directly at external scan inputs. Based on this extended circuit model, most of the aforementioned low-capture-power *X*-filling techniques, as well as test relaxation, can be directly applied for broadcast-based test compression, with little or no change.

3.5 Low-Power Test Ordering

During testing, fully-specified test vectors are applied to the CUT. Since the test vector application order also affects test power, properly ordering test vectors can also reduce test-induced switching activity. Several typical low-power test ordering techniques are described below.

3.5.1 Internal-Transition-Based Ordering

Transitions at nodes (scan FFs and gates) in a circuit can be used to guide test vector ordering (Chakravarty and Dabholkar 1994). In this method, information on transitions is represented by a complete directed graph, called a *transition graph* (TG). In a TG, a node represents a test vector, and the weight on an edge from node i to node j is the sum of shift-in and shift-out transitions for all nodes when test vector v_j is applied after test vector v_i. An example is shown in Fig. 3.33, where v_1, v_2, and v_3 are three test vectors. In addition, s and t represent the start and the end of scan testing, respectively. The time complexity of constructing a TG for n test vectors is $O(n^2)$, for which 2-value logic simulation needs to be conducted to compute the number of transitions during scan test operations. Timing-based logic simulation is required if greater accuracy is needed for test power estimation.

With a TG, the problem of finding the test vector order with a minimum test power dissipation can be solved by finding the test vector order with the smallest edge-weight sum. Obviously, this task is equivalent to the NP-complete traveling salesman problem. In practice, a greedy algorithm can be used to find a Hamiltonian path of minimum cost (i.e., the sum of edge-weights) in a TG. Its time complexity is $O(n^2 \log n)$ for a TG of n nodes or test vectors.

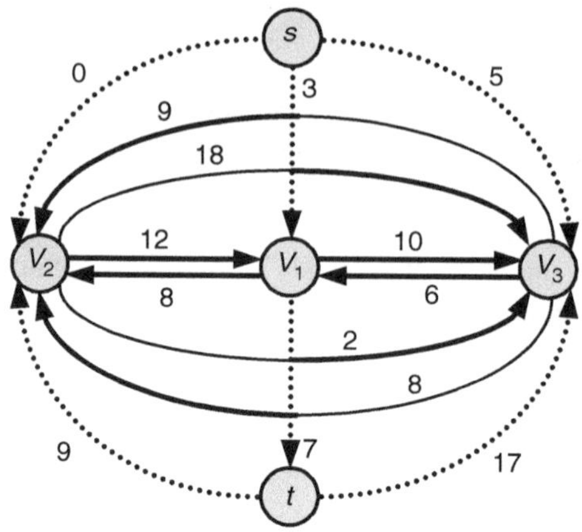

Fig. 3.33 Transition graph

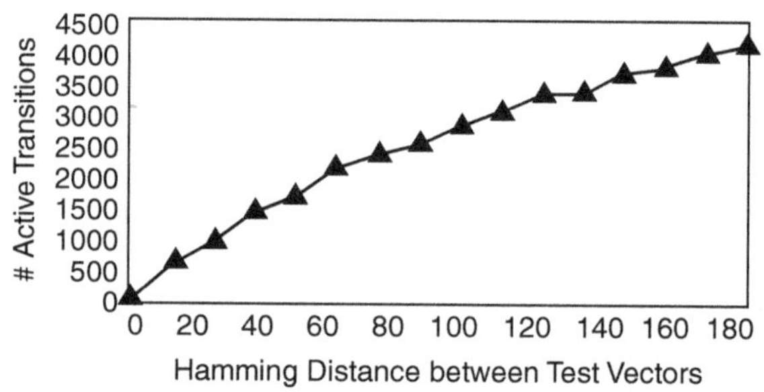

Fig. 3.34 Correlation between Hamming distance and transition activity

3.5.2 Inter-Vector-Hamming-Distance-Based Ordering

Constructing a TG for n test vectors requires $n \times (n - 1)$ logic simulation in order to obtain weights for all edges. This might be too time-consuming when a large number of test vectors are needed due to the circuit scale and/or a high fault coverage requirement. A method for solving this problem uses the Hamming distance between a pair of test vectors, instead of the number of transitions in the entire circuit, to estimate the switching activity caused by applying a pair of test vectors (Girard et al. 1998).

Given two test vectors $v_i = <v_{i_1} v_{i_2} \ldots v_{i_m}>$ and $v_j = <v_{j_1} v_{j_2} \ldots v_{j_m}>$, the Hamming distance between v_i and v_j is $\sum_{k=1}^{m} (v_{i_k} \oplus v_{j_k})$. Experimental results demonstrated a strong correlation between the Hamming distance and the transition activity in the combinational logic. An example is shown in Fig. 3.34. Based on this observation, it is reasonable to use Hamming distances, instead of transitions at nodes, as edge-weights in a TG. This method significantly speeds up TG construction, making it applicable to large circuits and/or large test sets.

3.5.3 Input-Transition-Density-Based Ordering

Hamming-distance-based ordering uses the number of transitions at circuit inputs to estimate circuit switching activity, without taking circuit characteristics into consideration. A more accurate method for estimating circuit switching activity considers not only whether an input transition occurs, but also its impact on circuit switching activity (Girard et al. 1999). The impact of a transition at primary input p_i can be expressed using the *induced activity function*, denoted by Φp_i, as follows:

$$\Phi p_i = \sum_{\forall x} D_{pi}(x) \times Fan(x)$$

where X is the output of a gate, $Dp_i(x)$ is the transition density of X due to a transition at the input p_i, and $Fan(x)$ is the number of fanout branches of X. Φp_i can be expanded as follows:

$$\Phi p_i = \sum_{\forall x} P\left(\frac{\partial val(x)}{\partial p_i}\right) \times f_{clock} \times P_t(p_i) \times Fan(x)$$

where $val(x)$ is the logic function of x, $P(\partial val(x)/\partial p_i)$ is the probability that the Boolean difference of $val(x)$ with respect to p_i evaluates to 1, f_{clock} is the clock frequency, and $P_t(p_i)$ is the transition probability of p_i. $P(\partial val(x)/\partial p_i)$ can be derived from the signal probability of each node using a procedure similar to the one for calculating detection probability (Bardell et al. 1987; Wang and Gupta 1997b). $P_t(p_i)$ can be calculated from the signal probability of p_i, denoted by $P_s(p_i)$, since $P_t(p_i) = 2 \times P_s(p_i) \times (1 - P_s(p_i))$. Note that $P_s(p_i)$ is simply the percentage of test vectors among the total test vectors whose $p_i = 1$.

Once the induced activity function of each input is obtained, a complete undirected graph $G = (V, E)$ can be constructed, with each edge corresponding to test vectors v_a and v_b having a weight defined as

$$weight(v_a, v_b) = \sum_{k=1}^{m} (\Phi p_i \times t_i(v_a, v_b))$$

where Φp_i is the induced activity function of input p_i, and $t_i(v_a, v_b)$ is 1 (0) if v_a and v_b have opposite (identical) logic values at input p_i, and m is the number of primary inputs. An order of test vectors that causes minimal test power can then be determined using heuristics, such as a greedy algorithm (Girard et al. 1998), to find a Hamiltonian path of minimum cost (i.e., the sum of edge-weights) in a TG. Compared with the method that must simulate the entire circuit for every pair of test vectors in order to build a TG (Chakravarty and Dabholkar 1994), the input-transition-density-based method is faster. Compared with the method that uses only Hamming distances as edge-weights (Girard et al. 1998), the input-transition-density-based method takes into account dependencies between internal nodes and circuit inputs, and tends to result in more effective test power reduction.

3.6 Low-Power Memory Test Generation

A system-on-chip circuit generally contains a large number of memory blocks, and each block is usually divided into a number of banks in order to increase access speed and optimize system costs (Cheung and Gupta 1996). In functional operations, only a few memory blocks, and one bank in such a block, are accessed at any time. In testing (especially *built-in self-test* (BIST)), however, concurrently testing multiple memory blocks or multiple banks is highly desirable for the purpose of reducing test time and simplify BIST control circuitry. This results in much higher power during testing than in functional operations. Therefore, power-aware memory test scheduling for multiple blocks and low-power memory test generation for each block are required. Typical methods for low-power memory test generation are described below.

3.6.1 Address Switching Activity Reduction

Low-power *random access memory* (RAM) testing can be realized by modifying a common test algorithm (e.g., *Zero-One*, *Checker Board*, *March B*, *Walking-0-1*, *SNP 2-Group*, etc.) so that test power is reduced (Cheung and Gupta 1996). The idea is to reorder the original test patterns to minimize switching activity on address lines without losing fault coverage. The number of transitions on an address line depends on the address counting method (i.e., the order in which addresses are enumerated during a read or write loop of a memory test algorithm) as well as the address bit position. In binary counting, for example, the LSB (MSB) address line has the largest (smallest) number of transitions. Table 3.6 shows the original and low-power versions of two memory test algorithms, *Zero-One* and *Checker Board*, where W0 (W1) represents writing a 0 (1) to an address location and R0 (R1) represents reading a 0 (1) from an address location. The symbol $\updownarrow$ represents a sequential access in any addressing order (increasing or decreasing) to all memory cells for which binary address counting is originally used. The low-power version uses *single bit change* counting, represented by the symbol $\updownarrow_s$. For example, the counting sequence of a two-bit single bit change code is $00 \rightarrow 01 \rightarrow 11 \rightarrow 10$.

Each low-power version of the memory test algorithm has the same fault coverage and time complexity as the original version, but reduces test power dissipation by a factor of 2 to 16 as a result of the modified addressing sequence.

Table 3.6 Original and low-power memory test algorithms

	Original test	Low-power test
Zero-One	$\updownarrow$(W0); $\updownarrow$(R0); $\updownarrow$(W1); $\updownarrow$(R1)	$\updownarrow_s$ (W0, R0, W1, R1)
Checker Board	$\updownarrow$ (W($1_{odd}/0_{even}$)); $\updownarrow$ (R($1_{odd}/0_{even}$)); $\updownarrow$ (W($0_{odd}/1_{even}$)); $\updownarrow$ (R($0_{odd}/1_{even}$))	$\updownarrow_s$ (W($1_{odd}/0_{even}$), R($1_{odd}/0_{even}$), W($0_{odd}/1_{even}$), R($0_{odd}/1_{even}$))

3.6.2 Precharge Restriction

Precharge circuits in *static random access memory* (SRAM) play the role of precharging and equalizing long and high capacitive bit lines, which is essential to ensure correct SRAM operations. It is well known that precharge circuitry is the principal contributor to power dissipation in SRAM. Experimental results have shown that it may represent up to 70% of overall power dissipation in an SRAM block (Liu and Svensson 1994).

A method for low-power SRAM testing exploits the predictability of the addressing sequence (Dilillo et al. 2006). In functional mode, all precharge circuits must constantly be active, since memory cells are selected randomly. In test mode, however, the access sequence is known and fixed. It is therefore possible to precharge only the columns that are to be selected according to the specific memory test algorithm during memory testing, resulting in reduced precharge activity. To implement this idea, one can use modified precharge control circuitry and exploit the first degree of freedom of March tests (i.e., any specific addressing sequence can be chosen). The modified precharge control logic contains an additional element for each column, as shown in Fig. 3.35. This element consists of one multiplexer and one NAND gate. LP_{test} selects between functional mode and test mode. The addressing sequence is fixed to *word line after word line* in test mode, and precharge activity is restricted to two columns (i.e., the selected column and the one subsequent to it) for each clock cycle. Pr_i is the precharge signal originally used, while CS_i' is the complement of the column selection signal. The multiplexer is for mode selection, and the NAND gate is used to force functional mode for a given column when it is selected for a read/write operation during the test. When LP_{test} is ON, CS_i' of column i drives the precharge of the next column $i+1$. Note that the precharge is active

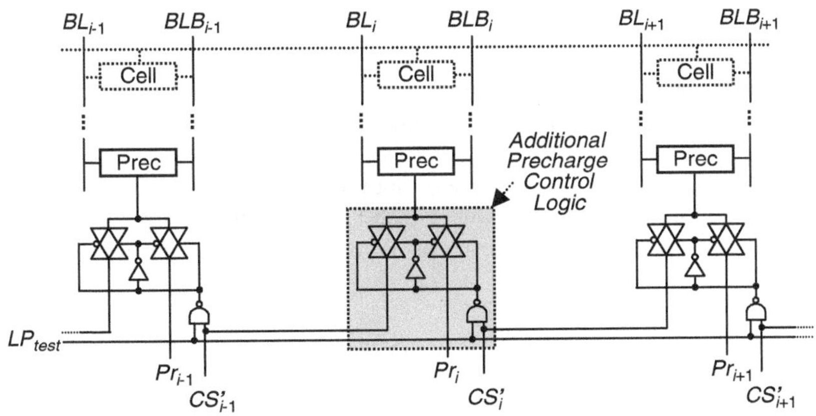

Fig. 3.35 A precharge control logic for low-power static random access memory (*SRAM*) testing

with the input signal at 0. Experiments used to validate this method have shown a significant test power reduction (~50%) with negligible impact on area overhead and memory performance.

3.7 Summary and Conclusions

The challenge of reducing test power adds a new dimension to test pattern generation, which is one of the most important tasks in VLSI testing. Various stages in test pattern generation can be explored for the purpose of reducing various types of test power. The major advantage of low-power test generation is that it causes neither area overhead nor performance degradation. Research in this field has yielded a considerable number of approaches and techniques in terms of low-power ATPG, low-power test compaction, low-power *X*-filling, and low-power test vector ordering for logic circuits under conventional (noncompressed) and advanced (compressed) scan testing, as well as low-power algorithms for memory testing.

This chapter has provided a comprehensive overview of the basic principals and fundamental approaches to low-power test generation. Detailed descriptions of typical low-power test generation methods have also been provided. As previously stated, the objective of this chapter is to help researchers devise more innovative solutions and practitioners build better low-power test generation flows in order to effectively and efficiently solve the problem of excessive test power.

There are four important issues that need to be further addressed in the future with regards to low-power test generation:

1. More effective and efficient flows for low-power test generation need to be developed by using the best combination of individual techniques in low-power test generation and low-power *design for testability* (DFT).
2. Faster and more accurate techniques need to be developed for analyzing the impact of test power instead of test power itself. For capture power, this means researchers should look beyond numerical switching activity and IR-drop to direct investigation of the impact of test power on timing.
3. More sophisticated power reduction techniques capable of focusing on regions that really need test power reduction should be developed.
4. *Low-power* testing needs to evolve into *power-aware* testing that has the following two characteristics: (1) capable of not reducing test power too far below its functional limit; and (2) if possible, capable of increasing test power in order to improve test quality (e.g., in terms of the capability of testing for small-delay defects).

Acknowledgments The authors wish to thank Dr. P. Girard of LIRMM, Prof. N. Nicolici of McMaster University, Prof. K. K. Saluja of University of Wisconsin – Madison, Prof. S. M. Reddy of University of Iowa, Dr. L.-T. Wang of SynTest Technologies, Inc., Prof. M. Tehranipoor of University of Connecticut, Prof. S. Kajihara and Prof. K. Miyase of Kyushu Institute of Technology,

Prof. K. Kinoshita of Osaka Gakuin University, Prof. X. Li and Prof. Y. Hu of Institute of Computing Technology of Chinese Academy of Sciences, Prof. Q. Xu of Chinese University of Hong Kong, Dr. K. Hatayama and Dr. T. Aikyo of STARC, and Prof. J.-L. Huang of National Taiwan University for reviewing this chapter and providing valuable comments.

References

M. Abramovici, M. Breuer, and A. Friedman, *Digital Systems Testing and Testable Design*. New York: Wiley-IEEE Press, revised edition, 1994.

N. Ahmed, M. Tehranipoor, and Y. Jayaram, "Transition Delay Fault Test Pattern Generation Considering Supply Voltage Noise in a SOC Design," in *Proc. of the Design Automation Conf.*, Jun. 2007a, pp. 553–538.

N. Ahmed, M. Tehranipoor, and V. Jayaram, "Supply Voltage Noise Aware ATPG for Transition Delay Faults," in *Proc. of the VLSI Test Symp.*, May 2007b, pp. 179–186.

P. H. Bardell, W. H. McAnney, and J. Savir, *Built-In Test for VLSI: Pseudo-Random Techniques*. London: John Wiley & Sons, 1987.

M. Bushnell and V. Agrawal, *Essentials of Electronic Testing for Digital, Memory & Mixed-Signal VLSI Circuits*. Boston: Springer, first edition, 2000.

K. M. Butler, J. Saxena, T. Fryars, G. Hetherington, A. Jain, and J. Lewis, "Minimizing Power Consumption in Scan Testing: Pattern Generation and DFT Techniques," in *Proc. of the International Test Conf.*, Oct. 2004, pp. 355–364.

S. Chakravarty and V. Dabholkar, "Two Techniques for Minimizing Power Dissipation in Scan Circuits during Test Application," in *Proc. of Asian Test Symp.*, Nov. 1994, 324–329.

A. Chandra and K. Chakrabarty, "System-on-a-Chip Test Data Compression and Decompression Architectures Based on Golomb Codes," *IEEE Trans. on Computer-Aided Design*, vol. 20, no. 3, pp. 355–368, Mar. 2001a.

A. Chandra and K. Chakrabarty, "Combining Low-Power Scan Testing and Test Data Compression for System-on-a-Chip," in *Proc. of the Design Automation Conf.*, Jun. 2001b, pp. 166–169.

A. Chandra and R. Kapur, "Bounded Adjacent Fill for Low Capture Power Scan Testing," in *Proc. of the VLSI Test Symp.*, Apr. 2008, pp. 131–138.

H. Cheung and S. Gupta, "A BIST Methodology for Comprehensive Testing of RAM with Reduced Heat Dissipation," in *Proc. of the International Test Conf.*, Oct. 1996, pp. 22–32.

F. Corno, P. Prinetto, M. Rebaudengo, and M. S. Reorda, "A Test Pattern Generation Methodology for Low Power Consumption," in *Proc. of the VLSI Test Symp.*, Apr. 1998, pp. 453–459.

D. Czysz, M. Kassab, X. Lin, G. Mrugalski, J. Rajski, and J. Tyszer, "Low Power Scan Shift and Capture in the EDT Environment," in *Proc. of the International Test Conf.*, Oct. 2008, Paper 13.2.

V. R. Devanathan, C. P. Ravikumar, and V. Kamakoti, "A Stochastic Pattern Generation and Optimization Framework for Variation-Tolerant, Power-Safe Scan Test," in *Proc. of the International Test Conf.*, Oct. 2007a, Paper 13.1.

V. R. Devanathan, C. P. Ravikumar, and V. Kamakoti "On Power-Profiling and Pattern Generation for Power-Safe Scan Tests," in *Proc. of the Design, Automation, and Test in Europe Conf.*, Apr. 2007b, pp. 534–539.

L. Dilillo, P. Rosinger, P. Girard, and B. M. Al-Hashimi, "Minimizing Test Power in SRAM Through Pre-Charge Activity Reduction," in *Proc. of the Design, Automation and Test in Europe*, Mar. 2006, pp. 1159–1165.

A. H. El-Maleh and A. Al-Suwaiyan, "An Efficient Test Relaxation Technique for Combinational & Full-Scan Sequential Circuits," in *Proc. of the VLSI Test Symp.*, Apr. 2002, pp. 53–59.

A. H. El-Maleh and K. Al-Utaibi, "An Efficient Test Relaxation Technique for Synchronous Sequential Circuits," *IEEE Trans. on Computer-Aided Design*, vol. 23, no. 6, pp. 933–940, June 2004.

H. Furukawa, X. Wen, K. Miyase, Y. Yamato, S. Kajihara, P. Girard, L.-T. Wang, and M. Tehranipoor, "CTX: A Clock-Gating-Based Test Relaxation and *X*-Filling Scheme for Reducing Yield Loss Risk in At-Speed Scan Testing," in *Proc. of the Asian Test Symp.*, Nov. 2008, pp. 397–402.

N. K. Jha and S. K. Gupta, *Testing of Digital Systems*. London: Cambridge University Press, first edition, 2003.

P. Girard, C. Landrault, S. Pravossoudovitch, and D. Severac, "Reducing Power Consumption during Test Application by Test Vector Ordering," in *Proc. of the International Symp. on Circuits and Systems*, May 1998, pp. 296–299.

P. Girard, L. Guiller, C. Landrault, and S. Pravossoudovitch, "A Test Vector Ordering Technique for Switching Activity Reduction during Test Operation," in *Proc. of 9th Great Lakes Symp. on VLSI*, Mar. 1999, pp. 24–27.

P. Girard, "Survey of Low-Power Testing of VLSI Circuits," *IEEE Design & Test of Computers*, vol. 19, no. 3, pp. 82–92, May-June 2002.

P. Girard, X. Wen, and N. A. Touba, *Low-Power Testing* (Chapter 7) in *Advanced SOC Test Architectures – Towards Nanometer Designs*. San Francisco: Morgan Kaufmann, first edition, 2007.

L. H. Goldstein and E. L. Thigpen, "SCOAP: Sandia Controllability/Observability Analysis Program," in *Proc. of the Design Automation Conf.*, June 1980, pp. 190–196.

P. Goel, "An Implicit Enumeration Algorithm to Generate Tests for Combinational Logic Circuits," *IEEE Trans. on Computers*, vol. C-30, no. 3, pp. 215–222, Mar. 1981.

I. Hamzaoglu and J. H. Patel, "Reducing Test Application Time fro Full Scan Embedded Cores," *in Proc. of the International Symp. on Fault-Tolerant Computing*, July 1999, pp. 260–267.

T. Hiraide, K. O. Boateng, H. Konishi, K. Itaya, M. Emori, H. Yamanaka, and T. Mochiyama, "BIST-Aided Scan Test - A New Method for Test Cost Reduction," in *Proc. of VLSI Test Symp.*, May 2003, pp. 359–364.

T.-C. Huang and K.-J. Lee, "An Input Control Technique for Power Reduction in Scan Circuits during Test Application," in *Proc. of the Asian Test Symp.*, Nov. 1999, pp. 315–320.

D. A. Huffman, "A Method for the Construction of Minimum Redundancy Codes," *Proc. of the Institute of Radio Engineers*, vol. 40, no. 9, pp. 1098–1101, Sept. 1952.

S. Kajihara, S. Morishima, A. Takuma, X. Wen, T. Maeda, S. Hamada, and Y. Sato, "A Framework of High-Quality Transition Fault ATPG for Scan Circuits," in *Proc. of the International Test Conf.*, Oct. 2006, Paper 2.1.

B. Keller, T. Jackson, and A. Uzzaman, "A Review of Power Strategies for DFT and ATPG," in *Proc. of the Asian Test Symp.*, Oct. 2007, pp. 213.

B. W. Kernighan and S. Lin, "An Efficient Heuristic Procedure for Partitioning Graphs," *The Bell System Technical Journal*, vol. 49, no. 2, 291–307, Feb. 1970.

A. Kokrady and C. P. Ravikumar, "Fast, Layout-Aware Validation of Test Vectors for Nanometer-Related Timing Failures," in *Proc. of the International Conf. on VLSI Design*, Jan. 2004, pp. 597–602.

L. Lee and M. Tehranipoor, "LS-TDF: Low Switching Transition Delay Fault Test Pattern Generation," in *Proc. of the VLSI Test Symp.*, Apr. 2008, pp. 227–232.

K.-J. Lee, J.-J. Chen, and C.-H. Huang, "Using a Single Input to Support Multiple Scan Chains," *in Proc. of the International Conf. on Computer-Aided Design*, Nov. 1998, pp. 74–78.

L. Lee, S. Narayan, M. Kapralos, and M. Tehranipoor, "Layout-Aware, IR-Drop Tolerant Transition Fault Pattern Generation," in *Proc. of the Design, Automation, and Test in Europe Conf.*, Mar. 2008, pp. 1172–1177.

W. Li, S. M. Reddy, and I. Pomeranz, "On Reducing Peak Current and Power during Test," in *Proc. of IEEE Computer Society Annual Symp. on VLSI*, May 2005, pp. 156–161.

X. Li, K.-J. Lee, and N. A. Touba, Test Compression (Chapter 6) in *VLSI Test Principles and Architectures: Design for Testability*. San Francisco: Morgan Kaufmann, first edition, 2006.

J. Li, Q. Xu, Y. Hu, and X. Li, "iFill: An Impact-Oriented *X*-Filling Method for Shift- and Capture-Power Reduction in At-Speed Scan-Based Testing," in *Proc. of Design, Automation, and Test in Europe*, Mar. 2008a, pp. 1184–1189.

J. Li, X. Liu, Y. Zhang, Y. Hu, X. Li, and Q. Xu, "On Capture Power-Aware Test Data Compression for Scan-Based Testing," in *Proc. of the International Conf. on Computer-Aided Design*, Nov. 2008b, pp. 67–72.

X. Lin, K.-H. Tsai, C. Wang, M. Kassab, J. Rajski, T. Kobayashi, R. Klingenberg, Y. Sato, S. Hamada, and T. Aikyo, "Timing-Aware ATPG for High Quality At-Speed Testing of Small Delay Defects," in *Proc. of the Asian Test Symp.*, Nov. 2006, pp. 139–146.

Y.-T. Lin, M.-F. Wu, and J.-L. Huang, "PHS-Fill: A Low Power Supply Noise Test Pattern Generation Technique for At-Speed Scan Testing in Huffman Coding Test Compression Environment," in *Proc. of the Asian Test Symp.*, Nov. 2008, pp. 391–396.

D. Liu and C. Svensson, "Power Consumption Estimation in CMOS VLSI Chips," *IEEE Journal of Solid-State Circuits*, vol. 29, no. 6, pp. 663–670, June 1994.

K. Miyase and K. Kajihara, "XID: Don't Care Identification of Test Patterns for Combinational Circuits," *IEEE Trans. Computer-Aided Design*, vol. 23, no. 2, pp. 321–326, Feb. 2004.

K. Miyase, K. Noda, H. Ito, K. Hatayama, T. Aikyo, Y. Yamato, H. Furukawa, X. Wen, and S. Kajihara, "Effective IR-Drop Reduction in At-Speed Scan Testing Using Distribution-Controlling *X*-Identification," in *Proc. of the International Conf. on Computer-Aided Design*, Nov. 2008, pp. 52–58.

N. Nicolici and B. M. Al-Hashimi, *Power-Constrained Testing of VLSI Circuits*. Boston: Springer, first edition, 2003.

N. Nicolici and X. Wen, "Embedded Tutorial on Low Power Test," in *Proc. of the European Test Symp.*, May 2007, pp. 202–207.

N. Nicolici, B. M. Al-Hashimi, and A. C. Williams, "Minimization of Power Dissipation during Test Application in Full-Scan Sequential Circuits Using Primary Input Freezing," *IEE Proceedings - Computers and Digital Techniques*, vol. 147, no. 5, pp. 313–322, Sept. 2000.

A. Papoulis, *Probability, Random variables and Stochastic Process*. New York: McGraw-Hill, 3rd edition, 1991.

K. P. Parker and E. J. McCluskey, "Probability Treatment of General Combinational Networks," *IEEE Trans. on Computers*, vol. C-24, no. 6, pp. 668–670, Jun. 1975.

I. Pomeranz, "On the Generation of Scan-Based Test Sets with Reachable States for Testing under Functional Operation Conditions," in *Proc. of the Design Automation Conf.*, Jun. 2004, pp. 928–933.

J. Rajski, J. Tsyzer, M. Kassab, and N. Mukherjee, "Embedded Deterministic Test," *IEEE Trans. on Computer-Aided Design*, vol. 23, no. 5, pp. 776–792, May 2004.

S. Ravi, "Power-Aware Test: Challenges and Solutions," in *Proc. of the International Test Conf.*, Oct. 2007, Lecture 2.2.

S. Ravi, V. R. Devanathan, and R. Parekhji, "Methodology for Low Power Test Pattern Generation Using Activity Threshold Control Logic," in *Proc. of the International Conf. on Computer-Aided Design*, Nov. 2007, pp. 526–529.

C. P. Ravikumar, M. Hirech, and X. Wen, "Test Strategies for Low-Power Devices," *Journal of Low Power Electronics*, vol. 4, no. 2, pp. 127–138, Aug. 2008.

S. Remersaro, X. Lin, Z. Zhang, S. M. Reddy, I. Pomeranz, and J. Rajski, "Preferred Fill: A Scalable Method to Reduce Capture Power for Scan Based Designs," in *Proc. of the International Test Conf.*, Oct. 2006, Paper 32.2.

S. Remersaro, X. Lin, S. M. Reddy, I. Pomeranz, and Y. Rajski, "Low Shift and Capture Power Scan Tests," in *Proc. of the International Conf. on VLSI Design*, Jan. 2007, pp. 793–798.

J. P. Roth, "Diagnosis of Automata Failures: A Calculus and A Method," *IBM Journal Research and Development*, vol. 10, no. 4, pp. 278–291, Apr. 1966.

R. Sankaralingam, R. R. Oruganti, and N. A. Touba, "Static Compaction Techniques to Control Scan Vector Power Dissipation," in *Proc. of the VLSI Test Symp.*, Apr. 2000, pp. 35–40.

R. Sankaralingam and N. A. Touba, "Controlling Peak Power during Scan Testing," in *Proc. of the VLSI Test Symp.*, Apr. 2002, pp. 153–159.

Y. Sato, S. Hamada, T. Maeda, A. Takatori, Y. Nozuyama, and S. Kajihara, "Invisible Delay Quality - SDQM Model Lights Up What Could Not Be Seen," in *Proc. of the International Test Conf.*, Nov. 2005, Paper 47.1.

S. Savir and S. Patil, "On Broad-Side Delay Test," in *Proc. of the VLSI Test Symp.*, Apr. 1994, pp. 284–290.

J. Saxena, K. Butler, V. Jayaram, and S. Hundu, "A Case Study of IR-Drop in Structured At-Speed Testing," in *Proc. of the International Test Conf.*, Sept. 2003, pp. 1098–1104.

N. Sitchinava, S. Samaranayake, R. Kapur, E. Gizdarski, F. Neuveux, and T. W. Williams, "Changing the Scan Enable during Scan Shift," *in Proc. of the VLSI Test Symp.*, Apr. 2004, pp. 73–78.

D.-S. Song, J.-H. Ahn, T.-J. Kim, and S.-H. Kang, "MTR-Fill: A Simulated Annealing-Based *X*-Filling Technique to Reduce Test Power Dissipation for Scan-Based Designs," *IEICE Trans. on Information & System, vol. E91-D, no. 4*, pp. 1197–1200, Apr. 2008.

N. A. Touba, "Survey of Test Vector Compression Techniques," *IEEE Design and Test of Computers*, vol. 23, no. 6, pp. 294–303, Apr. 2006.

S. Wang and W. Wei, "A Technique to Reduce Peak Current and Average Power Dissipation in Scan Designs by Limited Capture," in *Proc. of the Asian and South Pacific Design Automation Conf.*, Jan. 2005, pp. 810–816.

S. Wang and S. K. Gupta, "ATPG for Heat Dissipation Minimization during Test Application," in *Proc. of the International Test Conf.*, Oct. 1994, pp. 250–258.

S. Wang and S. K. Gupta, "ATPG for Heat Dissipation Minimization during Scan Testing," in *Proc. of the Design Automation Conf.*, Jun. 1997a, pp. 614–619.

S. Wang and S. Gupta, "DS-LFSR: A New BIST TPG for Low Heat Dissipation," in *Proc. of the International Test Conf.*, Nov. 1997b, pp. 848–857.

S. Wang and S. K. Gupta, "ATPG for Heat Dissipation Minimization during Test Application," *IEEE Trans. on Computers*, vol. 47, no. 2, pp. 256–262, Feb. 1994.

L.-T. Wang, X. Wen, H. Furukawa, F. Hsu, S. Lin, S. Tsai, K. S. Abdel-Hafez, and S. Wu, "VirtualScan: A New Compressed Scan Technology for Test Cost Reduction," in *Proc. of the International Test Conf.*, Oct. 2004, pp. 916–925.

J. Wang, X. Lu, W. Qiu, Z. Yue, S. Fancler, W. Shi, and D. M. H. Walker, "Static Compaction of Delay Tests Considering Power Supply Noise," in *Proc. of the VLSI Test Symp.*, May 2005a, pp. 235–240.

J. Wang, Z. Yue, X. Lu, W. Qiu, W. Shi, and D. M. H. Walker, "A Vector-Based Approach for Power Supply Noise Analysis in Test Compaction," in *Proc. of the International Test Conf.*, Oct. 2005b, Paper 22.2.

L.-T. Wang, C.-W. Wu, and X. Wen, editors, *VLSI Test Principles and Architectures: Design for Testability*. San Francisco: Morgan Kaufmann, first edition, 2006a.

J. Wang, D. M. H Walker, A. Majhi, B. Kruseman, G. Gronthoud, L. E. Villagra, P. van de Wiel, and S. Eichenberger, "Power Supply Noise in Delay Testing," in *Proc. of the International Test Conf.*, Oct. 2006b, pp. 1–10.

Z. Wang and K. Chakrabarty, "Test Data Compression for IP Embedded Cores Using Selective Encoding of Scan Slices," in *Proc. of the International Test Conf.*, Nov. 2005, pp. 581–590.

X. Wen, Y. Yamashita, K. Kajihara, L.-T. Wang, K. K. Saluja, and K. Kinoshita, "On Low-Capture-Power Test Generation for Scan Testing," in *Proc. of the VLSI Test Symp.*, May 2005, pp. 265–270.

X. Wen, S. Kajihara, K. Miyase, T. Suzuki, K. K. Saluja, L.-T. Wang, K. S. Abdel-Hafez, and K. Kinoshita, "A New ATPG Method for Efficient Capture Power Reduction during Scan Testing," in *Proc. of the VLSI Test Symp.*, May 2006, pp. 58–63.

X. Wen, K. Miyase, T. Suzuki, S. Kajihara, Y. Ohsumi, and K. K. Saluja, "Critical-Path-Aware *X*-Filling for Effective IR-Drop Reduction in At-Speed Scan Testing," in *Proc. of the Design Automation Conf.*, Jun. 2007a, pp. 527–532.

X. Wen, K. Miyase, S. Kajihara, T. Suzuki, Y. Yamato, P. Girard, Y. Ohsumi, and L.-T. Wang, "A Novel Scheme to Reduce Power Supply Noise for High-Quality At-Speed Scan Testing," in *Proc. of the International Test Conf.*, Oct. 2007b, Paper 25.1.

X. Wen, K. Miyase, T. Suzuki, S. Kajihara, L.-T Wang, K. K. Saluja, and K. Kinoshita, "Low Capture Switching Activity Test Generation for Reducing IR-Drop in At-Speed Scan Testing," *Journal of Electronic Testing: Theory and Applications, Special Issue on Low Power Testing*, vol. 24, no. 4, pp. 379–391, Aug. 2008a.

X. Wen, K. Miyase, S. Kajihara, H. Furukawa, Y. Yamato, A. Takashima, K. Noda, H. Ito, K. Hatayama, T. Aikyo, and K. K. Saluja, "A Capture-Safe Test Generation Scheme for At-Speed Scan Testing," in *Proc. of the European Test Symp.*, May 2008b, pp. 55–60.

P. Wohl, J. A. Waicukauski, S. Patel, and M. B. Amin, "Efficient Compression and Application of Deterministic Patterns in a Logic BIST Architecture," in *Proc. of the Design Automation Conf.*, Jun. 2003, pp. 566–569.

M.-F. Wu, J.-L. Huang, X. Wen, and K. Miyase, "Reducing Power Supply Noise in Linear-Decompressor-Based Test Data Compression Environment for At-Speed Scan Testing," in *Proc. of the International Test Conf.*, Oct. 2008, Paper 13.1.

J.-L. Yang and Q. Xu, "State-Sensitive *X*-Filling Scheme for Scan Capture Power Reduction," *IEEE Trans. on Computer-Aided Design of Integrated Circuits & Systems*, vol. 27, no. 7, pp. 1338–1343, July 2008.

M. Yilmaz, K. Chakrabarty, and M. Tehranipoor, "Interconnect-Aware and Layout-Oriented Test-Pattern Selection for Small-Delay Defects," in *Proc. of the International Test Conf.*, Oct. 2008, Paper 28.3.

Y. Yamato, X. Wen, K. Miyase, H. Furukawa, and S. Kajihara, "GA-Based *X*-Filling for Reducing Launch Switching Activity in At-Speed Scan Testing," in *Digest of IEEE Workshop on Defect and Data Driven Testing*, Oct. 2008.

T. Yoshida and M. Watari, "A New Approach for Low Power Scan Testing," in *Proc. of the International Test Conf.*, Sept. 2003, pp. 480–487.

Y. Zorian, "A Distributed BIST Control Scheme for Complex VLSI Devices," in *Proc. of the VLSI Test Symp.*, Apr. 1993, pp. 4–9.

Chapter 4
Power-Aware Design-for-Test

Hans-Joachim Wunderlich and Christian G. Zoellin

Abstract This chapter describes Design-for-Test (DfT) techniques that allow for controlling the power consumption and reduce the overall energy consumed during a test. While some of the techniques described elsewhere in this book may also involve special DfT, the topics discussed here are orthogonal to those techniques and may be implemented independently.

4.1 Introduction

The focus of this chapter is on techniques for circuits that implement scan design to improve testability. This applies to all current VLSI designs. The first part of the chapter deals with the design of the scan cells. Here, unnecessary switching activity is avoided by preventing the scan cells to switch during scan. This is achieved by gating either the functional output of a scan cell during shifting or by clock gating of the scan cell. Through careful test planning, clock gating can be employed to reduce test power without impacting fault coverage. The second part of the chapter deals with the scan paths in the circuit. Here, the segmentation of the scan path reduces the test power without increasing test time. Special clustering and ordering of the scan cells improves the effectiveness of power reduction techniques based on test planning and test generation. Finally, circuit partitioning techniques are the basis for test-scheduling methods. Three partitioning techniques are discussed. Circuits with parallel scan chains may be partitioned using gating of the scan clocks. In core-based designs, the test wrappers provide the DfT to partition the circuit effectively. Combinational logic may be partitioned at the gate level.

H.-J. Wunderlich (✉) and C.G. Zoellin
University of Stuttgart, Stuttgart, Germany
e-mail: wu@informatik.uni-stuttgart.de

P. Girard et al. (eds.), *Power-Aware Testing and Test Strategies for Low Power Devices*,
DOI 10.1007/978-1-4419-0928-2_4,

4.2 Power Consumption in Scan Design

This section discusses the power consumption in circuit designs that implement one or more scan paths. The phases of a scan test and their implications on power are described, so that the techniques described in the rest of this chapter can be evaluated. The scan test consists of shifting, launch, and capture cycles and many techniques that only reduce the power consumption for a subset of these three phases.

4.2.1 Power Consumption of the Circuit Under Test

Power consumption is categorized into static and dynamic power consumption. Dynamic power is consumed by the movement of charge whenever a circuit node incurs a switching event (i.e., a logic transition $0 \rightarrow 1$ or $1 \rightarrow 0$). Figure 4.1 outlines the typical waveform of the current I(t) during the clock cycles of a synchronous sequential circuit.

In synchronous sequential circuits, the memory elements update their state at a well-defined point in time. After the state of the memory elements has changed, the associated switching events propagate through the combinational logic gates. The gates at the end of the longest circuit paths are the last gates to receive switching events. At the same time, the longest paths usually determine the clock cycle. In Fig. 4.1, the circuit contains edge-triggered memory elements (e.g., flip-flops), so the highest current during the clock cycle is typically encountered at the rising clock edge. Subsequently, the switching events propagate through the combinational logic and the current decreases. The clock network itself also includes significant capacitance, and both clock edges contribute to the dynamic power consumption as well.

The peak single-cycle power is the maximum power consumed during a single clock-cycle, that is, when the circuit makes a transition from a state s_1 to a state s_2. Iterative (unrolled) representations of a sequential circuit are a common method to visualize the sequential behavior. Figure 4.2 shows a sequential circuit that makes a transition from state s_1 with input vector v_1 to state s_2 with input vector v_2.

If the peak single-cycle power exceeds a certain threshold power, the circuit can be subject to IR-drop. This may result in erroneous behavior of otherwise good

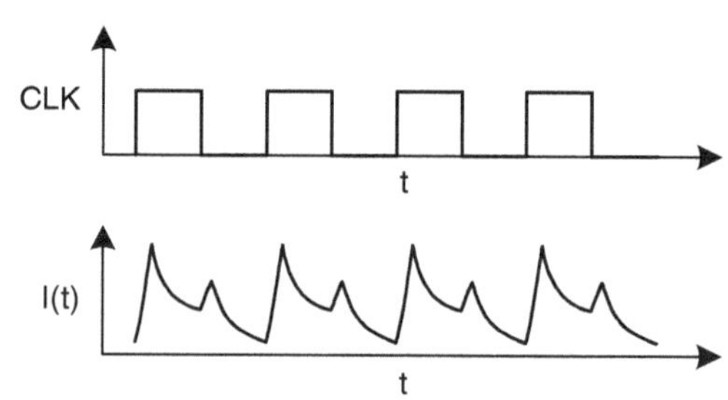

Fig. 4.1 Power consumption during a clock cycle

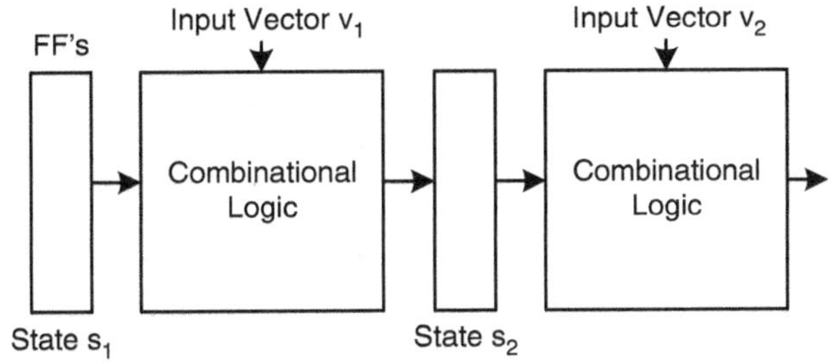

Fig. 4.2 Peak single-cycle power

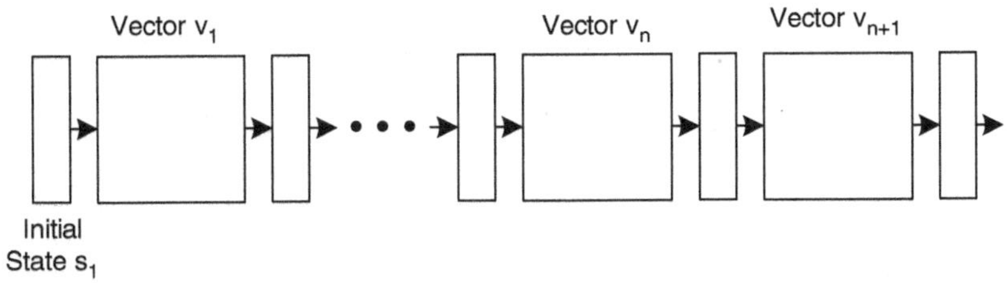

Fig. 4.3 Peak *n*-cycle power

chips, which are subsequently rejected (so-called yield-loss). In the remainder of this chapter, whenever the term peak power is used, it refers specifically to the peak single-cycle power.

The peak *n*-cycle power is the maximum of the power averaged over n clock cycles (Fig. 4.3) (Hsiao et al. 2000). The peak *n*-cycle power is used to determine local thermal stress as well as the required external cooling capability.

The average power is the energy consumed during the test over the total test time. Average power and total energy are important measures when considering battery life in an online test environment. Most of the common literature does not distinguish between average power and peak *n*-cycle power and often includes *n*-cycle averages under the term average power. When the term average power is used in this chapter, it refers to the average of the power consumption over a time frame long enough to include thermal effects.

4.2.2 Types of Power Consumption in Scan Testing

Scan design is the most important technique in design for testability. It significantly increases the testability of sequential circuits and enables the automation of structural testing with a fault-model. Figure 4.4 shows the general principle of scan design. When signal *scan enable* is set to "1," the circuit is operated in a way that all of the memory elements form a single shift register, the scan path. Multiple, parallel scan shift registers are called scan chains. During shifting, the circuit goes through numerous, possibly nonfunctional states and state transitions.

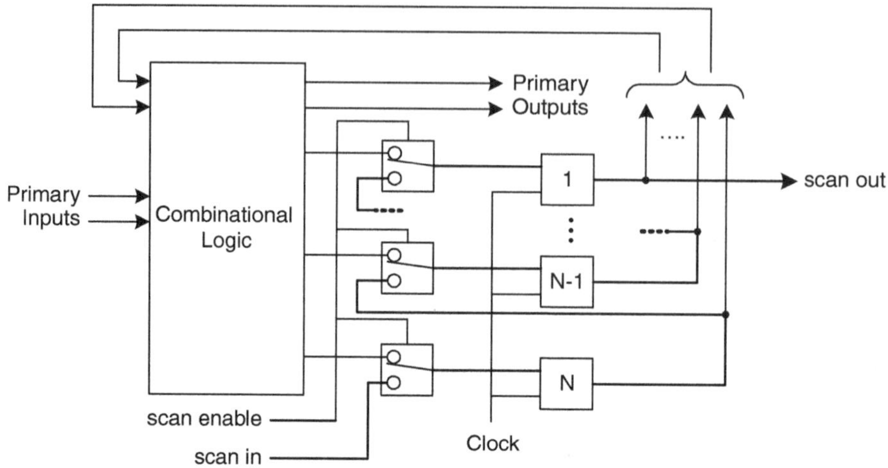

Fig. 4.4 Principle of the scan path

Fig. 4.5 Circuit states and transitions during shifting

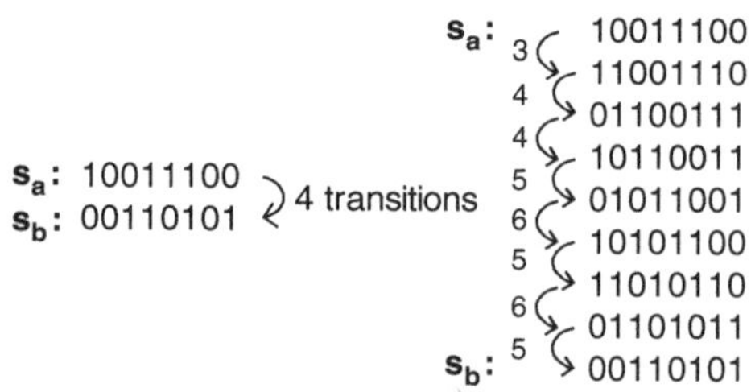

For scan-based tests, power is divided into shift power and capture power. *Shift power* is the power consumed while using the scan path to bring a circuit from state s_a (e.g., a circuit response) to s_b (e.g., a new test pattern) (Fig. 4.5). Since most of the test time is spent shifting patterns, shift power is the largest contributor to overall energy consumption. Excessive peak power during shifting may cause *scan cell failures* corrupting the test patterns or the corruption of state machines like BIST controllers.

Capture power is the power consumed during the cycles that capture the circuit responses. For stuck-at tests, there is just one capture cycle per pattern. For transition tests, there may be two or more launch or capture cycles. In launch-off-shift transition tests, the last shift cycle (launch cycle) is directly followed by a functional cycle that captures the circuit response. In launch-off-capture transition tests, the transition is launched by a functional cycle followed by a functional cycle to capture the response. Excessive peak power during the launch and capture cycles may increase the delay of the circuit and cause erroneous responses from otherwise good circuits. This is especially true for at-speed transition testing, since capture cycles occur with the functional frequency.

Power may also be distinguished according to the structure where it is consumed. A large part of the test power is consumed in the clock tree and the scan cells. In high-frequency designs, clocking and scan cells can consume as much power as the combinational logic. Usually, only a small part of the power is consumed in the control logic (such as BIST controllers), the pattern generators, and the signature registers. A detailed analysis of the contributors to test power may be found in Gerstendörfer and Wunderlich (2000).

4.3 Low-Power Scan Cells

Scan cells are the primary means of implementing a scan path. A plethora of scan cell designs have been proposed. This chapter discusses the power implications of the two most common designs and describes techniques to reduce both the power consumed in the scan cell as well as the power consumed by the combinational logic driven by the scan cell.

4.3.1 Power Considerations of Standard Scan Cells

The most common types of scan cells have been discussed in Sect. 1.3.2. Scan based on *muxed-D* cells requires only a single clock signal to be routed to the scan cell and any type of flip-flop may be used as its basis. Hence, muxed-D can take advantage of a number of low-power flip-flop designs such as double-edge triggered flip-flops (Chung et al. 2002).

For *LSSD*, the shift operation is exercised using two separate clock signals A and B. These clock signals are driven by two nonoverlapping clock waveforms, which provide increased robustness against variation and shift-power–related IR-drop events. Figure 4.6 shows an LSSD scan cell implemented with transmission gates. The transmission gate design has very low overall power consumption (Stojanovic and Oklobdzija 1999).

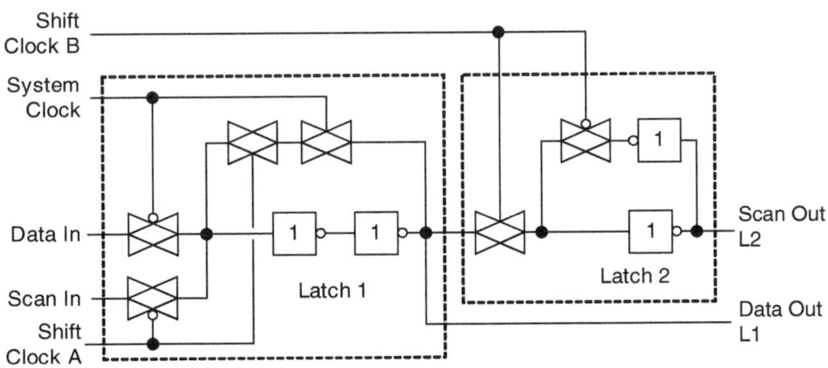

Fig. 4.6 LSSD scan cell using transmission gate latches

For designs such as Fig. 4.6, a significant portion of the power is consumed in the clock buffers driving the transmission gates. Hence, clock gating of the local clock buffers is an important technique to further reduce power.

4.3.2 Scan Clock Gating

Clock gating is an important technique for power reduction during functional mode. Clock gating reduces the switching activity by two means: first by preventing memory elements from creating switching events in the combinational logic, and second by preventing the clock transitions in the leaves of the clock tree. A common application of clock gating during scan testing is to deactivate the scan clock during useless patterns application (Fig. 4.7). *Useless patterns* do not detect additional faults that are not already detected by other patterns.

During scan-based BIST, a sequence of pseudorandom test patterns is applied to the circuit. Fault simulation is used to determine the useless patterns and resimulating the patterns in reverse and permuted order may uncover additional useless patterns.

The *pattern suppression* of Gerstendörfer and Wunderlich (1999) employs a simple controller that deactivates the scan clock during useless patterns. Girard et al. (1999) present a similar technique of suppressing useless patterns in nonscan circuits.

Figure 4.8 shows the DfT architecture for clock gating during the test. The circuit has the common self-test DfT of a scan path with a test pattern generator and a signature analyzer. The test controller generates the scan clocks and contains a pattern counter. A simple decoder generates the clock gating signal from the pattern count.

Using the information obtained from fault simulation, a simple table is constructed. For example, the result of the fault simulation may look as listed in Fig. 4.9.

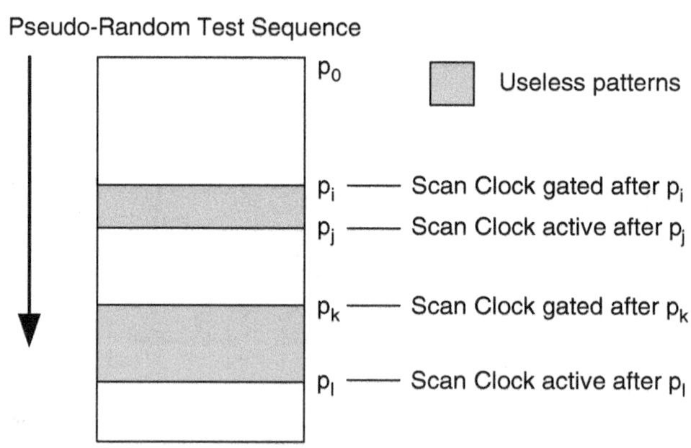

Fig. 4.7 Scan clock gating of useless patterns

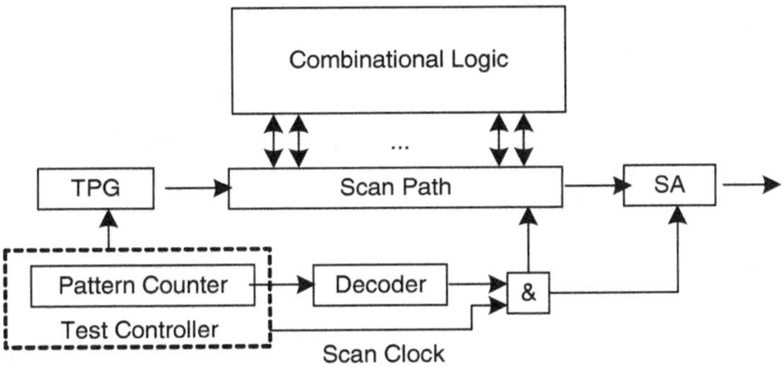

Fig. 4.8 Design for test with scan clock gating during useless patterns

index	binary	# faults
0	0000	17
1	0001	9
2	0010	4
3	0011	0
4	0100	5
5	0101	2
6	0110	3
7	0111	0

index	binary	#faults
8	1000	2
9	1001	0
10	1010	0
11	1011	1
12	1100	0
13	1101	0
14	1110	0
15	1111	0

Fig. 4.9 Fault simulation result for a test set with 16 patterns

Fig. 4.10 Boolean function of the decoder

{ 0000, 0001, 0010, 0100, 0101, 0110, 1000, 1011, 1100 } on-set

{ 0011, 0111, 1001, 1010 } off-set

{ 1101, 1110, 1111 } dc-set

Fig. 4.11 Decoder for pattern suppression for the example

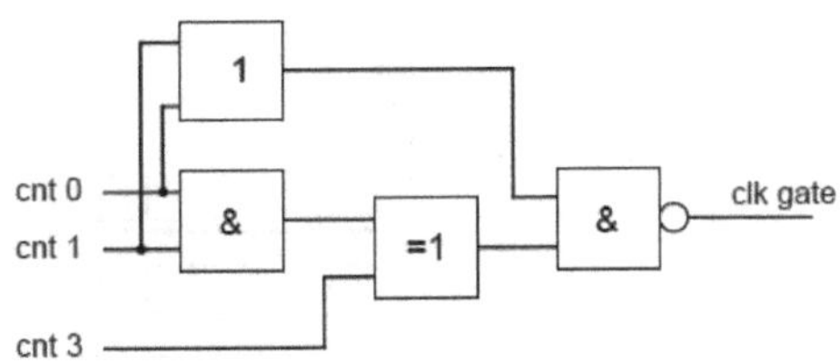

The first three patterns detect new faults and the pattern with index 3 does not. Shifting is suspended during patterns 3, 7, 9, and 10, and enabled during patterns 0, 1, 2, 4, 5, 6, 8, 11, and 12. The clock is enabled for pattern 12 to shift out the circuit response of pattern 11. The test controller stops the BIST after pattern 12. Now, the resulting Boolean function is shown in Fig. 4.10.

This function is minimized and synthesized using a standard tool flow. Figure 4.11 shows the decoder for the example.

For larger circuits, the overhead for the decoder is just a few percent as reported in Gerstendörfer and Wunderlich (1999). It has been shown that pattern suppression reduces the average power consumption by approximately 10%. However, the reduction may be significantly higher if the test length is very high or if the circuit has been designed for random testability.

For pattern suppression, the scan clocks can be gated at the root of the clock tree. The general idea of avoiding useless activity during shifting is common to most of the techniques presented in this chapter. In most cases, they rely on DfT that allows disabling scan clocks.

To achieve improved granularity of the clock gating, the clocks may be gated closer to the memory cells (Wagner 1988). However, the savings obtained using clock gating diminish if it is applied to individual cells. In functional design, clocks are usually gated at the register granularity (e.g., of 32 or 64 bits). During test, an acceptable granularity is to deactivate a scan chain, a group of scan chains, or a test partition. Figure 4.12 shows a commonplace design for test architectures that employ parallel scan chains such as the *STUMPS* design (Self-Test Using MISR and Parallel SRSG). Here, the scan clock of every scan chain may be disabled individually by setting a *Test Hold* register.

In order to implement the scan clock gating, all of the clock gating functionality can be added to the *local clock buffers*. Figure 4.13 shows an example of a local clock buffer that allows for clock gating during functional mode and during scan as well. If signal *Testmode* is set to "0," then the clock gating is controlled by *Activate* and the outputs *Load Clock B* and *System Clock* operate an associated LSSD cell in a master/slave mode. If *Testmode* is set to "1," then the *Scan Clock A* and the *Load Clock B* operate the LSSD cell in scan mode. The signal *Test Hold* deactivates the clock during both scan and capture clocks of the test.

The clock buffer employs a dynamic logic gate for the clock gating. The dynamic logic style allows to design the clock buffer in such a way that the clocks stay off

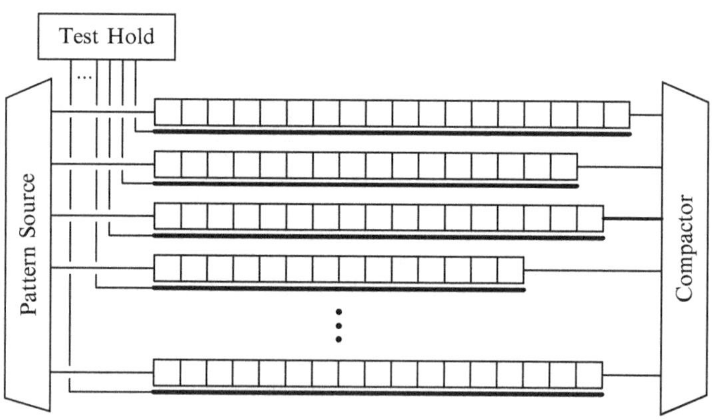

Fig. 4.12 DfT with parallel scan chains and clock gating per chain

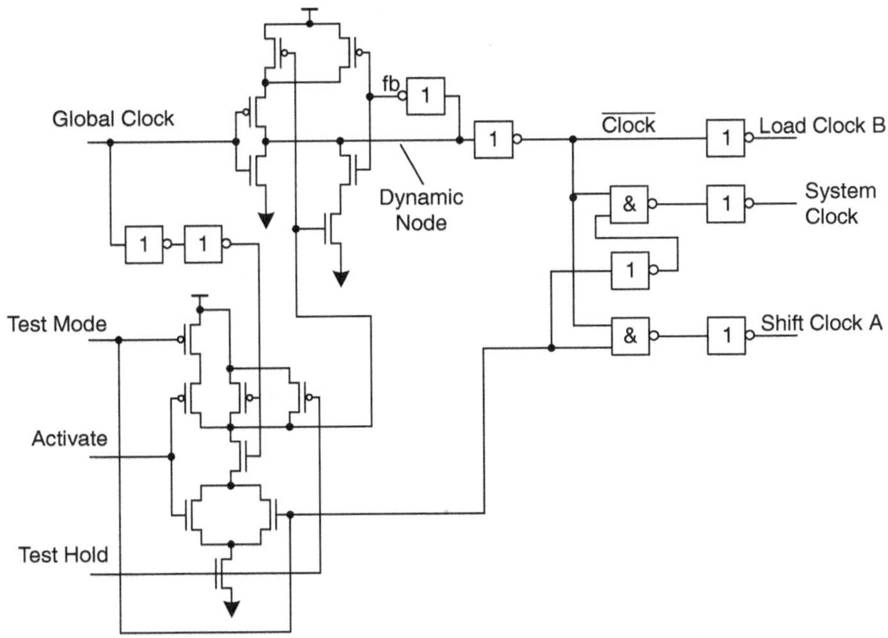

Fig. 4.13 Local Clock Buffer for functional and scan clock gating (Pham et al. 2006)

during the complete clock cycle, even if one of the clock gating signals exhibits glitches. This way race conditions are avoided. The precharge of the dynamic logic is controlled by the logic function:
$\neg$(ScanEnable $\wedge$ (Testmode $\vee$ Activate)).

In the *partial-scan* design only a subset of all memory elements of a circuit can be scanned. In this case, it is highly beneficial to disable the clocks of the nonscan elements during shifting. This avoids the power consumption in the nonscan cells and the associated clock buffers. It also blocks the switching events of the combinational logic attached to the scan cells from propagating further through the circuit. Figure 4.14 outlines the principle.

4.3.3 *Test Planning for Scan Clock Gating*

If a circuit is designed with parallel scan chains that can be deactivated as in Fig. 4.12, the shifting of a scan chain may be avoided completely if the values controlled and observed by that scan chain do not contribute to the fault coverage of the test. In other words, turning off the clocks of the scan chain does not alter the fault coverage. In Zoellin et al. (2006), it was shown that the power consumed during the BIST of large industrial circuits, like the Cell processor^{TM}, can be reduced significantly without impairing fault coverage.

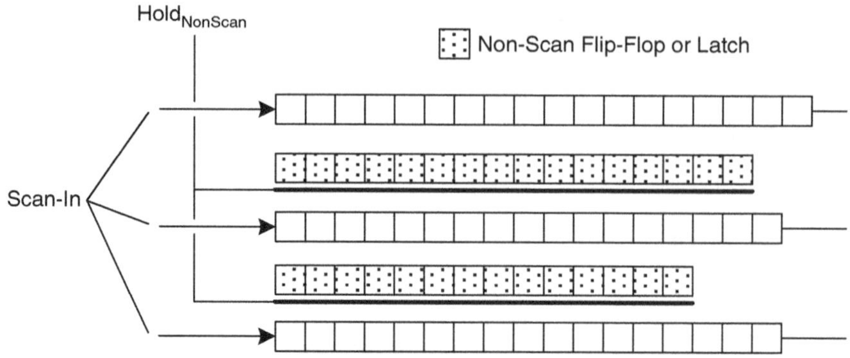

Fig. 4.14 Using nonscan cells to block the propagation of switching activity

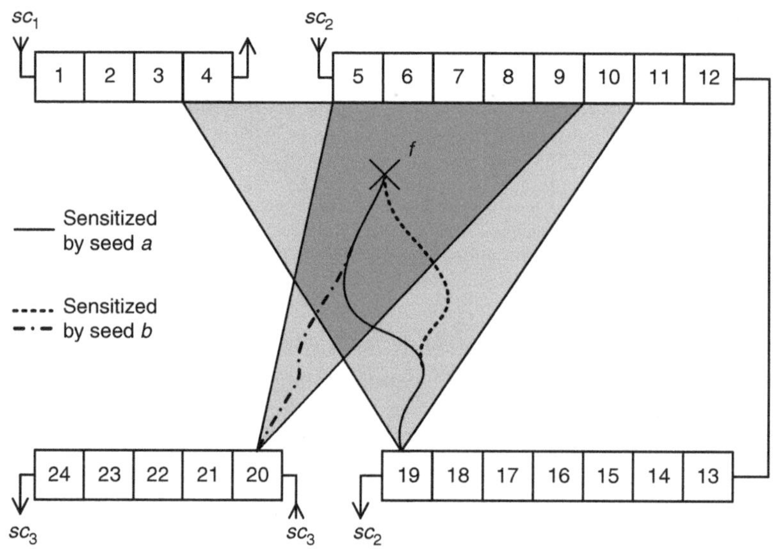

Fig. 4.15 Example of detecting fault f

Test planning is the process of assigning configurations of the scan clock gating for each session of a test such that a certain set of faults is detected. For example, in a BIST based on the *STUMPS* design (Self-Test Using MISR and Parallel SRSG), the BIST consists of several sessions and each session is started by a seed of the linear feedback shift register. For every seed, test planning computes a configuration of the scan chains such that fault coverage is not impaired.

Most faults detected by the complete test can be detected by several sessions and may often be observed in several scan cells. In the example of Fig. 4.15, the fault f may be detected by a test session started using seed a and a test session started by seed b. In the case of seed a, the fault is detected in scan cell 19 and in the case of seed b the fault is detected in cells 19 and 20. Only one of these combinations is required.

To ensure that the path that detects the fault is completely sensitized, it is sufficient to activate all of the scan cells in the input cone together with the scan cell observing the fault effect. For example, to detect the fault in cell 19 of Fig. 4.15, it is sufficient to activate the scan cells {4, 5, 6, 7, 8, 9, 10, 19}. Since the clocks cannot be activated individually, this is mapped to the scan chains $\{sc_1, sc_2\}$.

These degrees of freedom are now encoded into constraints for a *set covering* problem. In the example of Fig. 4.15, the constraints are $\{a, \{sc_1, sc_2\}\}$, $\{b, \{sc_1, sc_2\}\}$, and $\{b, \{sc_{2,3}\}\}$. For the optimization of the test plan, the constraints for all of the faults to be detected have to be generated. The set covering is then solved by a branch & bound method. The cost function for the minimization is an estimate of the power consumption.

Imhof et al. (2007) report that the power reduction obtained by test planning of a pseudorandom BIST is approximately 40–60%. The larger the number of test sessions, the higher the power reduction. Sankaralingam and Touba (2002) show that even for deterministic tests, a careful combination of scan cell clustering, scan cell ordering, test generation, and test planning can obtain a power reduction of approximately 20%.

4.3.4 Toggle Suppression

During shifting, the functional outputs of the scan cells continue to drive the combinational logic. Hence, the combinational logic is subject to high switching activity during the scanning of a new pattern. Except for the launch cycle in launch-off-shift transition tests, the shifting does not contribute to the result of the test.

Hence a very effective method to reduce the shift power is to *gate the functional output* of the scan cell during shifting. The gating can be achieved by just inserting an AND or OR gate after the functional output of the scan cell. However, in this case the entire delay of the AND or OR gate will impact the circuit delay. Instead, it is more desirable to integrate the gating functionality into the scan cell itself.

Figure 4.16 shows a *muxed-D* scan cell based on a master-slave flip-flop. The NAND gate employed to gate the functional output incurs only a very small delay

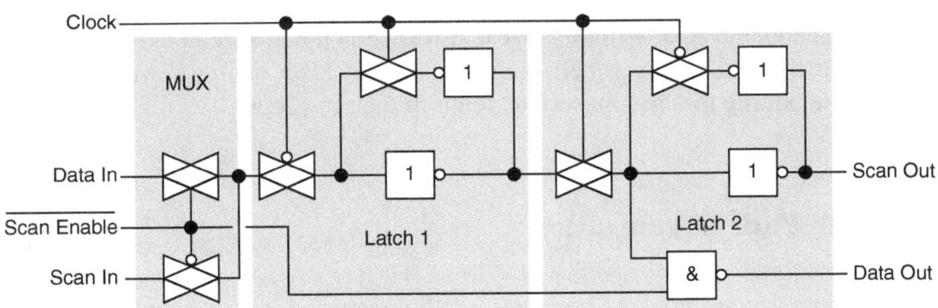

Fig. 4.16 Master-slave muxed-D cell with toggle suppression

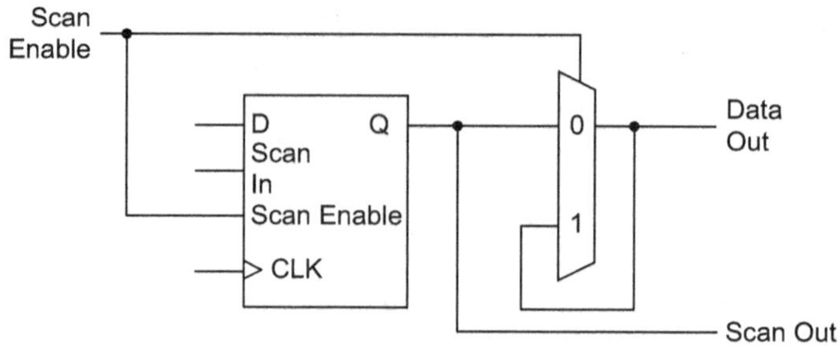

Fig. 4.17 Toggle suppression implemented with multiplexer

overhead, since the NAND-input can be driven by the QN node of the slave latch. Hertwig and Wunderlich (1998) have reported that *toggle suppression* reduces average shift power by almost 80% on average. But the switching activity during the capture cycle is not reduced and overall peak power consumption is almost unaffected.

In order to use toggle suppression with launch-off-shift transition tests, the control signal for the output gating has to be separated from the scan enable signal. However, this increases the wiring overhead of the scheme significantly.

The techniques described above reduce the peak power during shifting since all of the scan cells are forced to a specific value, and the application of the test pattern to the combinational logic may incur switching in up to 50% of the scan cells. To provide additional control over the peak power for launch and capture cycles, the functional output of the scan cell can be gated using a memory element. The memory element then stores the circuit response of the preceeding pattern, and by appropriately ordering the test patterns, the peak power can be reduced. For example, Zhang and Roy (2000) have proposed the structure in Fig. 4.17, which uses an asynchronous feedback loop across a multiplexer to implement a simple latch. Similar to the NAND-based approach, the impact on the circuit delay can be reduced by integrating the gating functionality into the scan cell. Parimi and Sun (2004) use a master-slave edge-triggered scan cell and duplicate the slave latch.

It may be sufficient to apply toggle suppression to a subset of the scan cells. ElShoukry et al. (2007) use a simple heuristic to select scan cells to be gated. The cost function is based on a scan cell's contribution to the power consumption and takes into account available timing slack. It was shown that adding toggle suppression to just 50% of the scan cells achieves almost 80% of the power reduction compared to adding toggle suppression to all of the scan cells.

4.4 Scan Path Organization

This section discusses how the scan path can be organized in a way that the shifting process uses less power and that it assists other techniques for power reduction.

Fig. 4.18 General scan insertion flow

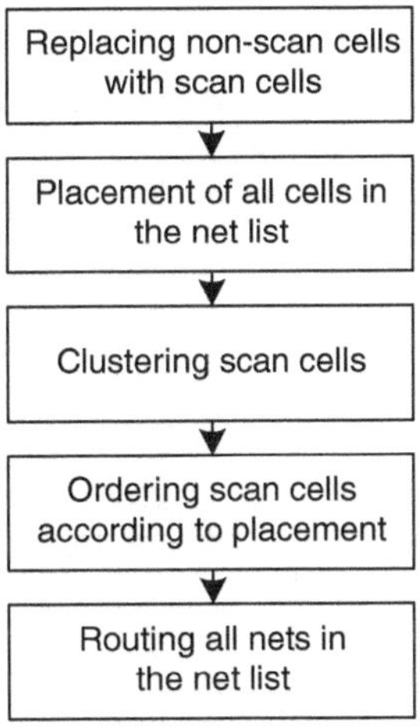

Figure 4.18 shows the general flow of *scan insertion* into a design. Commercial tools support all of these steps and the techniques discussed in this section may be used to extend or replace some of the steps in Fig. 4.18.

4.4.1 Scan Path Segmentation

A common method to reduce the excess switching activity during shifting is to split the scan path into several *segments*. Shifting is then done one segment after the other. The segments not currently active are not clocked and do not contribute to shift power. The technique reduces both peak and average power.

Figure 4.19 shows the structure proposed by Whetsel (2000). Here, a scan path of length t is split into 3 segments of length $1/3t$.

The activation of the segments is controlled using the scan clocks. Because the shift input is multiplexed using the clocks, only a multiplexer for the shift outputs is required. However, either the shift clocks for each segment have to be routed individually or scan clock gating is employed as described in Sect. 4.3.1. Figure 4.20 shows the clock sequence for the example above.

For launch-off-shift transition faults, in the clock sequence of Fig. 4.20, only the shift of the segment lastly activated launches a transition to be captured. In this case, it is possible to apply an additional launch shift cycle to all segments just before the capture cycle.

If the segmentation is done this way, the test time remains the same. For two segments, shift power is reduced by approximately 50%, for three segments the reduction is approximately 66%. Whetsel (2000) has reported that two or three segments have the best ratio of power reduction versus implementation overhead. The technique reduces both the peak power during shifting as well as the overall test

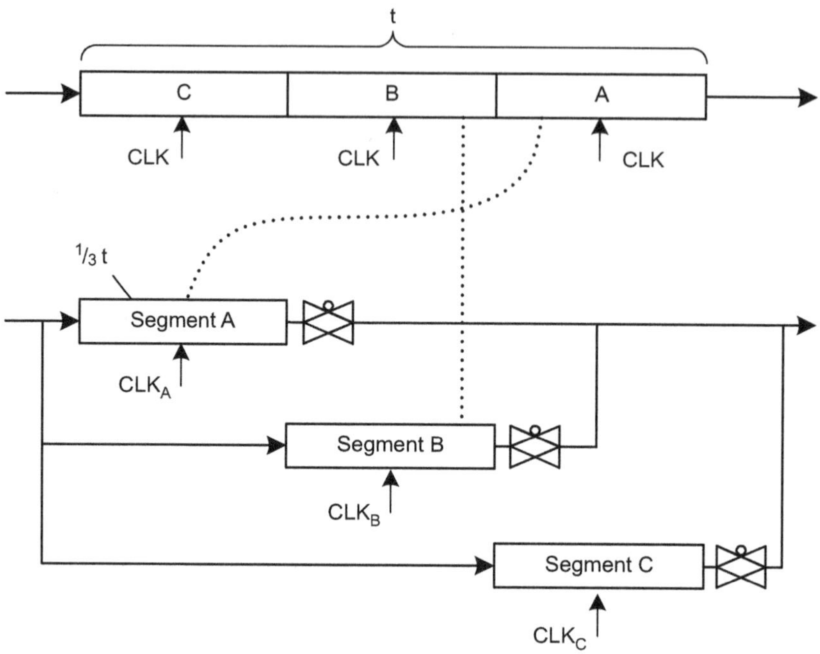

Fig. 4.19 Scan path segmentation

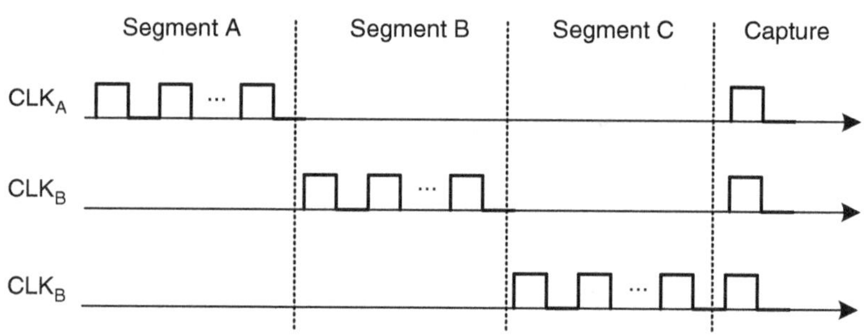

Fig. 4.20 Clock sequence for the scan segmentation in Fig. 4.19

energy. Since test time is kept, average power is reduced as well. However, the power consumption during the capture cycle is not reduced by the clock sequence above.

If the DfT architecture consists of multiple scan chains anyway, like in the STUMPS architecture, the technique can also be applied using just the scan clock gating from Fig. 4.12 from Sect. 4.3.1. In this case, the test time is increased compared to scanning all chains in parallel.

4.4.2 Extended Clock Schemes for Scan Segmentation

The *clock sequence* in Fig. 4.20 has two remaining drawbacks: first, the clock frequency used to shift the individual scan segment is not reduced and may be subject to local IR-drop effects. Second, the power of the capture cycle is not reduced, which is a significant issue especially in transition tests.

To solve the first problem, instead of shifting individual segments at full clock frequency the segments can be shifted in an alternating fashion. This technique is also often called *staggered clocking* or *skewed clocking* in low-power design. Figure 4.21 shows the clock sequence for the three scan segments A, B, and C of Fig. 4.19 as proposed by Bonhomme et al. (2001). Now, each individual clock has a lower frequency. This increases the robustness of the pattern shifting against IR-drop events. Girard et al. (2001) show how staggered clocking may be applied to the pattern generator as well.

The peak power of the launch and capture is not reduced with the previous clock sequence. The staggered clocking above can be done for the launch and capture clocks as well (Fig. 4.22).

In this case, only transition faults that are launched and captured by the same segment can be tested. Figure 4.23 shows an example where all of the flip-flops are contained in a single segment, which launches the transition, sensitizes the

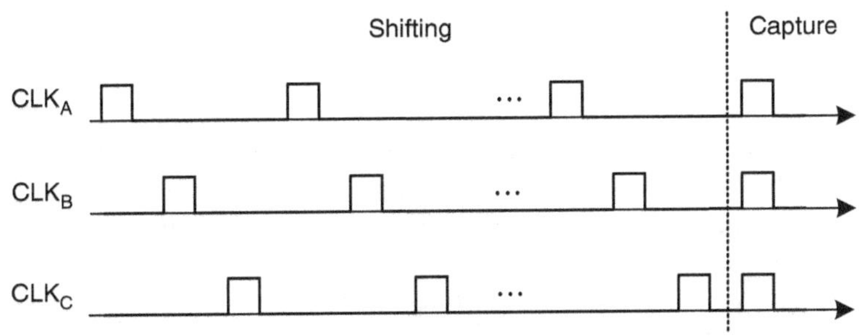

Fig. 4.21 Staggered clock sequence for shift peak power reduction

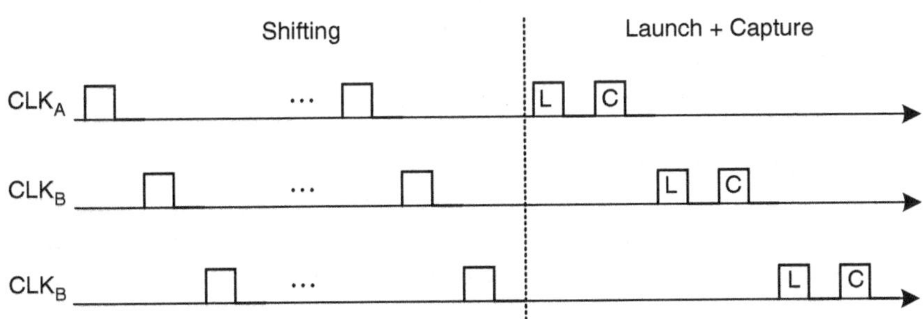

Fig. 4.22 Clock sequence for launch-capture peak power reduction

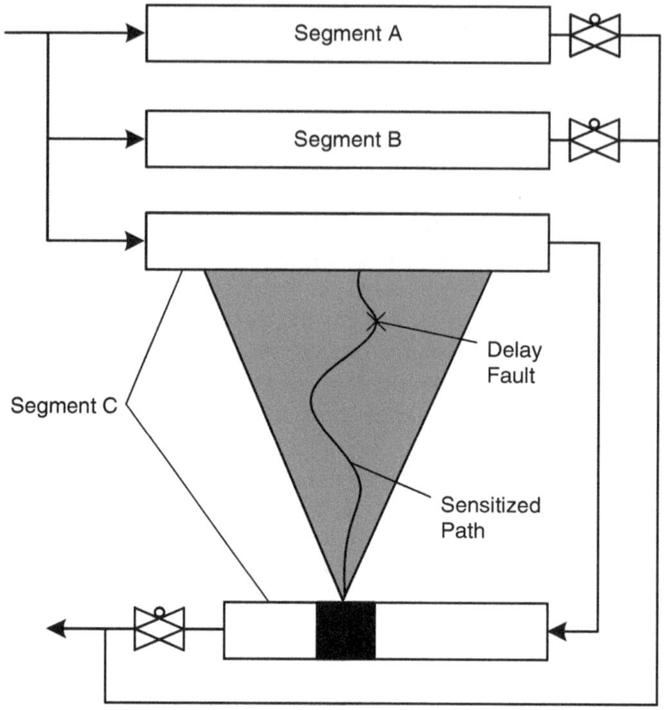

Fig. 4.23 Input cone that is contained in a single scan segment

propagation path, and observes the fault. In this case, the fault can be tested by only executing the launch and capture cycle for segment A. However, capturing segment A before segment B may change the justification for segment B and special care is required during test generation.

Most of the fault-coverage can be retained by using additional combinations of launch and capture clocks. The combinations of segments to be activated to detect a certain set of faults can be determined by solving the set covering problem discussed in Sect. 4.3.3. The set of faults that can be tested using just the launch and capture clocks of a single or a few segments can be increased by clustering the scan cells into scan segments appropriately. Rosinger et al. (2004) report that appropriate planning and clustering allow to reduce the peak power during capture by approximately 30–50%.

Yoshida and Watari (2002) use even more fine-grained clock staggering by manipulating the duty cycles of each scan clock. This allows to more closely interleave the shifting of several segments and can improve the shift frequency. However, the modification of the clock duty cycle requires a significantly higher design and verification effort.

4.4.3 Scan Cell Clustering

In many power reduction techniques, the effectiveness of the method is influenced by the organization of the scan chain. For example, in the scan path segmentation presented above, the shifting of a segment may be avoided completely if there is no fault observed in the segment and if the segment contains no care bit of the pattern. In the example in Fig. 4.24, a path is sensitized by a test pattern and the response is captured in scan cell 11. If all of these flip-flops are in the same scan segment like in Fig. 4.23 of Sect. 4.4.1, only that segment has to be activated to test all of the faults along the path. In fact, similar relations hold for many other, more advanced test generation and test planning techniques.

The goal of *scan clustering* is to cluster the scan cells of a circuit into k segments or parallel scan chains, where each segment contains at most t scan cells (Fig. 4.25). The clustering tries to increase the likelihood that the scan cells with care bits and the observing scan cells are in the same segment. Since it is undesirable to synthesize DfT hardware based on a specific test set, the optimization is based on the circuit

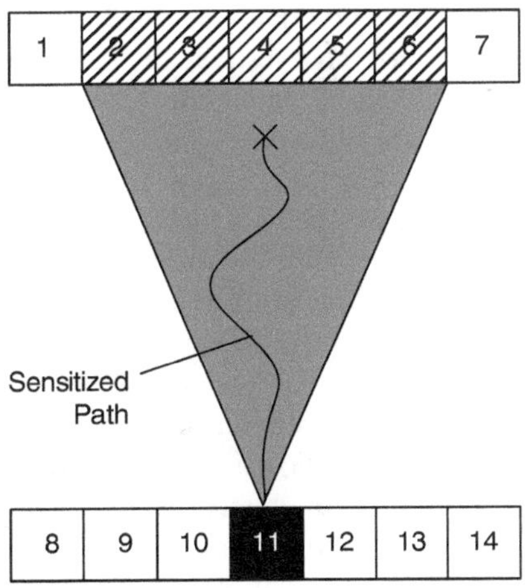

Fig. 4.24 Scan cell observing a fault with input cone

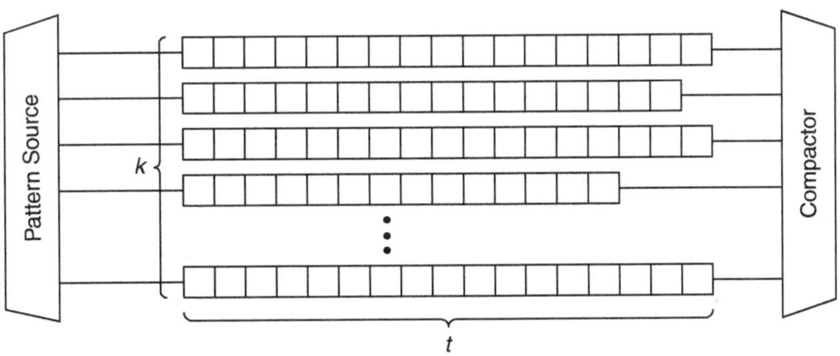

Fig. 4.25 Parameters k and t in scan chain clustering

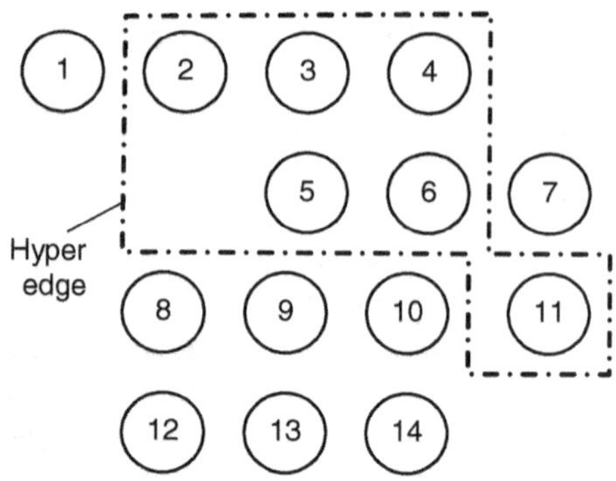

Fig. 4.26 Hyper graph and hyper edge for Fig. 4.24

structure. In the DfT insertion process, the scan clustering is followed by layout-based scan cell ordering that tries to minimize the routing overhead of the scan design.

The problem of clustering the scan cells is mapped to a *graph partitioning* problem. The technique described here uses a *hyper graph* representation of all the constraints. The vertices of the hyper graph are the scan cells of the circuit. The scan cells in the input cone of a given scan cell are sufficient to sensitize all of the paths that can be observed. The hyper graph contains one *hyper edge* for each input cone to a scan cell. Figure 4.26 shows the hyper edge for the example of Fig. 4.24. For example, the hyper edge for cell 11 is {2,3,4,5,6,11}.

Now, the optimized clustering is the partitioning of the vertices of the hyper graph into k partitions of up to t scan cells such that the global edge cut is minimized. The *hyper graph partitioning* problem is NP-complete and a large number of heuristics exist (Karypis and Kumar 1999).

For scan clustering, a *problem-specific heuristic* such as the one proposed by Elm et al. (2008) can achieve favorable results with very low computation time (linear-time complexity) even for multimillion gate designs. This kind of clustering can improve the effectiveness of power reduction techniques by approximately 40% compared to regular scan insertion techniques.

4.4.4 Scan Cell Ordering

For a given test set, the order of the scan cells determines where transitions occur during shifting. Figure 4.27 shows a rather extreme case of this. The first ordering in the example has the worst case-switching activity, whereas in the second ordering only two transitions occur in the test pattern and test response.

Most current test generation tools have the capability to provide partially specified test sets. The ordering-aware filling method "repeat fill" will cause the test patterns to have very few transitions already, and the gains possible by scan cell

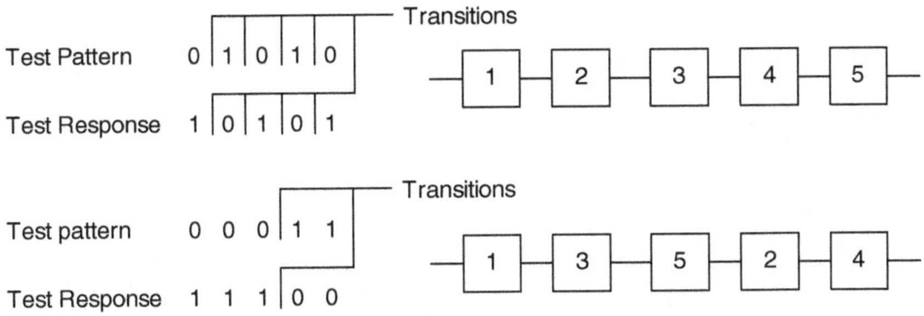

Fig. 4.27 Influence of scan cell order on switching activity during shift

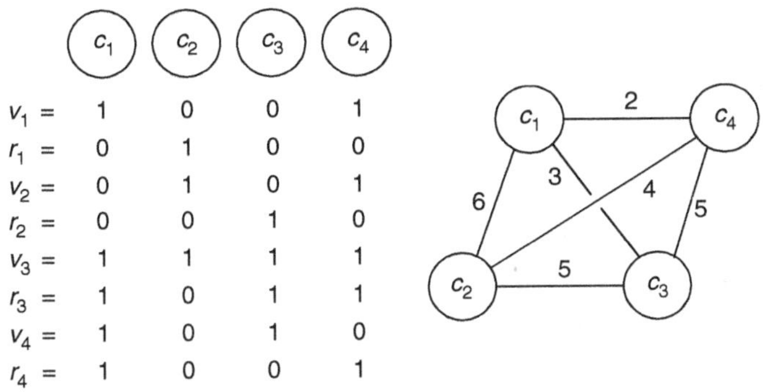

Fig. 4.28 Test and response vectors used to compute edge weights

ordering are rather low. Scan cell ordering is effective if the test set is randomly filled or if the test set is highly compacted. However, even slight changes in the test generation procedure can cancel out any improvements by the scan ordering of already-existing hardware. Furthermore, the hardware overhead for *scan wiring* can be a substantial contributor to the overhead for DfT and power-aware scan cell clustering with regular, layout-aware ordering may be preferable.

The problem of finding the optimal order of a set of scan cells C with respect to a given test set is translated into finding a Hamiltonian path in a weighted, undirected graph $G(C, E)$. The weight of an edge between two scan cells c_i and c_j is the number of transitions that would occur if c_i were followed by c_j (or c_j by c_i). In the example in Fig. 4.28, the weight of the edge between c_1 and c_2 is 6, since the sequence $\{v_1, r_1, v_2, r_3, v_4, r_4\}$ would result in 6 transitions if c_1 were followed by c_2 or vice versa.

In the example described above, the optimum solution is c_1-c_4-c_2-c_3-c_1. This solution is found by solving the *traveling salesman problem* (TSP) for the graph. TSP is a well-known NP-hard problem. Bonhomme et al. (2003) have reported good results with a $O(n^2)$ greedy heuristic. However, an ordering based solely on solving the TSP above results in significant routing overhead. Bonhomme et al. (2003) propose to trade off power reduction and routing overhead. For this, the chip area is

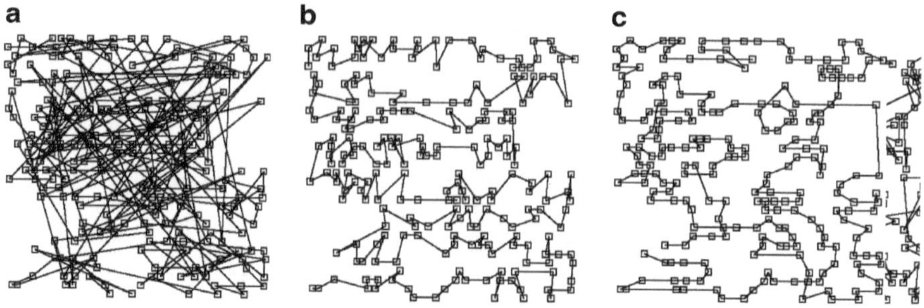

Fig. 4.29 Wiring for power-aware order (**a**), power aware routing-constrained order (**b**), and commercial tool (**c**)

divided into several tiles to which the partial solutions are constrained, such that no ordering decisions with high overhead are taken. In this case, scan cell ordering can provide approximately 20% power reduction when compared with scan cell ordering that optimizes for routing overhead (Fig. 4.29).

4.4.5 Scan Tree and Scan Forest

The *scan tree* is a generalization of the scan path. In a scan tree, a scan cell's output may be connected to several other scan cells as seen in Fig. 4.30. *Scan forests* are the extension of this concept to parallel scan chains.

Scan cells connected to the same fanout will receive the same test stimuli. And to avoid any impact on fault coverage, special care must be taken. For example, for stuck-at faults, it is sufficient to ensure that all the scan cells in the input cone can be controlled independently (cf. Fig. 4.24). Two scan cells are called compatible if they are not part of any common input cone. For the scan tree, scan cells should be ordered by their compatibility. Chen and Gupta (1995) use a graph-based approach to find pseudoprimary inputs that may receive the same test vectors.

Scan trees are often used to reduce test time and test data volume. The *Illinois scan architecture* by Hamzaoglu and Patel (1999) is a special case of the scan tree in which only the scan-in has a fanout larger than one. Hellebrand et al. (2000) combine the scan tree with a test pattern decompression technique to improve the compression efficiency.

The scan tree may also be combined with a regular scan path (Fig. 4.31). This mode of operation is often called "*serial mode*" and is used to provide conventional scan access to the circuit for debugging as well as for cases where some of the scan cells are incompatible.

The principle may also be applied to the fan-in of scan cells. For example, in the *double-tree design* of Fig. 4.32, the scan cells 8, 9, and 10 are computed as the XOR of the two predecessors. Alternatively, additional control signals are used to select a predecessor with a multiplexer as suggested by Bhattacharya et al. (2003).

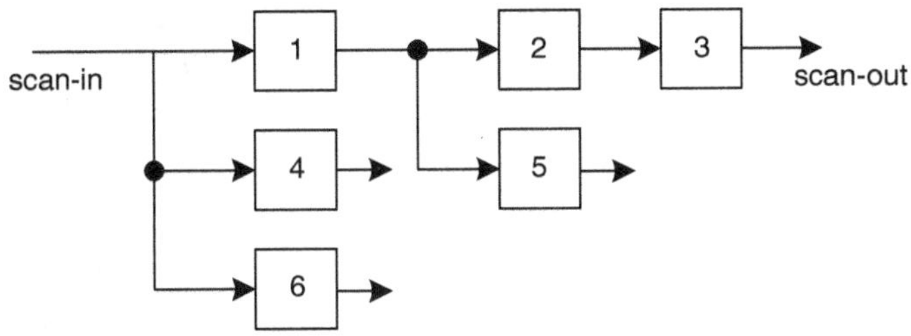

Fig. 4.30 Example of a scan tree

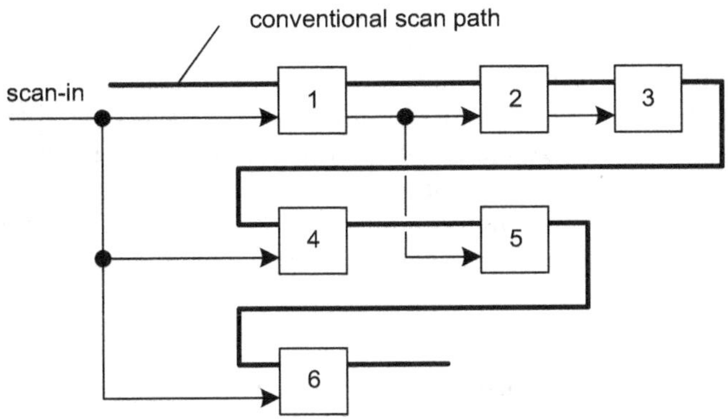

Fig. 4.31 Scan tree combined with conventional scan path

Fig. 4.32 Double scan-tree

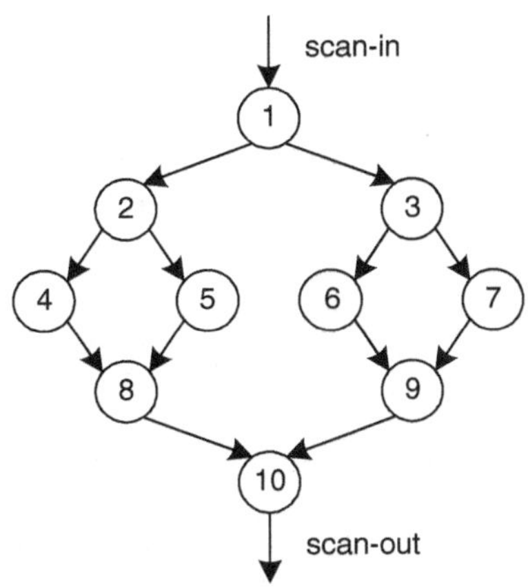

The scan segmentation of Sect. 4.4.1 is a special case of the double scan-tree with multiplexers in Fig. 4.32. And in fact, power reduction for the more general structure works in a similar way. Here, scan clock gating is used to reconfigure the double tree according to the care bits in a test pattern. The scan gating is implemented such that any complete path through the double tree can be activated at a time. If a test pattern has care bits in scan cells 1, 5, and 8, it is sufficient to scan just the cells in the path 1–2–5–8–10 (Path-1 in Fig. 4.33).

Fig. 4.33 Scan path configurations for Fig. 4.32

Select = 00 ➔ Path-0: 1→2→4→8→10
Select = 01 ➔ Path-1: 1→2→5→8→10
Select = 10 ➔ Path-2: 1→3→6→9→10
Select = 11 ➔ Path-3: 1→2→7→8→10

In most test sets, care bits are rather sparse and often only a few paths have to be scanned for a complete pattern. When constructing the scan tree of Fig. 4.32, the scan cells that are most likely to contain a care bit should be closer to the root of the tree. The problem of clustering and ordering scan cells in this way can be mapped to the algorithms presented in Sects. 4.4.3 and 4.4.4. Xiang et al. (2007) have presented such a technique for constructing forests of scan trees.

For the double scan tree with clock gating, Bhattacharya et al. (2003) report a reduction in shift power consumption of up to 90%. Similar to the scan segmentation in Sect. 4.4.1, special attention is required for the peak power consumption during launch and capture cycles of transition tests. Also, the routing overhead must be taken into account when constructing the scan tree.

4.4.6 Inserting Logic into the Scan Path

Combinational logic can be inserted to apply certain patterns with lower shift power consumption. In most ATPG test sets, many patterns have similar assignments to the pseudoprimary inputs because of common path sensitization criteria between faults.

If the test set for a circuit is known, the probability of the assignment in each scan cell can be computed. This prediction is subsequently used to select the optimal *polarity* of the scan cells. This reduces the number of transitions during shifting, but not during the capture cycle. Figure 4.34 shows a single test cube that is filled using the repeat fill method. The test pattern has two transitions and the associated test response has one transition. By using the inverted output of the second scan cell in the example, the number of transitions in the final pattern and response is reduced to just one.

However, often it is highly undesirable to have DfT structures that rely on a specific test set, since even a slight change in the test generation process may change the test set. Instead, more general measures from *testability analysis* can be employed, for example, the methods COP by Brglez et al. (1984) or PROTEST by Wunderlich (1985).

Correlation between the assignments to certain pseudoprimary inputs can be exploited to improve the prediction of the scan cell value and further reduce the number of transitions in the pattern. Sinanoglu et al. (2002) embed a linear function into the scan path as depicted in Fig. 4.35. Here issues of routing overhead and computational complexity mandate that the linear function is implemented over just a short segment of the scan path. The algorithm proposed by Sinanoglu et al. (2002) works by the divide-and-conquer paradigm and uses a given test set as the input.

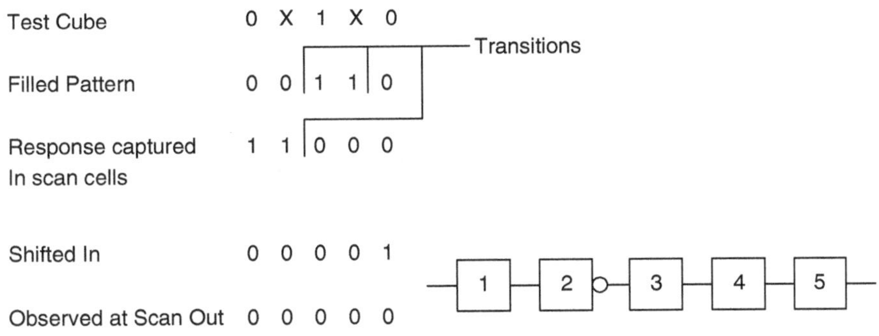

Fig. 4.34 Example of scan segment inversion

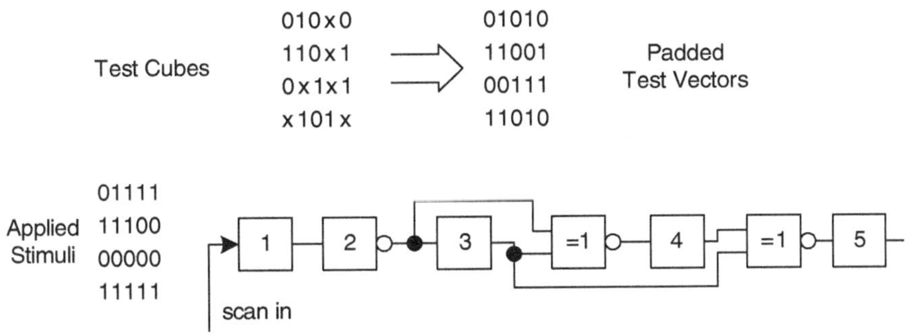

Fig. 4.35 Scan segment inversion by embedding a linear function

This technique provides a 10–20% reduction of the shift power. But other than the selection of the scan cell polarity, the approach causes an area overhead of approximately 5%.

Inversion of certain scan segments can also improve the efficiency of pseudorandom BIST. Here, the goal is to increase the detection probability for test patterns with low weight (i.e., probability of a "1" <50%). This is directly related to the estimation of the weights of the random pattern test such as by Wunderlich (1987). To further increase the effectiveness of the test, Lai et al. (2004) make the segment inversion configurable (Fig. 4.36).

The BIST is split into several sessions with different configurations. In this case, the reduction of the energy consumption of the entire test is achieved by reducing the test time to achieve a certain fault coverage similar to deterministic BIST.

4.5 Partitioning for Low Power

Partitioning allows for splitting the test of an entire chip into several independent tests. These independent tests are then scheduled using one of the techniques in

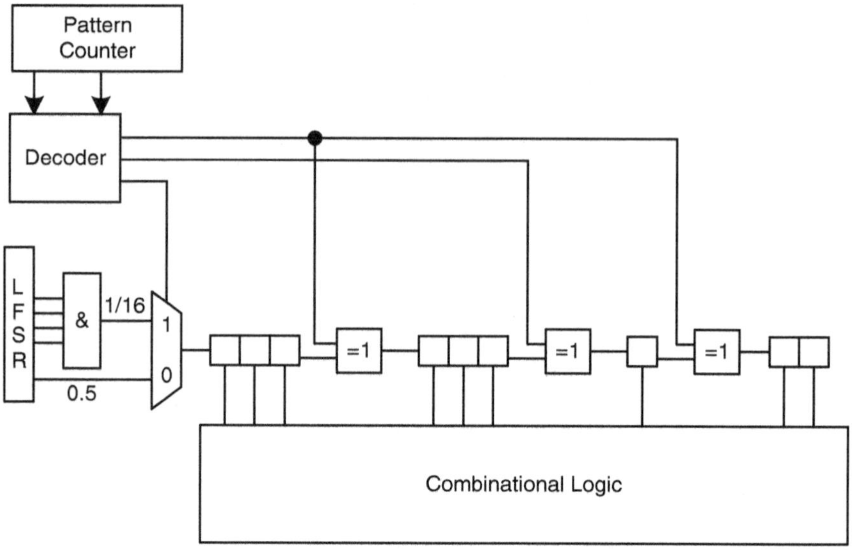

Fig. 4.36 Pseudorandom BIST with scan segment inversion

Chap. 6 to optimize for test time, bandwidth, and power. The common techniques for partitioning are discussed in the fifth section of this chapter. In designs with parallel scan chains, clock gating of the scan chains is an effective means of partitioning. If a core-based design implements *test wrappers* around cores, for example, using *IEEE Std. 1500*, partitioning is part of the general test strategy. And finally, under certain circumstances, the combinational logic may be partitioned at the gate level.

Franch et al. (2007) have shown that the capacitance in the idle partitions of a circuit is fully available to support the power supply of the active partitions. Hence, partitioning together with test scheduling and planning effectively reduce the likelihood of IR-drop during the peak power cycles. Sehgal et al. (2007) report how partitioning can be implemented for high-end microprocessors and how it effects the overall test cost.

4.5.1 Partitioning by Clock Gating

The individual scan chains of a DfT with parallel scan chains can be used for partitioning the circuit using the *scan clock gating* of Sect. 4.3.1. In some cases, if the scan chains have been organized according to Sect. 4.4.3, the partitioning may be static. Figure 4.37 shows an example of this. Here the tests for each partition may be applied individually. The partition that is deactivated does not contribute to the power consumption. If the clustering of Sect. 4.4.3 was not completely successful, a small number of tests may be required to recover lost fault coverage.

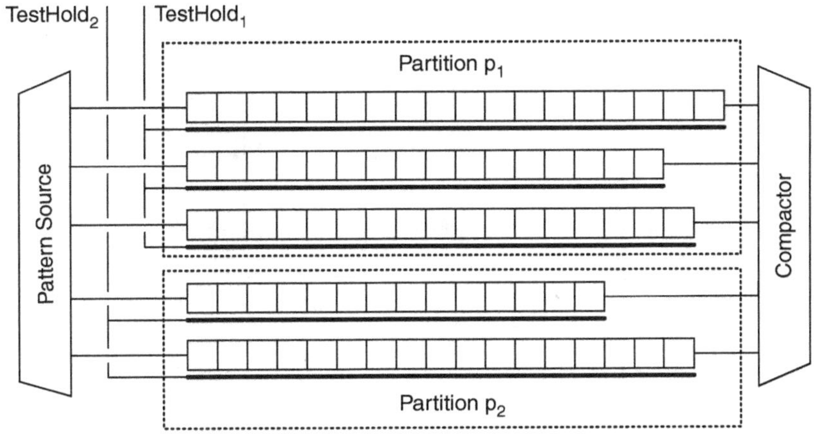

Fig. 4.37 Circuit partitioning using scan gating

Partitioning a circuit this way should be distinguished from the test planning in Sect. 4.3.3. The partitioning described here exposes test sessions and configurations of the clock gating to the scheduling process. The configurations of the scan clocks are predetermined and the test sets are generated for these configurations. The test scheduling then determines a schedule that suits the given power envelope. In test planning, the test set is given and the configuration of the scan clocks is determined by the process of test planning.

4.5.2 Partitioning in Core-Based Design

At higher granularity, Systems-on-Chip designer use hierarchical and reuse-oriented design methods such as core-based design. For example, the circuit in Fig. 4.38 contains two cores. The test of the entire circuit may be executed by testing one core after the other.

Under some circumstances, it may prove difficult to achieve the desired fault coverage in the glue logic between the partitions. The standard solution to this problem is to insert a *boundary register* as shown in Fig. 4.39. If the boundary register cannot be inserted for timing reasons, additional patterns may be added to recover some of the fault-coverage (e.g., Xu et al. 2007).

This type of partitioning in core-based design is a common industry practice and provides the basis for the scheduling techniques in Chap. 6. For example, IP cores wrapped according to IEEE Std. 1500 already provide the boundary register to support power-aware test scheduling.

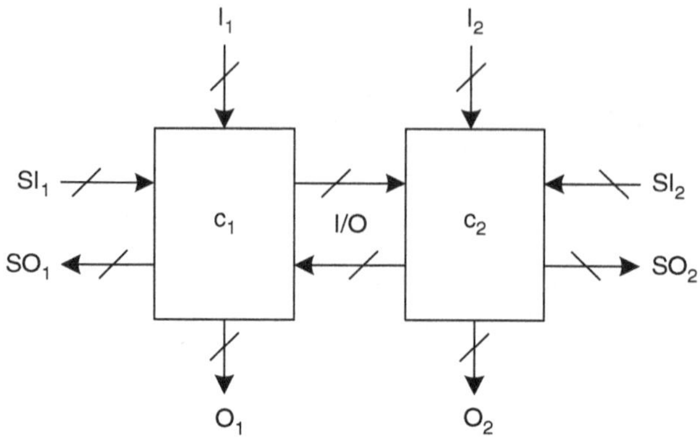

Fig. 4.38 A circuit with two cores

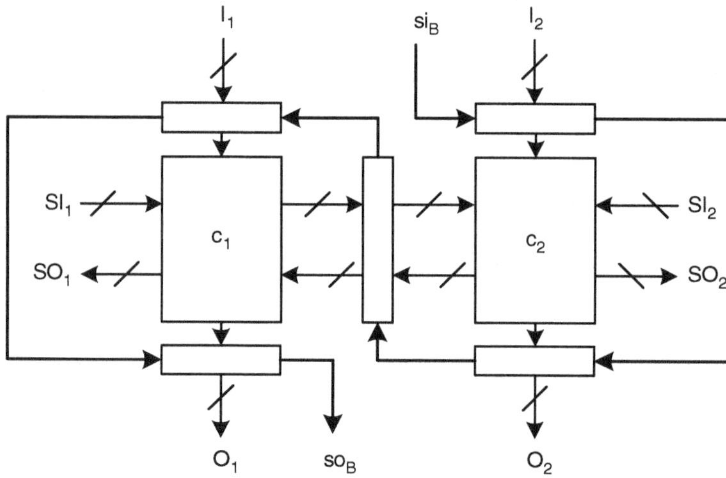

Fig. 4.39 Partition isolation of two cores using boundary registers

4.5.3 Partitioning of the Combinational Logic

If a more fine-grained partitioning is required, the combinational logic may be partitioned at the gate level. Here the partition isolation is implemented using multiplexers as shown in Fig. 4.40. In the example, A/B and C/D are the primary input vectors of the partitions p_1 and p_2 and Q and R are the output vectors. When testing partition p_1, the multiplexers are used to reroute part of the primary inputs and outputs of p_2 such that the boundary between p_1 and p_2 may be tested as well (and vice versa when testing p_2).

Girard et al. (2000) partition the circuit with the h-Metis *hyper graph partitioning* tool by Karypis and Kumar (1999), such that only a small number of multiplexers is

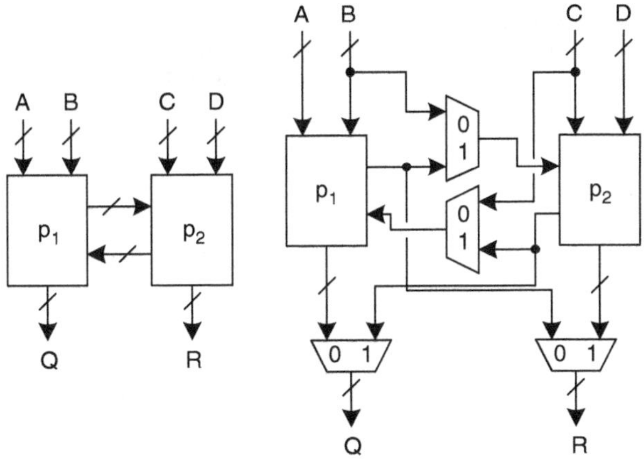

Fig. 4.40 Partitioning of the combinational logic by multiplexers

required. The resulting partitioning requires only a few percent overhead due to the multiplexers added to the circuit. During formulation of the hyper graph instance, a penalty is given to edges that are on the critical path of the circuit. Consequently, multiplexers are only inserted on paths with available timing slack.

4.6 Summary and Conclusions

This chapter has provided a short overview of the most important techniques for power-aware DfT.

Scan cells are the first targets when trying to avoid unnecessary switching activity. Gating the functional output of a scan cell during shifting is a suitable technique when very low power consumption is targeted. Clock gating of the scan cell is easily implemented as part of functional clock gating. This clock gating can be employed to reduce test power without impacting fault coverage by careful test planning. The segmentation, clustering, and ordering of scan cells reduce the power consumption and improve the effectiveness of power reduction techniques such as test planning and power-aware test generation. Finally, the partitioning techniques discussed in this chapter provide the basis for the test scheduling described in Chap. 6.

The discussed techniques are able to reduce the power consumption by an order of magnitude, especially when applying all of the techniques in a single DfT. However, it should be pointed out that test power should not be reduced below functional power to avoid the concerns related to undertesting of the circuit. For this reason, programmable techniques such as clock gating may be preferable over static techniques since they allow to adapt the power consumption to the test quality requirements observed in production.

Acknowledgments We would like to thank Frederik Heinrich for his indispensable support in creating the figures for this chapter. Further gratitude goes to our colleague Michael Kochte for thorough proofing and reviewing the content of this chapter as well as providing helpful comments. We would also like to thank the coeditors of this book for reviewing the chapter.

References

B. B. Bhattacharya, S. C. Seth, and S. Zhang, "Double-tree scan: A novel low-power scan-path architecture," in *Proceedings International Test Conference (ITC 2003), September 28–October 3, 2003, Charlotte, NC, USA*, pp. 470–479.

Y. Bonhomme, P. Girard, L. Guiller, C. Landrault, and S. Pravossoudovitch, "A gated clock scheme for low power scan testing of logic ICs or embedded cores," in *Proceedings 10th Asian Test Symposium (ATS 2001), November 19–21, 2001, Kyoto, Japan*, pp. 253–258.

Y. Bonhomme, P. Girard, L. Guiller, C. Landrault, and S. Pravossoudovitch, "Efficient scan chain design for power minimization during scan testing under routing constraint," in *Proceedings 2003 International Test Conference (ITC 2003), September 28–October 3, 2003, Charlotte, NC, USA*, pp. 488–493.

F. Brglez, P. Pownall, and R. Hum, "Applications of testability analysis: From ATPG to critical delay path tracing," in *Proceedings International Test Conference (ITC 1984), October 1984, Philadelphia, PA, USA*, pp. 705–712.

C.-A. Chen and S. K. Gupta, "A methodology to design efficient BIST test pattern generators," in *Proceedings IEEE International Test Conference (ITC 1995), October 21–25, 1995, Washington, DC, USA*, pp. 814–823.

W. Chung, T. Lo, and M. Sachdev, "A comparative analysis of low-power low-voltage dual-edge-triggered flip-flops," *IEEE Transactions on VLSI Systems*, vol. 10, no. 6, pp. 913–918, 2002.

M. Elm, H.-J. Wunderlich, M. E. Imhof, C. G. Zoellin, J. Leenstra, and N. Mäding, "Scan chain clustering for test power reduction," in *Proceedings 45th Design Automation Conference (DAC 2008), June 8–13, 2008, Anaheim, CA, USA*, pp. 828–833.

M. ElShoukry, M. Tehranipoor, and C. P. Ravikumar, "A critical-path-aware partial gating approach for test power reduction," *ACM Transactions on Design Automation of Electronic Systems (TODAES)*, vol. 12, no. 2, 2007, pp. 242–247.

R. Franch, P. Restle, N. James, W. Huott, J. Friedrich, R. Dixon, S. Weitzel, K. Van Goor, and G. Salem, "On-chip timing uncertainty measurements on IBM microprocessors," in *Proceedings IEEE International Test Conference (ITC 2007), October 23–25, 2007, Santa Clara, CA, USA*, pp. 1–7.

S. Gerstendörfer and H.-J. Wunderlich, "Minimized power consumption for scan-based BIST," in *Proceedings IEEE International Test Conference (ITC 1999), September 28–30, 1999 Atlantic City, NJ, USA*, pp. 77–84.

S. Gerstendörfer and H.-J. Wunderlich, "Minimized power consumption for scan-based BIST," *Journal of Electronic Testing*, vol. 16, no. 3, pp. 203–212, 2000.

P. Girard, L. Guiller, C. Landrault, and S. Pravossoudovitch, "A test vector inhibiting technique for low energy BIST design," in *Proceedings 17th IEEE VLSI Test Symposium (VTS 1999), April 25–30, 1999, San Diego, CA, USA*, pp. 407–412.

P. Girard, C. Landrault, L. Guiller, and S. Pravossoudovitch, "Low power BIST design by hypergraph partitioning: methodology and architectures," in *Proceedings IEEE International Test Conference (ITC 2000), October 2000, Atlantic City, NJ, USA*, pp. 652–661.

P. Girard, L. Guiller, C. Landrault, S. Pravossoudovitch, and H.-J. Wunderlich, "A modified clock scheme for a low power BIST test pattern generator," in *Proceedings 19th IEEE VLSI Test Symposium (VTS 2001), April 29–May 3, 2001, Marina Del Rey, CA, USA*, pp. 306–311.

I. Hamzaoglu and J. H. Patel, "Reducing test application time for full scan embedded cores," in *Proceedings International Symposium on Fault-Tolerant Computing (FTCS 1999), June 15–18, 1999, Madison, Wisconsin, USA*, pp. 260–267.

S. Hellebrand, H.-J. Wunderlich, and H. Liang, "A mixed mode BIST scheme based on reseeding of folding counters," in *Proceedings IEEE International Test Conference (ITC 2000), October 2000, Atlantic City, NJ, USA*, pp. 778–784.

A. Hertwig and H.-J. Wunderlich, "Low power serial built-in self-test," in *IEEE European Test Workshop (ETW 1998), May 27–29, 1998, Sitges, Barcelona, Spain*, pp. 49–53.

M. S. Hsiao, E. M. Rudnick, and J. H. Patel, "Peak power estimation of VLSI circuits: New peak power measures," *IEEE Transactions on VLSI Systems*, vol. 8, no. 4, pp. 435–439, 2000.

M. E. Imhof, C. G. Zoellin, H.-J. Wunderlich, N. Mäding, and J. Leenstra, "Scan test planning for power reduction," in *Proceedings 44th Design Automation Conference (DAC 2007), June 4–8, 2007, San Diego, CA, USA*, pp. 521–526.

G. Karypis and V. Kumar, "Multilevel k-way hypergraph partitioning," in *Proceedings 36th Conference on Design Automation (DAC 1999), June 21–25, 1999, New Orleans, LA, USA*, pp. 343–348.

L. Lai, J. H. Patel, T. Rinderknecht, and W.-T. Cheng, "Logic BIST with scan chain segmentation," in *Proceedings IEEE International Test Conference (ITC 2004), October 26–28, 2004, Charlotte, NC, USA*, pp. 57–66.

N. Parimi and X. Sun, "Toggle-masking for test-per-scan VLSI circuits," in *Proceedings 19th IEEE International Symposium on Defect and Fault-Tolerance in VLSI Systems (DFT 2004), October 10–13, 2004, Cannes, France*, pp. 332–338.

D. Pham, T. Aipperspach, D. Boerstler, M. Bolliger, R. Chaudhry, D. Cox, P. Harvey, P. Harvey, H. Hofstee, C. Johns, J. Kahle, et al., "Overview of the architecture, circuit design, and physical implementation of a first-generation Cell processor," *IEEE Journal of Solid-State Circuits*, vol. 41, no. 1, pp. 179–196, 2006.

P. M. Rosinger, B. M. Al-Hashimi, and N. Nicolici, "Scan architecture with mutually exclusive scan segment activation for shift- and capture-power reduction," *IEEE Transactions on CAD of Integrated Circuits and Systems (TCAD)*, vol. 23, no. 7, pp. 1142–1153, 2004.

R. Sankaralingam and N. A. Touba, "Reducing test power during test using programmable scan chain disable," in *Proceedings 1st IEEE International Workshop on Electronic Design, Test and Applications (DELTA 2002), January 29–31, 2002, Christchurch, New Zealand*, pp. 159–166.

A. Sehgal, J. Fitzgerald, and J. Rearick, "Test cost reduction for the AMD Athlon processor using test partitioning," in *Proceedings IEEE International Test Conference (ITC 2007), October 23–25, 2007, Santa Clara, CA, USA*, pp. 1–10.

O. Sinanoglu, I. Bayraktaroglu, and A. Orailoglu, "Test power reduction through minimization of scan chain transitions," in *Proceedings 20th IEEE VLSI Test Symposium (VTS 2002), April 28–May 2, 2002, Monterey, CA, USA*, pp. 166–172.

V. Stojanovic and V. Oklobdzija, "Comparative analysis of master-slave latches and flip-flops for high-performance and low-power systems," *IEEE Journal of Solid-State Circuits*, vol. 34, no. 4, pp. 536–548, 1999.

K. D. Wagner, "Clock system design," *IEEE Design & Test of Computers*, vol. 5, no. 5, pp. 9–27, 1988.

L. Whetsel, "Adapting scan architectures for low power operation," in *Proceedings IEEE International Test Conference (ITC 2000), October 2000, Atlantic City, NJ, USA*, pp. 863–872.

H.-J. Wunderlich, "PROTEST: A tool for probabilistic testability analysis," in *Proceedings 22nd ACM/IEEE Conference on Design Automation (DAC 1985), June 23–26, 1985, Las Vegas, Nevada, USA*, pp. 204–211.

H.-J. Wunderlich, "On computing optimized input probabilities for random tests," in *Proceedings 24th ACM/IEEE Design Automation Conference (DAC 1987), June 28–July 1, 1987, Miami Beach, FL*, pp. 392–398.

D. Xiang, K. Li, J. Sun, and H. Fujiwara, "Reconfigured scan forest for test application cost, test data volume, and test power reduction," *IEEE Transactions on Computers*, vol. 56, no. 4, pp. 557–562, 2007.

Q. Xu, D. Hu, and D. Xiang, "Pattern-directed circuit virtual partitioning for test power reduction," in *Proceedings IEEE International Test Conference (ITC 2007), October 23–25, 2007, Santa Clara, CA, USA*, pp. 1–10.

T. Yoshida and M. Watari, "MD-SCAN method for low power scan testing," in *Proceedings 11th Asian Test Symposium (ATS 2002), November 18–20, 2002, Guam, USA*, pp. 80–85.

X. Zhang and K. Roy, "Power reduction in test-per-scan BIST," in *6th IEEE International On-Line Testing Workshop (IOLTW 2000), July 3–5, 2000, Palma de Mallorca, Spain*, pp. 133–138.

C. Zoellin, H.-J. Wunderlich, N. Maeding, and J. Leenstra, "BIST power reduction using scan-chain disable in the Cell processor," in *Proceedings IEEE International Test Conference (ITC 2006), October 24–26, 2006, Santa Clara, CA, USA*, pp. 1–8.

Chapter 5
Power-Aware Test Data Compression and BIST

Sandeep Kumar Goel and Krishnendu Chakrabarty

Abstract The test data volume for manufacturing test of modern devices is increasing rapidly. This is due to the facts that the transistor count for these chips is increasing exponentially and the use of advanced technology introduces new physical and timing-related defects, which require new types of test. It is well known that power consumption during test is much higher than in the functional mode due to increased switching activity in test mode. Therefore, efficient techniques that minimize both test data volume and test power consumption are required. Techniques such as test data compression and built-in-self-test (BIST) are used commonly to handle the problem of increased test data volume. In this chapter, several low-power state-of-the-art test data compression and BIST techniques are discussed. Their advantages and disadvantages are discussed from area, performance, and power point of view.

5.1 Introduction

Predesigned *intellectual property* (IP) cores are now routinely used in large *system-on-a-chip* (SOC) designs. An SOC design integrates multiple cores (e.g., microprocessor, memory, DSPs, and I/O controllers) on a single piece of silicon. Despite these benefits, IP cores continue to pose several difficult test challenges. Two problems that are becoming increasingly important are power consumption during manufacturing test and test data volume. The precomputed test patterns provided by the core vendor must be applied to each core within the power constraints of the SOC. In addition, test data compression is necessary to overcome the limitations of the *automatic test equipment* (ATE), for example, tester data memory and I/O channel capacity.

S.K. Goel (✉)
LSI Corporation, Milpitas, CA, USA
e-mail: skgoel@ee.duke.edu

K. Chakrabarty
Duke University, Dhuram, NC, USA
e-mail: krish@ee.duke.edu

P. Girard et al. (eds.), *Power-Aware Testing and Test Strategies for Low Power Devices*,
DOI 10.1007/978-1-4419-0928-2_5,

A number of fault models such as stuck-at, transition delay, and shorts/opens are typically used today to achieve high defect coverage (Bushnell and Agrawal 2000). Often, n-detect stuck-at test sets are used with a large value of n to detect unmodeled faults. As a result, the test-data volumes for today's integrated circuits are prohibitively high. For example, the test-data volume for transition-delay faults is two to five times higher than that for stuck-at faults (Keller et al. 2004), and test patterns for such sequence- and timing-dependent faults are more important for newer technologies (Ferhani and McCluskey 2006). Moreover, due to shrinking process technologies, the physical limits of photolithographic processing and new materials such as copper interconnects, many new types of manufacturing defects cannot be accurately modeled using known fault models (Vermeulen et al. 2004). Although efficient methods exist today for fault-model-oriented test generation (Cox and Rajski 1994) (Yang et al. 2004), there is a lack of understanding on how best to combine the test sets thus obtained, that is, derive the most effective union of the individual test sets, without increasing the test-data volume excessively by simply using all the patterns for each fault model. As a result, the 2007 International Technology Roadmap for Semiconductors predicted that the test-data volume for integrated circuits will be as much as 38 times larger and the test-application time will be approximately 17 times larger in 2015 than it was in 2007 (ITRS 2007). Test compression is therefore essential to reduce the test-data volume and testing time.

Power consumption during testing is also a paramount consideration since excessive heat dissipation can damage the circuit under test. Since power consumption in test mode is higher than during normal operation, special care must be taken to ensure that the power rating of the SOC is not exceeded during test application (Zorian 1993). A number of techniques to control power consumption in test mode have been presented in the literature. These include test scheduling algorithms under power constraints (Chou et al. 1997), low-power built-in self-test (BIST) (Wang and Gupta 1999; Gerstendörfer and Wunderlich 1999; Girard et al. 1999; Corno et al. 2000), and techniques for minimizing power during scan testing (Wang and Gupta 1997a; Dabholkar et al. 1998; Sankaralingam et al. 2000). Power consumption and the resulting heat dissipation are especially important for SOCs since test scheduling techniques and test access architectures for system integration attempt to reduce testing time by applying scan/BIST vectors to several cores simultaneously (Chakrabarty 2000; Sugihara et al. 1998). Therefore, it is extremely important to decrease power consumption while testing the IP cores in an SOC.

In this chapter, we describe low-power *test compression* and *BIST* techniques. First, we present an overview of various low-power test techniques. A number of techniques to control power consumption in test mode have also been presented in the literature. These can be broadly classified as (a) structural, (b) algorithmic, and (c) tester based.

1. *Structural methods*: These methods, which do not address test data volume or testing time, are based on the following design techniques.

- *Gated scan chains*: These refer to schemes that use gating techniques to clock portions of the scan chain during scan operation. These techniques are discussed in Chap. 4 of this book.
- *Modified test pattern generator (TPG)*: Test generation circuits can be tailored to yield low-power vectors without significantly affecting the fault coverage and testing time (Girard et al. 2001; Corno et al. 2000). The method presented in Girard et al. (2001) is based on gated clock scheme for the TPG. The TPG is divided into two groups of flip-flops, and each group is activated by a clock running at half the speed of the normal clock. Another TPG based on cellular automata is presented in Corno et al. (2000).
- *Modified scan latch and vector inhibition*: Scan power can also be reduced by modifying the scan cell and adding gating logic to mask the scan path activity during shifting (Gerstendörfer and Wunderlich 1999). This approach has already been discussed in Chap. 4. The *vector-inhibiting* technique presented in Girard et al. (1999) provides a hardware solution to the power minimization problem and is shown to significantly decrease power consumption during BIST sessions. The method decreases the switching activity in the internal nodes of the circuit under test by avoiding useless patterns (nondetecting patterns) to be applied to thc circuit.
- *Scan chain organization*: The switching activity in the scan chain can be reduced by shortening and reorganizing the scan chains. The scan array solution presented in Xu et al. (2001) reduces power dissipation by using two-dimensional scan arrays, which reduce switching activity and allow the use of a slower scan clock. For more information on this approach, please refer to Chap. 4.

2. *Algorithmic methods*: These include *automatic test pattern generation* (ATPG) under power constraints, techniques based on test data compression and test scheduling algorithms.

   - *ATPG techniques*: ATPG techniques for generating vectors that lead to low-power testing are described in Chap. 3. However, while these techniques provide reduction in power consumption, they do not lead to any appreciable decrease in test-data volume.
   - *Test-data compression*: Test generation for low-power scan testing usually leads to an increase in the number of test vectors (Wang and Gupta 1997a). On the contrary, static compaction of scan vectors causes significant increase in power consumption during testing (Sankaralingam et al. 2000). While compacted vectors are useless if they exceed power constraints, uncompacted vectors cannot be used as they require excessive tester memory. Power minimization based on test-data compression was first presented in Chandra and Chakrabarty (2002).
   - *Test scheduling*: Test scheduling techniques for system integration attempt to reduce testing time by applying scan/BIST vectors to several cores simultaneously (Chou et al. 1997; Chakrabarty 2000). Test scheduling is typically carried out under power constraints since multiple cores are tested in parallel. For more information about this approach, please refer to Chap. 6.

3. *Tester frequency*: Reduction in power dissipation can be achieved by running the tester at a slower frequency. Although this method offers the simplest way to reduce power consumption, it leads to unacceptable testing times and is, therefore, impractical. We note that structural methods for reducing test power in SOCs require modification to the embedded cores, for example, via scan latch reordering (Dabholkar et al. 1998), scan chain, and scan cell redesign. This is usually not feasible for IP cores. ATPG techniques are also unfeasible for IP cores, since they require gate-level structural models (Pomeranz et al. 1999). Moreover, ATPG techniques that address test power do not directly consider test data volume and testing time issues. We, therefore, focus in this chapter on test-data compression that can reduce test power, test-data volume, and testing time simultaneously.

5.2 Coding-Based Compression Methods

In this section, we review various coding-based low-power compression methods. Coding methods typically target runs of 1s and 0s in the test data. These runs can be compressed into shorter code words.

5.2.1 Golomb Code

It was first shown by Chandra and Chakrabarty (2002) that scan vector compaction does not always leads to higher power consumption. In particular, it was demonstrated that *Golomb coding* can be used to decrease both peak and average power for IP cores. In this way, there is no need to either reduce the scan clock rate for low-power or add blocking logic to the scan cells. The use of a low-cost on-chip decoder allows us to achieve significant test-data compression, and the decompressed scan vectors cause very little switching activity in the scan chains during test application. In order to reduce test-data volume, Golomb coding maps don't care bits to 0s, so as to generate long runs of 0s. Such a strategy is also beneficial for reducing shift power for scan testing, since long runs of 0s can be scanned in without causing any toggling in the scan cells.

We first review Golomb coding and its application to test data compression in Chandra and Chakrabarty (2001). The major advantages of Golomb coding of test data include high compression, analytically predictable compression results, and a low-cost and scalable on-chip decoder. In addition, an interleaved decompression architecture allows multiple cores in an SOC to be tested concurrently using a single ATE I/O channel.

The first step in the encoding procedure presented in Chandra and Chakrabarty (2001) is to map all don't-care bits in the test set to zero. However, such a mapping of don't-cares to 0s is not the most effective strategy for low-power test compression. We explain later how the don't-cares must be mapped to minimize power consumption during testing.

The next step is to select the Golomb code parameter m referred to as the group size. Once m is determined, for example, using the methods described in Chandra and Chakrabarty (2001), the runs of zeros in the test data stream are mapped to groups of size m (each group corresponding to a run length). The number of such groups is determined by the length of the longest run of zeros in the test set. The set of run-lengths $\{0, 1, 2, \ldots, m-1\}$ forms group A_1; the set $\{m, m+1, \ldots, 2m-1\}$, group A_2; etc. In general, the set of run-lengths $\{(k-1)m, (k-1)m+1, \ldots, km-1\}$ comprises group A_k. To each group A_k, we assign a group prefix of $(k-1)$ ones followed by a zero. We denote this by $1^{(k-1)0}$. If m is chosen to be a power of 2, that is, $m = 2^N$, each group contains 2^N members and a $\log_2 m$-bit sequence (tail) uniquely identifies each member within the group. Thus, the final code word for a run-length L that belongs to A_k group is composed of two parts: a group prefix and tail. The prefix is $1^{(k-1)0}$ and the tail is a sequence of $\log_2 m$ bits. The encoding process is illustrated in Fig. 5.1 for $m = 4$.

Since the decoder for Golomb coding needs to communicate with the tester, and both the codewords and the decompressed data can be of variable length, proper synchronization must be ensured through careful design. In particular, the decoder must communicate with the tester to signal the end of a block of variable-length decompressed data. These and other related decompression issues are discussed in detail in Chandra and Chakrabarty (2001).

For scan vectors, the dynamic power consumption during testing depends on the number of transitions that occur in the scan chain as well as on the number of circuit elements that switch during the scan-in and scan-out operations. Power estimation models based on the switching activity of circuits have been presented in the literature (Girard et al. 1999; Sankaralingam et al. 2000). Low-power test compression methods often use the *weighted transitions metric* (WTM), introduced and validated in Sankaralingam et al. (2000), to estimate the power consumption due to scan vectors. The WTM models the fact that the scan in power for a given vector depends not only on the number of transitions in it but also on their relative positions. The weighted transitions count metric is also strongly correlated to the switching

Group	Run-Length	Group Prefix	Tail	Codeword
A_1	0	0	00	000
	1		01	001
	2		10	010
	3		11	011
A_2	4	10	00	1000
	5		01	1001
	6		10	1010
	7		11	1011
A_3	8	110	00	11000
	9		01	11001
	10		10	11010
	11		11	11011

Fig. 5.1 An illustration of the Golomb coding procedure

activity in the internal nodes of the core under test during the scan-in operation. It was shown experimentally by Sankaralingam et al. (2000) that scan vectors that have higher WTM dissipate more power in the core under test. Therefore, the low-power compression method in Chandra and Chakrabarty (2002) reorders the test patterns such that the WTM measure is minimized. This problem is mapped to the traveling salesman problem and solved using a heuristic method.

5.2.2 Alternating Run-Length Code

Another low-power test compression method is based on the use of the *alternating run-length code*, which is based on the *frequency-directed run-length* (FDR) code (Chandra and Chakrabarty 2003b). We first review FDR coding and its application to test-data compression (Chandra and Chakrabarty 2003a).

The FDR code is a data compression code that maps variable-length runs of 0s to variable-length code words. The encoding procedure is illustrated in Fig. 5.2. As an example, consider a run of six 0s (0000001) in the input stream. This run belongs to group A3 and it is mapped to the code word 110000. The reader is referred to Chandra and Chakrabarty (2003a) for a detailed discussion and motivation for the FDR code.

It was shown by Chandra and Chakrabarty (2003a) that the FDR code is very efficient for compressing data that has few 1s and long runs of 0s. However, for data streams that are composed of both runs of 0s and runs of 1s, the FDR code is rather inefficient. A different code is presented by Chandra and Chakrabarty (2003b) to efficiently compress both runs of 0s and 1s.

Figure 5.3 illustrates the encoding procedure for the new alternating run-length code. The alternating run-length code is also a variable-to-variable-length code and consists of two parts – group prefix and tail. The prefix identifies the group in which

Group	Run Length	Group Prefix	Tail	Codeword
A_1	0	0	0	00
	1		1	01
A_2	2	10	00	1000
	3		01	1001
	4		10	1010
	5		11	1011
A_3	6	110	000	110000
	7		001	110001
	8		010	110010
	9		011	110011
	10		100	110100
	11		101	110101
	12		110	110110
	13		111	110111

Fig. 5.2 Illustration of the FDR code

Group	a=0 Run-Length of 0s	a=1 Run-Length of 1s	Group Prefix	Tail	Codeword
A_1	0	0	0	0	00
	1	1		1	01
A_2	2	2	10	00	1000
	3	3		01	1001
	4	4		10	1010
	5	5		11	1011
A_3	6	6	110	000	110000
	7	7		001	110001
	8	8		010	110010
	9	9		011	110011
	10	10		100	110100
	11	11		101	110101
	12	12		110	110110
	13	13		111	110111

Fig. 5.3 The alternating run-length code

Input data stream: 000000111111000001 (18-bits)

FDR encoded data: 1100000000000000011010 (22-bits)

Alternating run-length encoded data:- 11000010111010 (14-bits)

|a = 0|a = 1|a = 0|

Fig. 5.4 An example of FDR and alternating run-length coding

the run-length lies and the tail identifies the member within the group. An additional parameter associated with this code is the alternating binary variable a. The encoding produced by the alternating run-length code for a given run-length depends on the value of a. If $a = 0$, the run-length is treated as a run of 0s. On the other hand, if $a = 1$, the run-length is treated as a run of 1s. Note that the value of a for the different runs are not added to the encoded data stream.

Figure 5.4 shows the encoded data obtained using the two codes for a data stream composed of interleaved runs of 0s and 1s. We observe that the size of the FDR-encoded data set (22 bits) is larger than the size of the input data set (18 bits); hence, the FDR code provides no compression for this case. On the other hand, the size of the alternating run-length-encoded data set (14 bits) is smaller than the size of the input data set. Therefore, we are able to achieve compression with the new code. We also note that $a = 0$ is used for compressing a runs of 0s, and $a = 1$ is used for compressing a runs of 1s, and $a = 0$ is then used for compressing the next run of 0s. Hence, a is inverted after each run is encoded and it keeps alternating between 0 and 1 thereafter.

A low-cost decompression architecture is presented in Chandra and Chakrabarty (2003a), and it is demonstrated that the mapping of don't-cares to 1s and 0s can be done in the same way as in Chandra and Chakrabarty (2002) to minimize the WTM measure. A careful mapping of the don't-cares to 0s and 1s, followed by alternating run-length coding of the resulting test data, not only provides reduction in test-data volume but also minimizes the scan power dissipation.

5.2.3 Recent Advances in Coding-Based Compression Methods

Since early work on using Golomb and alternating run-length coding, several other coding methods have been reported for low-power test-data compression. We first review compression based on a combination of run-length and *Huffman coding* referred to as *RL-Huffman encoding* (Nourani and Tehranipoor 2005). The key idea here is to perform run-length coding in the first step, where the mapping of don't-cares to 1s and 0s is based on the strategy adopted for run-length coding, that is, maximize the length of a run (a contiguous stream of 1s or 0s). The run-lengths are referred to as block lengths. In the next step, Huffman coding is used to encode the block length values. As an important benefit, this approach allows the compression to be traded-off with the decompressor cost. Figure 5.5 shows an example of run-length coding applied to a set of scan vectors.

Figure 5.6 highlights the subsequent encoding of block lengths using Huffman coding. The Huffman tree for generating the code words is shown in Fig. 5.6a, whereas the code words and the reduction in data volume ("savings," in number of bits) is shown in Fig. 5.6b.

In another low-power compression method, run-length coding is combined with *scan-latch ordering* (Rosinger et al. 2001). The coding method is based on the encoding of runs of both 1s and 0s using the same code word, but distinguishing these two runs using an extra "sign bit." This compression technique is referred to as symmetric coding in Rosinger et al. (2002a), where techniques are described to achieve

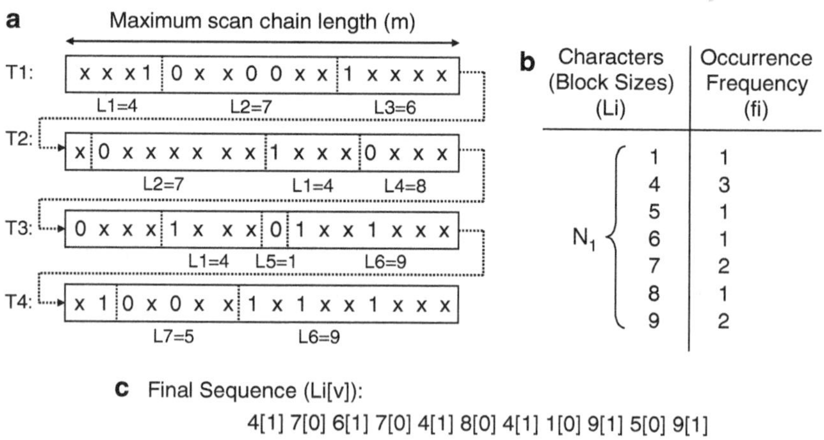

Fig. 5.5 The application of run-length coding to a small example. (**a**) Set of scan vectors. (**b**) Block sizes and frequency. (**c**) Final sequence

a

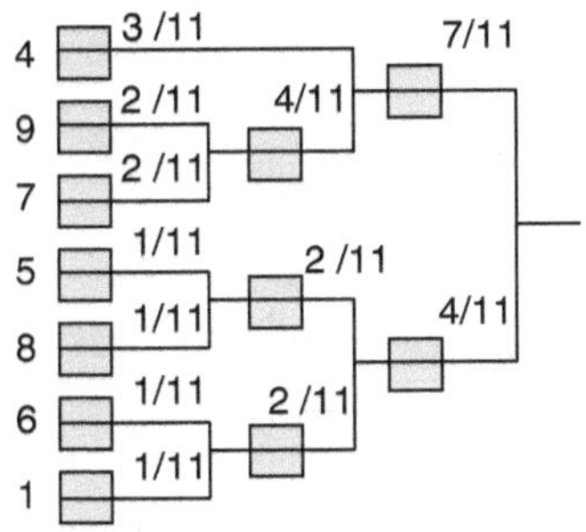

b

	Characters (Li)	Occurrence Frequency (fi)	Huffman Code (Ci)	Saving (Si)
N_1	4	3	00	+6
	9	2	010	+12
	7	2	011	+8
	5	1	100	+2
	8	1	101	+5
	6	1	110	+3
	1	1	111	−2
				S=+34

Fig. 5.6 Huffman coding applied to the example of Fig. 5.5. (**a**) Huffman tree. (**b**) Huffman codes and resulted savings

trade-offs between scan power and test-data compression. Since test application at the SOC level must consider system-level issues arising from the integration of other cores, the compression strategy for a core must also consider the compression/decompression solution used for other cores in the system. This problem is addressed in Gonciari et al. (2003), where test-data compression and test power are addressed from a system integrator's perspective.

Low-power test compression based on *selective encoding* (Wang and Chakrabarty 2005; Wang and Chakrabarty 2008) is described in Li et al. (2008). The key idea here is to utilize selective encoding to reduce capture power, which has recently emerged as a serious concern due to high clock frequencies in functional mode. The selective encoding method is illustrated in Fig. 5.7. Scan slices are mapped to code words by assigning 1s and 0s appropriately to their don't-cares. The mapping of don't-cares in each scan slice is not unique; different mappings can lead to the same reduction in test-data volume. It is shown in Li et al. (2008) that this flexibility in don't-care mapping can be exploited to reduce the capture power or eliminate capture violations during scan testing.

In order to reduce shift power for scan testing with selective encoding, several techniques for mapping don't-care bits are presented in Badereddine et al. (2008). All 0-filling (all don't-cares in a scan slice are mapped to zero before compression), all 1-filling (all don't-cares are mapped to 1), and adjacent filling (referred to as MT-filling) are considered. It is shown that while all these mapping techniques lead to high compression, the MT-filling strategy reduces peak and power consumption the most, since it reduces switching activity in the scan chains.

In more recent work, the entropy of the test data is used to guide the filling of don't-care bits for a compression based on fixed-length symbols. The filling is accompanied with a check for capture-power violations, thereby test compression is achieved without any capture power violations. The *entropy* is a measure for the disorder of a data set and it presents a fundamental limit on the amount of compression that can be achieved, and it is independent of the compression scheme (Balakrishnan and Touba 2007). Consider the test data after don't-care-filling as shown in Table 5.1.

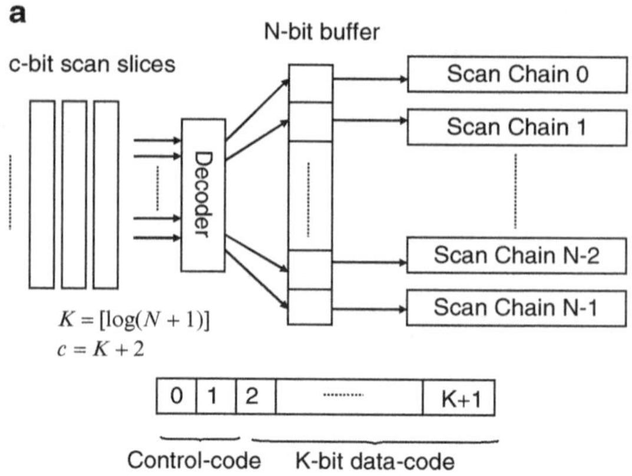

b

Slice	Slice code: Control code	Slice code: Data code	Description
XX00 010X	00	0101	Start a new slice, map X to 0, set bit 5 to 1
1110 0001	00	0111	Start a new slice, map X to 0, set bit 7 to 1
	11	0000	Enter group-copy-mode, starting from bit 0 (i.e., group 0)
	11	1110	The data is 1110
XXXX XX11	01	1000	Start a new slice, map X to 1, no bits are set to 0

Fig. 5.7 Illustration of the selective encoding method. (**a**) Concept. (**b**) Example

Table 5.1 Illustration of entropy computation for test-data compression

Vector 1	0001	0011	0001	1100	0000
Vector 2	0000	1001	0000	0010	1000
Vector 3	0000	0000	0000	0011	0010
Vector 4	1000	0001	1001	0010	0000

i	Symbol x_i	Frequency	Probability p_i	Huffman code
1	0000	7	0.35	10
2	0001	3	0.15	110
3	0010	3	0.15	111
4	1000	2	0.10	010
5	0011	2	0.10	000
6	1001	2	0.10	011
7	1100	1	0.05	0010

The number of occurrences for each type of symbol is shown in column "frequency," and the "probability p_i" is calculated by dividing the frequency with the number of symbols (n) in the data set. On the basis of the above table, its entropy can be calculated as $-\Sigma_i \ p_i \log p_i = 2.564$. The maximum compression is given by (*symbol length–entropy*)/*symbol length* = 35.9%. Higher the entropy, the less the potential test compression that can be achieved. In Liu and Xu (2009), the entropy of the test set is used to evaluate the impact of don't-care-filling on compression, whereas the scan-capture power is evaluated by the Hamming distance between each test pattern and its response (i.e., capture transitions in the state elements).

5.3 LFSR-Decompressor-Based Compression Methods

LFSR-based decompressors have been widely adopted in industry because they can be used to drive many scan chains in parallel, they are easy to implement, and the associated compression methods provide high compression. Over the past few years, these techniques have been enhanced to address the need for low power as well as low test-data volume. The key idea behind these methods is to generate seeds for LFSR reseeding such that the don't-cares in test cubes are filled to reduce the transition count.

One of the first techniques for reducing test power for *LFSR reseeding* was presented in Rosinger et al. (2002b). Two LFSRs are used in the underlying test architecture. The main LFSR generates the test cube through conventional reseeding. An extra "masking" LFSR is used to generate a set of mask bits. If the number of 1s in a test cube is less than the number of 0s, then the outputs of the two LFSRs are ANDed together. The mask cube has a 1 for each specified 1 in the test cube and a don't-care for each specified 0 or don't-care in the test cube. If the number of 0s in the test cube is less than the number of 1s, the outputs of the two LFSRs are ORed together. In this case, the mask cube has a 0 for each specified 0 in the test cube and a don't-care for each specified 1 or a don't-care in the test cube. A seed is computed for the extra masking LFSR so that it generates the mask cube. Test power is reduced because the outputs of the two LFSRs are ANDed or ORed, thus reducing the probability of transitions.

An improvement over Rosinger et al. (2002b) for low-power LFSR reseeding is presented in Lee and Touba (2007). The main idea behind this encoding algorithm is to exploit the fact that the number of transitions in a test cube is always less than the number of specified bits in it. Thus, rather than using LFSR reseeding to directly encode the specified bits as in conventional LFSR reseeding, the method in Lee and Touba (2007) divides the test cube into blocks and only uses LFSR reseeding to produce the blocks that contain transitions. For blocks that do not contain any transition, the logic value fed into the scan chain is simply held constant. This approach reduces the number of transitions in the scan chains and in most cases also reduces the total number of specified bits that must be generated by the LFSR as compared with conventional LFSR reseeding.

More recently, low-power test application has been integrated in an embedded test environment (Mrugalski et al. 2007; Czysz 2008a, b). The hardware decompressor for *Embedded Deterministic Test* (EDT) relies on a ring generator followed by a phase shifter. The low-power decompressor presented in Mrugalski et al. (2007) is shown in Fig. 5.8, where (a) and (b) refer to different implementations of the overall architecture. The same data can now be provided to the scan chains for a number

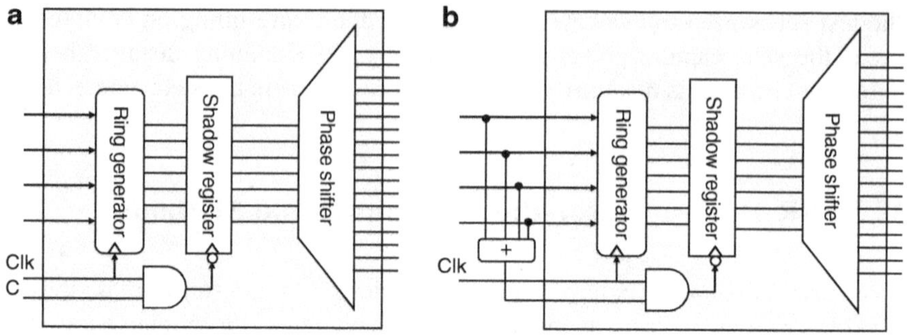

Fig. 5.8 The low-power decompressor from Mrugalski et al. (2007). (**a**) Implementation 1. (**b**) Implementation 2

of shift cycles through a shadow register placed between the ring generator and the phase shifter. The shadow register captures and saves, for a number of cycles, a desired state of the ring generator, whereas the generator itself keeps advancing to the next state needed to encode another group of specified bits. As a result, independent operations of the ring generator and its shadow register allow virtually any state that causes no conflicts with specified bits to reduce the transition count. A new test-cube encoding scheme is also presented to achieve high compression with this decompression architecture.

5.4 Broadcast-Scan-Based Compression Methods

In this section, we discuss broadcast methods and scan architectures that target low power in addition to the reduction of test-data volume and test time.

An alternative class of compression methods targeting low test power is based on broadcast scan. The segmented addressable scan architecture simultaneously leads to low test power and low test-data volume (Al-Yamani et al. 2005). This architecture combines the benefits of Illinois scan (Hmazaoglu 1999) (reduced test-data volume) and scan-segment decoding (Rosinger 2004) (low test power). It allows multiple segments to be loaded simultaneously without the use of any mapping logic. A multiple-hot decoder is used to allow the simultaneous loading of multiple segments. To reduce test power, the segments that are not loaded in a given round are not clocked. This approach also allows faster clocking of the test patterns within the same power budget.

Finally, recent work on progressive random access scan (PRAS) offers a promising alternative approach to reduce test-data volume, test time, and test power (Baik and Saluja 2005). The main idea here is to provide individual access (i.e., random access) to each scan cell. Such accessibility to every scan cell eliminates unnecessary switching activity during scan shifting, and it reduces test-data volume and test time. In the PRAS architecture, scan cells are configured as an SRAM-like grid structure. Some additional peripheral and test control logic is added. During test mode, scan cells in one of the rows is enabled, allowing a read/write operation. The SRAM-like read/write operation is achieved by modifying every regular flip-flop into a grid-accessible flip-flop.

5.5 Low-Power BIST Techniques

The traditional approach of manufacturing testing using only external automated test equipment (ATE) is becoming more and more difficult and costly. Factors that drive the test costs up are the increase in pin count, test-data volume, speed, and corresponding required ATE accuracy. Especially the test-data volume has risen dramatically due to a combination of growth in transistor count and new advanced test methods [such as small delay testing (Lin et al. 2006; Yilmaz et al. 2008; Goel et al. 2009)] that add significantly to the test set size. Furthermore, detection of small timing-related defects requires at-speed or faster than at-speed test application (Kruseman et al. 2004; Turakhia et al. 2007), which is becoming very difficult (if not impossible) to achieve with the conventional ATE for modern GHz chips.

Several test methods are applied to handle the problem of increased test-data volume. Various test-data compression techniques described in the previous sections reduce the demands on both vector memory and test application time but still require the presence of an ATE. Another effective approach is *Built-In Self Test* (BIST) (Bardell et al. 1987), in which a circuit is equipped with an on-circuit stimuli generator and a response evaluator. A BIST-equipped circuit tests itself, thereby reducing the ATE storage requirement to almost nil. Furthermore, BIST provides at-speed test application and enables the use of low-cost ATE as requirements on timing accuracy, vector memory, and pin count are strongly reduced. Also, BIST offers superior test quality because a large number of patterns can be applied to the circuit using on-chip TPG, which increases the detection probability of unmodeled defects (Benware et al. 2003). Other advantages of BIST are board-level/system-level test and in-field test of critical applications.

BIST for embedded memories is mature and widely used in industry, whereas the industrial use of BIST for random logic also called "*Logic BIST*" is increasing. In BIST, pattern generation and evaluation is done on-chip and, therefore, the basic BIST architecture requires three additional components to be added to the circuit-under test. As shown in Fig. 5.9, these three components are (1) test stimuli generator, (2) test response analyzer, and (3) test controller or test scheduler.

A *test stimuli generator* generates stimuli bits that are required to test the circuit under test. Examples of test stimuli generators are ROM with stored patterns, a counter, a *linear feedback shift register* (LFSR), and *cellular automata* (CA). A *test response analyzer* monitors the response from the circuit and performs a comparison between the observed response and the expected response. On the basis of the comparison, it generates a pass/fail signature. Typical examples of response analyzer are

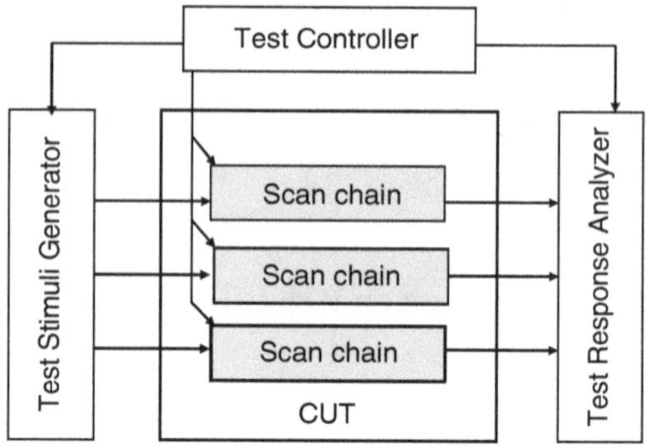

Fig. 5.9 Basic architecture of BIST

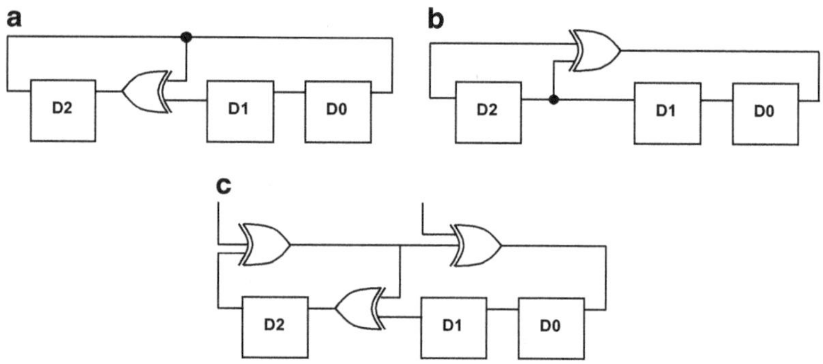

Fig. 5.10 Example of LFSR: (**a**) internal XOR, (**b**) external XOR, and (**c**) MISR

comparator, a parity tree, and a linear circuit with LFSR used as signature analyzer. The test controller is required to start and stop the test.

Depending on the TPG type, different BIST schemes provide a trade-off between test application time (also test quality) and the area overhead. In a ROM-based pattern generator BIST (Kuban and Bruce 1984), test patterns are generated using traditional automatic test pattern generation (ATPG) tool and fault simulated to achieve the maximum fault coverage. Later, these patterns are stored on-chip in the ROM to support BIST functionality. Depending on the number of patterns that need to be stored in the ROM, this approach requires a large area overhead.

Because of low area overhead and low design efforts, BIST schemes using pseudorandom pattern generators are popularly used. Using an LFSR, a large number of pseudorandom patterns can be easily generated. Another advantage of using LFSR is that it can be integrated with a linear circuit to obtain a *Multiple Input Signature Register* (MISR), which is very popularly used as a response analyzer. Typically a LFSR consists of D flip-flops and linear logic elements (XOR gates). Depending on the location of XOR gates, there are two types of LFSRs (1) Internal-XOR and (2) External XOR. Figure 5.10a,b show an example of both types of LFSRs, whereas Fig. 5.10c shows an example of a MISR. Every LFSR is associated with a polynomial, which can be used to understand the outputs from the LFSR. Detailed information about LFSR, MISR construction, and polynomial algebra can be found in Bardell et al. (1987).

From the test application point of view, most BIST schemes can be classified into two categories (Agrawal et al. 1993a, b): (1) *test-per-clock* and (2) *test-per-scan*. In a test-per-clock BIST, a test vector is applied and test responses are observed every single clock cycle. While, in the test-per-scan BIST, a test vector is applied and test responses are observed every single scan cycle. The length of a scan cycle depends on the length of the longest scan chain in the design. A scan cycle is equal to the length of the longest scan chain plus one or more capture cycles. Depending on the length of the scan cycle, usually test-per-scan BIST approaches are slower than test-per-clock approaches. As most modern designs contain multiple scan chains, test-per-scan BIST schemes are more widely used than test-per-clock schemes. An example of test-per-scan BIST scheme is STUMP (*self-test using MISR and parallel shift-register sequence generator*) (Bardell et al. 1987). In this scheme, outputs of pseudorandom pattern generator are fed to a linear network. The linear network consists of XOR gates and also known as phase shifter. Outputs of the linear network are directly connected to the scan chain inputs. The scan chain outputs are connected to the MISR inputs. The purpose of linear network is to minimize the correlation between the bits shifted into the scan chains to provide higher detection of faults.

As explained in Bardell et al. (1987), one of the drawbacks of *pseudorandom patterns* (LFSR) is that they provide low coverage for hard-to-detect or *random-pattern resistant* (RPR) faults. RPR faults have very low random detection probability, which is defined as the probability that a random pattern will detect the fault. To improve the fault coverage for RPR faults, efficient techniques such as *test point insertion* (TPI) (Tamarapalli and Rajski 1996; Touba and McCluskey 1996) and weighted pseudorandom patterns (Bardell et al. 1987; Waicukauski et al. 1989; Hartmann and Kenmnitz 1993; Pomeranz and Reddy 1993) are used. In TPI techniques, detection probabilities for RPR faults are improved by inserting control and/or observe points in the design. In the weighted pseudorandom pattern techniques, weights are assigned to the inputs such that outputs of a pseudorandom pattern generator have nonuniform signal probabilities instead of uniform signal probability of 0.5.

As mentioned in Sect. 1, power consumption during test is significantly higher than during normal operation (Zorian 1993). Especially for BIST, low correlation between successive vectors generated from LFSR results in very high switching activity that increases the power consumption during test. If the power consumption during any cycle (also referred to as *peak power consumption*) exceeds the maximum power budget, it can permanently damage the circuit under test. Higher power consumption combined with long test application time for BIST due to large number of patterns can also elevate the circuit temperature. In the worst case, excessive heat dissipation can results in hot spots in the design that can affect the reliability of the circuit. Another problem is the increased current density (di/dt) that can also affect the reliability of the circuit (Weste and Eshraghian 1992). Therefore, from the power point of view, it is important to minimize both the peak power consumption as well as the total energy consumption during test. Various efficient power minimization techniques for BIST have been proposed in literature that can be classified into four categories: (1) vector inhibition and selection, (2) modified TPG, (3) modified scan and reordering, and (4) circuit partitioning and test scheduling. In the next Sects. 5.5.1–5.5.4, we review some of the widely used techniques for each category.

5.5.1 Vector Inhibition and Selection

As not all vectors generated by the pseudorandom pattern generator detect faults, these vectors can be masked without impacting the fault coverage. Masking of these patterns reduces the switching activity in the circuit and hence minimizes the power consumption. However, masking and selection of vectors require additional circuitry to be added to the circuit.

In Corno et al. (1999a, b), *test vector selection* and masking scheme are presented. In the proposed scheme, the output vector sequence generated by the TPG is fault simulated and set-covering based algorithms are used to find the set of patterns that detects faults; these patterns are refereed to as *useful patterns*. The basic flow for this scheme is shown in Fig. 5.11. The Enable block generates a signal "1" for every useful vector and Mask circuit passes the vector generated by the LFSR to the circuit under test. For every *nondetecting vector*, Enable block generates a "0," indicating that the sequence generated by the LFSR should be masked.

In Girard et al. (1999) and Manich et al. (2000), a *test vector inhibiting* technique is proposed to mask sequences of test vectors generated by the LFSR. The motivation behind the proposed approach is that not all pseudorandom vectors generated from the LFSR detect fault, and the subsequences of consecutive nondetecting vectors are often of great length. Therefore, inhibiting the LFSR during the generation of nondetecting vectors can reduce the switching activity in the circuit without impacting the fault coverage and test application time. An example of vector inhibition scheme is shown in Fig. 5.12a. The output of LFSR is connected to a decoding logic that generates a "1" as soon as it detects a nondetecting vector. A D-flip-flop clocked by the decoding logic is used to control the passing or inhibition of generated vector to the circuit. Knowledge about nondetecting patterns can be obtained by fault simulating the patterns generated by the LFSR. Please note that to further decrease the switching activity, multiple subsequences can be inhibited. However, this will increase the size of decoding logic and associated circuitry.

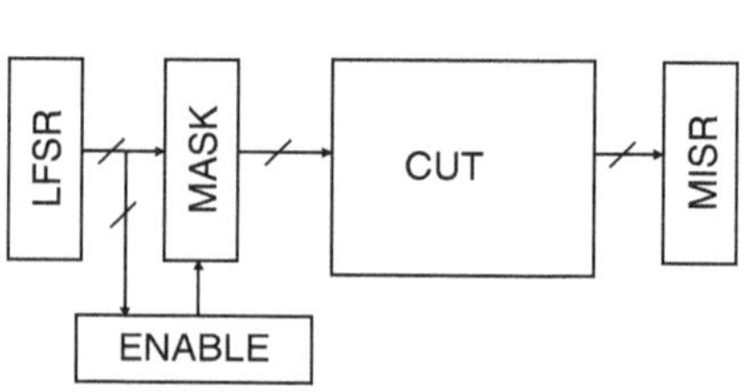

Fig. 5.11 Vector selection-based low-power BIST

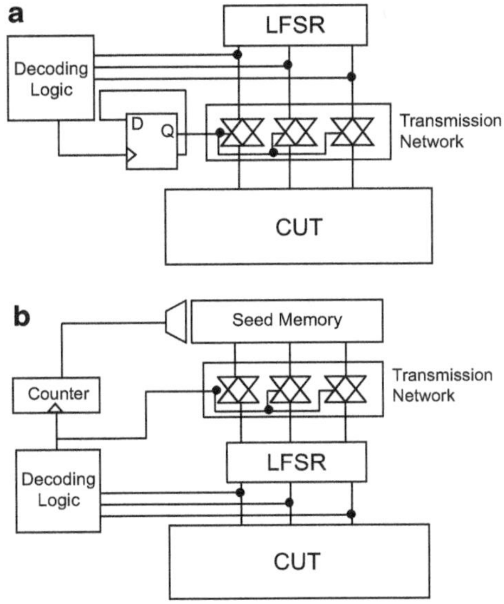

Fig. 5.12 Vector inhibition scheme for low-power BIST. (**a**) Original vector inhibition, (**b**) vector inhibition combined with LFSR reseeding technique

Vector inhibition technique can also be combined with LFSR reseeding (Hellebrand et al. 1992) technique (as shown in Fig. 5.12b) to minimize the switching activity while maximizing the coverage for RPR faults. In LFSR reseeding, a LFSR is loaded multiple times with different seeds to generate vectors that detect a large number of faults including RPR faults. In the proposed technique, seed memory consists of two parts: (1) one part contains seeds required for generating vectors for RPR faults and (2) second part contains seeds that are used to inhibit portions of pseudorandom sequences that do not detect any fault. Primarily, a seed in the second part corresponds to the last nondetecting vector in a test sequence.

5.5.2 Modified TPG

To minimize the switching activity and related power dissipation, concept of *dual-speed LFSR* (DS-LFSR) is proposed in (Wang and Gupta 1997b). A DS-LFSR consists of two LFSRs: (1) one *normal-speed LFSR* clocked by the normal clock, and (2) a *slow-speed LFSR* clocked by a clock whose speed is 1/dth of the normal clock. The proposed method reduces the switching activity at the circuit inputs connected to the slow-speed LFSR while achieving similar or higher fault coverage. To understand how a sequence generated by a normal LFSR can be generated by a DS-LFSR, let's consider the sequence shown in Fig. 5.13a.

The same sequence of 4-bit vectors can be partitioned into two groups: S and N as shown in Fig. 5.13b. The partitioned sequence shown in Fig. 5.13b can be reordered according to the sequence shown in Fig. 5.13c and can be generated by two independent LFSRs. For the N portion of the reordered sequence, a normal-speed LFSR

a: Original sequence	b: Partitioned sequence S	b: Partitioned sequence N	c: Reordered sequence S	c: Reordered sequence N
0001	00	01	01	01
1000	10	00	01	10
0100	01	00	01	01
0010	00	10	01	00
1001	10	01	10	01
1100	11	00	10	10
0110	01	10	10	11
1011	10	11	10	00
0101	01	01	11	01
1010	10	10	11	10
1101	11	01	11	11
1110	11	10	11	00
1111	11	11	00	01
0111	01	11	00	10
0011	00	11	00	11
0011	00	00	00	00

(a → b: Partitioned; b → c: Reordered)

Fig. 5.13 Vector reordering for DS-LFSR approach. (**a**) Original sequence, (**b**) partitioned sequence, and (**c**) reordered sequence

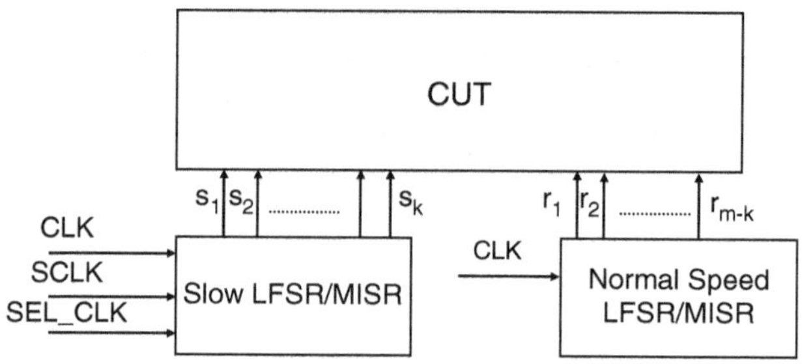

Fig. 5.14 DS-LFSR-based low-power BIST

can be used. However, for the S portion of the reordered sequence, a slow-speed LFSR can be used, as its output changes only every fifth cycle. For an m-bit original LFSR, the corresponding DS-LFSR will have a k-bit slow-speed LFSR driven by a clock with period $2^{(m-k)}$ times the period of the normal clock, and $(m - k)$ bit normal-speed LFSR driven by normal clock. Figure 5.14 shows the complete architecture of the DS-LFSR-based BIST.

The slow-speed LFSR is connected to both normal clock (CLK) and slow-clock (SCLK). The control signal SEL_CLK selects SCLK when slow LFSR is used as a TPG, whereas normal clock CLK is selected when the CUT is in normal mode or the slow LFSR is configured as a multiple input signature register (MISR). By

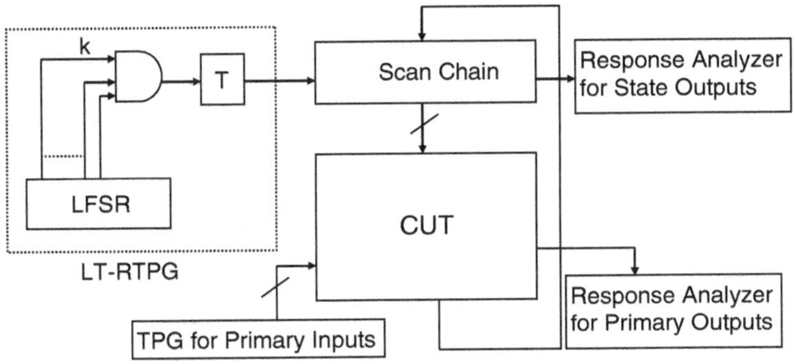

Fig. 5.15 LT-RTPG low-power BIST architecture

maximizing the number of inputs connected to the slow LFSR, significant reduction in power consumption can be obtained.

Another scheme called *low-transition random TPG* (LT-RTPG) for switching activity minimization is described in Wang and Gupta (1999) and Wang and Gupta (2006). The key idea behind LT-RTPG is that in order to minimize the transition activity at circuit inputs during shift cycles, neighboring scan flip-flops in a scan chain should be assigned identical values at most times. Reduced transition activity at the circuit inputs also reduces switching activity in the combinational logic connected to the circuit inputs. Figure 5.15 shows the basic architecture for an LR-RTPG-based BIST scheme; a LT-RPTG consists of a LFSR, a k-input AND gate, and a *T (toggle) flip-flop*.

As the probability that the T flip-flop toggles its state at cycle $t+1$ is independent of its state at cycle t, signal probability at the T flip-flop output that is connected to the scan chain is 0.5. A T flop-flop holds it value until a signal value "1" is assigned at its input. As the sequence generated by the LFSR holds randomness property, it is expected that for a large value of k, AND gate output will not change for a long period of time, thereby enabling shifting of identical values in the scan chain. The probability that the T flip-flop will toggle at any time t can be calculated as $1/2^k$, where k is the number of LFSR outputs connected to the AND gate. If the input of a scan chain is directly connected to the r-stage LFSR, then there is 2^{r-1} number of transitions at the scan chain input (Bardell et al. 1987). However, with LT-RTPG, the number of transition at the scan chain input is equal to 2^{r-k}. Therefore, LT-RTPG reduces the transition activity by a factor of 2^{k-1}. As the assignment of identical values in scan cells reduces the fault coverage, very large value of k is not recommended. In Wang and Gupta (1999), it is shown that for $k = 2$ or 3, the loss in fault coverage is very minimal, whereas significant reduction in power consumption can be obtained. To minimize the power consumption and increase the fault coverage for RPR faults, LT-RPTG is combined with 3-weight random TPG in Wang (2002).

The use of an *adjacency-based TPG* along with the conventional random LFSR is proposed in Girard et al. (2000). The basic concept of adjacency-based testing

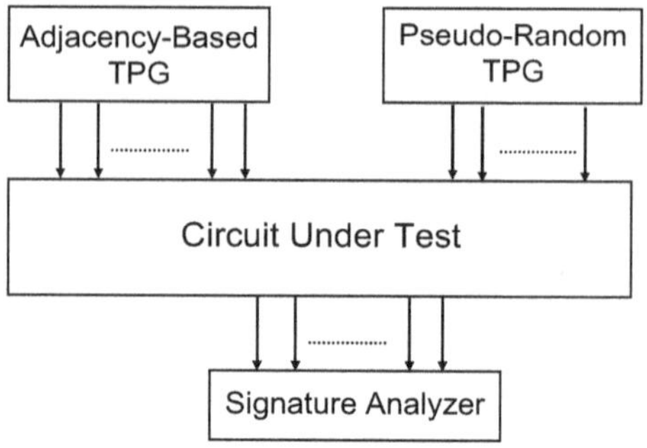

Fig. 5.16 Adjacency-based TPG for low-power BIST

was first proposed in Craig and Kime (1985). In adjacency test approach, only a single transition is applied to the circuit in each clock cycle, that is, the hamming distance between two successive patterns is always one. Please note that the use of adjacency-based test generator is not recommended for testing large circuits as prohibitive test length may result to achieve decent fault coverage. In the proposed low-power BIST scheme as shown in Fig. 5.16, a test pattern that is applied every cycle to the circuit consists of two parts. In the first part, which is generated by the adjacency-based TPG, only one bit is changed compared to the first part of the previous pattern. The second part is generated using the conventional LFSR. As the number of transition at the circuit inputs are greatly reduced (depending on the size of adjacency-based TPG), the proposed method results in low-power consumption. To further minimize power consumption, outputs of the adjacency-based TPG can be connected to those inputs in the circuit that have maximum influence on the internal switching activity. For example, an input with a large fan-out cone is an ideal candidate for this. A gain function called *induced activity function* is used to calculate which inputs should be connected to adjacency-based TPG. Remaining inputs are connected to conventional TPG.

Test power minimization technique based on *gated clock scheme* for TPG block has been proposed in Girard et al. (2001). In the proposed technique, a clock whose speed is half of the normal speed is used to activate one-half of the D flip-flops in the LFSR during one clock cycle of the test session. In the next clock cycle, the second half of the D flip-flops in the LFSR is activated using another clock. Both clocks are synchronized with master clock and have same period but shifted in time. Figure 5.17 shows the basic scheme of the proposed TPG. As only half of the circuit inputs changes every cycle, this scheme reduces the power consumption in the circuit. Also, as only half of the flip-flops in the LFSR are activated at a time, it also reduces power consumption in the LFSR. The clock divider circuit that can be used to generate two clocks required by this scheme and the associated timing waveform are shown in Fig. 5.18. It is important to note here is that as different clocks are used to feed flip-flops in the TPG, two different clock trees are used in the proposed scheme. This scheme does not require any major modification in the circuit and requires negligible area overhead.

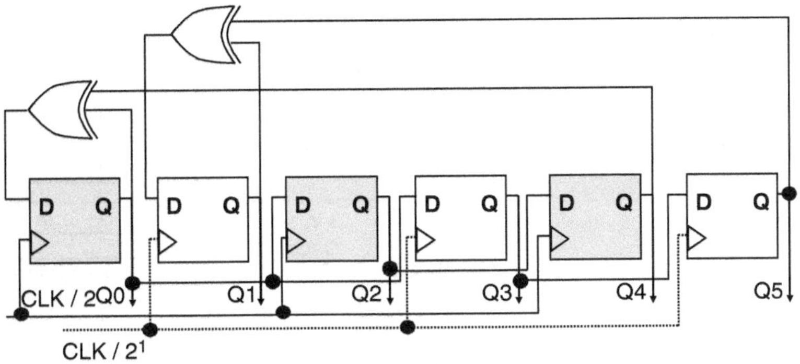

Fig. 5.17 Low-power TPG using gated clock scheme (Girard et al. 2001)

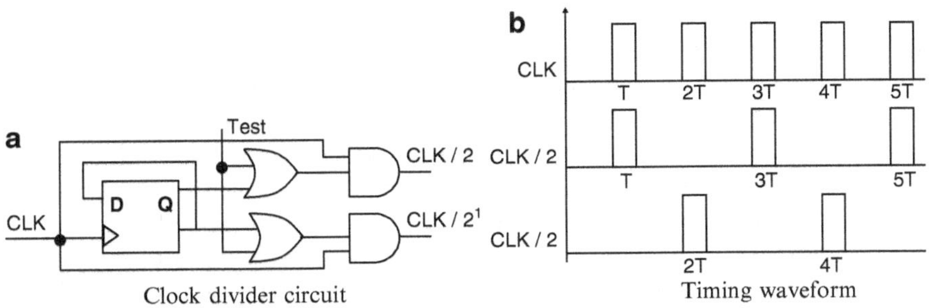

Fig. 5.18 Modified clock generator (Girard et al. 2001)

5.5.3 Modified Scan and Reordering

Transition frequency-based scan cell ordering to minimize total power consumption is proposed in Bellos et al. (2004). In the proposed technique, first a long sequence of pseudorandom patterns is applied to the circuit and transition frequency due to the scan-out of the test responses is calculated for each pair of internal scan cells. Next, scan cells are reordered in such a way that the total transition frequency is minimized. Finally, several different seeds are used for the TPG and resulting vectors are fault simulated. The seed that generates vectors with desired fault coverage and minimum number of test vectors is selected. For the scan cells connected to the inputs of the scan chains, cells with outputs that have minimum influence on the internal switching activity are selected. The disadvantages of *scan cell reordering*-based techniques are impact on timing closure and routing congestion due to very long scan paths.

The use of a *smoother* to minimize the switching activity is proposed in Lai et al. (2004). A smoother circuit modifies the bits generated by the LFSR such that the number of transitions is minimized. The resulting sequence has nonuniform *signal probabilities* instead of uniform signal probability of 0.5 like in the case of pseudorandom sequences. As the smoother technique does not differentiate between

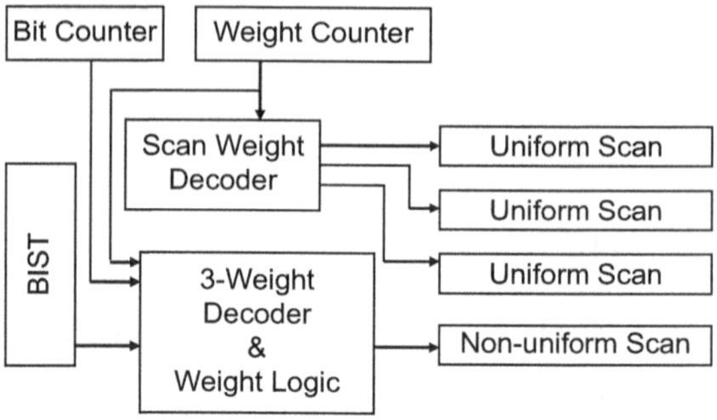

Fig. 5.19 Scan chain partitioning low-power BIST scheme

detecting and nondetecting patterns, it usually results in lower fault coverage. To minimize the fault coverage loss, scan cell reordering is used, which again suffers from the timing closure and routing congestion problems.

A *scan partitioning*-based low-power BIST scheme using *3-valued weighted* random pattern generation (Pomeranz and Reddy 1993) is described in Lee and Touba (2005). In this proposed scheme, random test is used for easy-to-detect faults, whereas for RPR faults, 3-valued weighted random pattern generation is used. In a 3-valued weighted pattern generation, each scan cell is weighted to one of the three values: 0, 1, or random. In the proposed scheme, two types of scan chains are defined: (1) *uniform scan* and (2) *nonuniform scan*. A uniform scan is defined as a scan chain in which all scan cells have the same weight in each weight set. A nonuniform scan is defined as a scan chain in which each scan cell has an individual weight as in the case of conventional weighting. Figure 5.19 shows the architecture for the proposed scheme. As a single weight decoder can be used for all uniform scans, maximizing the number of uniform scan partitions can minimize the total area overhead associated with the decoding logic as well as the power consumption during shift.

As this method only constraints scan cell assignment to scan partitions and not the scan cells order in each partition, this scheme has minimal impact on the routing congestion.

5.5.4 Test Scheduling

For large designs, instead of testing multiple blocks in parallel, *test scheduling* for BISTed blocks can be performed to minimize the overall power consumption (Zorian 1993). Different test scheduling algorithms that take various factors such as block type, test type, test time, power consumption per block into account can be used. To execute the test scheduling, different *BIST control* schemes such as *centralized* and *distributed* can be used. In a centralized approach, a single controller communicates with different BISTed blocks and schedules their tests such that there is no resource conflict and test cost under consideration is minimized. For

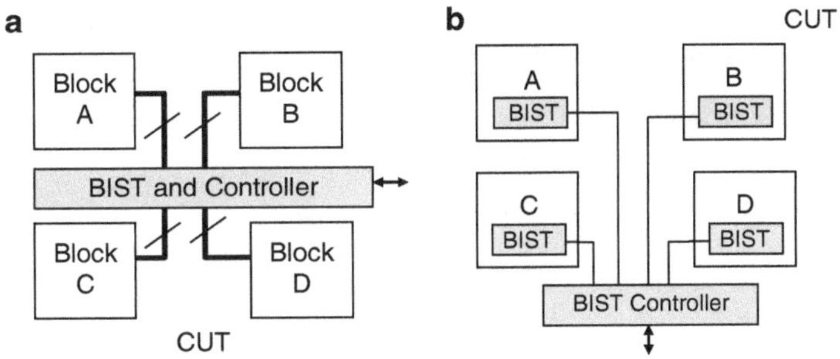

Fig. 5.20 (**a**) Centralized BIST scheme, (**b**) distributed BIST scheme

large designs, the connectivity between blocks and the BIST logic/controller can be a bottleneck; therefore, distributed control is recommended for these designs. In a distributed scheme, multiple BIST engines are used and control block controls the execution of the individual tests. Figure 5.20 shows an example of centralized and distributed control scheme. A distributive approach is more flexible but centralized scheme is more cost effective.

5.6 Summary and Conclusions

Low-power test compression has received attention for over a decade in the research community and commercial tools have now emerged. In this chapter, we have reviewed the several key low-power test compression techniques that have been presented in the literature and we have also described how these methods have evolved over the past decade. We have focused on coding techniques based on data compression, compression methods that rely on an LFSR decompressor, as well as various low-power BIST techniques. The DFT methods presented in this chapter will allow us to test next-generation integrated circuits without exceeding power limits and thereby reduce yield loss and test cost.

References

V. D. Agrawal, C. R. Kime and K. K. Saluja, "A tutorial on Built-In Self-Test, Part 1: Principles," *IEEE Design and Test of Computers*, vol. 10, no. 1, pp. 73–82, 1993a.

V. D. Agrawal, C. R. Kime and K. K. Saluja, "A tutorial on Built-In Self-Test, Part 2: Applications," *IEEE Design and Test of Computers*, vol. 10, no. 2, pp. 69–77, 1993b.

A. Al-Yamani, E. Chmeler, and M. Grinchuck, "Segmented addressable scan architecture," *Prof. IEEE VLSI Test Symposium*, pp. 405–411, May 2005.

N. Badereddine, Z. Wang, P. Girard, K. Chakrabarty, A. Virazel, S. Pravossoudovitch, and C. Landrault, "A selective scan slice encoding technique for test data volume and test power reduction," *Journal of Electronic Testing: Theory and Applications*, vol. 24, pp. 353–364, August 2008.

D.H. Baik and K.K. Saluja, "Progressive random access scan: A simultaneous solution to test power, test data volume and test time," *Proc. IEEE International Test Conference*, pp. 1–10, November 2005.

P. H. Bardell, W. H. McAnney, and J. Savir, "Built-In Test for VLSI: Pseudorandom techniques," *John Wiley & Sons*, New York, 1987.

M. Bellos, D. Bakalis and D. Nikolos, "Scan cell ordering for low power BIST," in *Proc. International Symposium on VLSI Emerging Trends in VLSI Systems Design*, 2004.

B. Benware, C. Schurmyer, N. Tamarapalli, K. –H Tsai, S. Ranganathan, R. Madge, and P. Krishnamurthy, "Impact of multiple-detect test patterns on product quality," in *Proc. International Test Conference*, October 2003, pp. 1031–1040.

M. L. Bushnell and V. D. Agrawal, "Essentials of electronic testing," *Norwell*, MA, Kluwer, 2000.

K. J. Balakrishman and A. Touba, "Relationship between entropy and test data compression," *IEEE Transactions on VLSI Systems*, pp. 386–395, 2007.

D. Czysz, G. Mrugalski, J. Rajski, and J. Tyszer, "Low-power test data application in EDT environment through decompressor freeze," *IEEE Transactions on Computer-Aided Design of Integrated Circuits and Systems*, vol. 27, pp. 1278–1290, July 2008a.

D. Czysz, M. Kassab, X. Lin, G. Mrugalski, J. Rajski and J. Tyszer, "Low power scan shift and capture in the EDT environment," in *Proc. IEEE International Test Conference*, October 2008b.

K. Chakrabarty, "Test scheduling for core-based systems using mixed-integer linear programming," *IEEE Transactions on Computer-Aided Design of Integrated Circuits and Systems*, vol. 19, pp. 1163–1174, October 2000.

A. Chandra and K. Chakrabarty, "Low-power scan testing and test data compression for system-on-a-chip," *IEEE Transactions on Computer-Aided Design of Integrated Circuits and Systems*, vol. 21, pp. 597–604, May 2002.

A. Chandra and K. Chakrabarty, "System-on-a-chip test data compression and decompression architectures based on Golomb codes," *IEEE Transactions on Computer-Aided Design of Integrated Circuits & Systems*, vol. 20, pp. 355–368, March 2001.

A. Chandra and K. Chakrabarty, "Test data compression and test resource partitioning for system-on-a-chip using frequency-directed run-length (FDR) codes," *IEEE Transactions on Computers*, vol. 52, pp. 1076–1088, August 2003a.

A. Chandra and K. Chakrabarty, "A unified approach to reduce SOC test data volume, scan power and testing time," *IEEE Transactions on Computer-Aided Design of Integrated Circuits & Systems*, vol. 22, pp. 352–362, March 2003b.

R. M. Chou, K. K. Saluja, and V. D. Agarwal, "Scheduling tests for VLSI systems under power constraints," *IEEE Transactions on VLSI Systems*, vol. 5, pp. 175–185, June 1997.

R. M. Chou, K. K. Saluja, and V. D. Agrawal, "Power Constraint Scheduling of Tests," in *Proc. International Conference on VLSI Design*, January 1994, pp. 271–274.

H. Cox and J. Rajski, "On necessary and nonconflicting assignments in algorithmic test pattern generation," *IEEE Transactions on Computer-Aided Design of Integrated Circuits and Systems*, vol. 13, no. 4, pp. 515–530, April 1994.

F. Corno, M. Rebaudengo, and M. S. Reorda, "Low power BIST via nonlinear hybrid cellular automata," in *Proc. IEEE VLSI Test Symposium*, pp. 29–34, April 2000.

F. Corno, M. Rebaudengo, M. S. Reorda, and M. Violante, "A new BIST architecture for low power circuits," in *Proc. European Test Workshop*, pp. 160–164, May 1999a.

F. Corno, M. Rebaudengo, M.S. Reorda, and M. Violante, "Optimal vector selection for low power BIST," in *Proc. International Symposium on Defect and Fault Tolerance in VLSI Systems*, November 1999b, pp. 219–226.

G. L. Craig and C. R. Kime, "Pseudo-exhaustive adjacency testing: A BIST approach for stuck-open faults," in *Proc. International Test Conference*, October 1985, pp. 126–137.

V. Dabholkar, S. Chakravarty, I. Pomeranz, and S. M. Reddy, "Techniques for minimizing power dissipation in scan and combinational circuits during test application," *IEEE Transactions on Computer-Aided Design of Integrated Circuits and Systems*, vol. 17, pp. 1325–1333, December 1998.

F.-F. Ferhani and E. J. McCluskey, "Classifying bad chips and ordering test sets," in *Proc. International Test Conference*, pp. 1–10, October 2006.

S. Gerstendorfer and H.-J Wunderlich, "Minimized power consumption for scan-based BIST," in *Proc. International Test Conference*, September 1999, pp. 77–84.

P. Girard, L. Guiller, C. Landrault, and S. Pravossoudovitch, "An adjacency-based test pattern generator for low power BIST design," in *Proc Asian Test Symposium*, December 2000, pp. 459–464.

P. Girard, L. Guiller, C. Landrault, S. Pravossoudovitch, and H. J. Wunderlich, "A modified clock scheme for a low power BIST test pattern generator," in *Proc. VLSI Test Symposium*, April 2001, pp. 306–311.

P. Girard, L. Guiller, C. Landrault, and S. Pravossoudovitch, "A test vector inhibiting technique for low energy BIST design," in *Proc. VLSI Test Symposium*, April 1999, pp. 407–412.

S. K. Goel, N. Devta-Prasanna, and R. Turakhia, "Effective and efficient test pattern generation for small delay defects," in *Proc. VLSI Test Symposium*, May 2009.

P. T. Gonciari, B. M. Al-Hashimi and N. Nicolici, "Test data compression: The system integrator's perspective," in *Proc. IEEE/ACM Design, Automation and Test in Europe (DATE) Conference*, March 2003, pp. 726–731.

B. Keller, M. Tegethoff, T. Bartenstein, and V. Chickermane, "An economic analysis and ROI model for nanometer test," in *Proc. International Test Conference*, October 2004, pp. 518–524.

I. Hamzaoglu and J. H. Patel, "Reducing test application time for full scan embedded cores," *Proc. IEEE International Symposium on Fault-Tolerant Computing*, June 1999, pp. 260–267.

J. Hartmann and G. Kenmnitz, "How to do weighted random testing for BIST," in *Proc. International Conference on Computer-Aided Design*, 1993.

S. Hellebrand, S. Tarnick, J. Rajski, and B. Courtois, "Generation of vector patterns through reseeding of multiple-polynomial linear feedback shift registers," in *Proc. International Test Conference*, October 1992, pp. 120–129.

Semiconductor Industry Association, International Technology Roadmap for Semiconductors (ITRS), 2007. [Online]. Available: http://www.itrs.net/Links/2007ITRS/Home2007.htm.

G. Karypis and V. Kumar, "A fast and high quality multilevel scheme for partitioning irregular graphs," Technical Report 95–035, *Department of Computer Science*, University of Minnesota, 1988.

B. Kruseman, A. K. Majhi, G. Gronthoud, and E. Eichenberger, "On hazard-free patterns for fine-delay testing," in *Proc. International Test Symposium*, October 2004, pp. 213–222.

J. Kuban and W. Bruce, "Self testing the Motorola MC6804P2," *IEEE Design and Test of Computers*, vol. 1, no. 2, 1984.

J. Lee and N. A. Touba, "LFSR-reseeding scheme achieving low-power dissipation during test," *IEEE Transactions on Computer-Aided Design of Integrated Circuits and Systems*, vol. 26, pp. 396–401, February 2007.

J. Li, X. Liu, Y. Zhang, Y. Hu, X. Li, and Q. Xu, "On capture power-aware test data compression for scan-based testing," in *Proc. IEEE/ACM International Conference on Computer-Aided Design (ICCAD)*, May 2008, pp. 67–72.

X. Liu and Q. Xu, "A generic framework for scan capture power reduction in fixed-length symbol-based test compression environment," in *Proc. IEEE/ACM Design, Automation and Test in Europe (DATE) Conference*, April 2009.

N. C. Lai, S. J. Wang, and Y. H. Fu, "Low power BIST with smoother and scanchain reorder," in *Proc. Asian Test Symposium*, November 2004, pp. 40–45.

J. Lee and N. A. Touba, "Low power BIST based on scan partitioning," in *Proc. International Symposium on Defect and Fault Tolerance in VLSI Systems*, October 2005, pp. 33–41.

X. Lin, K. H Tsai, C. Wang, M. Kassab, J. Rajski, T. Kobayashi, R. Klingenberg, Y. Sato, S. Hamada, and T. Aikyo, "Timing-aware ATPG for high quality at-speed testing of small delay defects," in *Proc. Asian Test Symposium*, November 2006, pp. 139–146.

S. Manich, A. Gabarro, M. Lopez, J. Figueras, P. Girard, L. Guiller, C. Landrault, S. Pravossoudovitch, P. Texieira, and M. Santos, "Low power BIST by filtering non-detecting vectors," *Journal of Electronic Testing: Theory and Applications*, vol. 16, issue 3, 2000.

G. Mrugalski J. Rajski, D. Czysz and J. Tyszer, "New test data decompressor for low power applications," in *Proc. IEEE/ACM Design Automation Conference*, June 2007, pp. 539–544.

N. Nicolici and B. M. Al-Hashimi, "Scan latch partitioning into multiple scan chains for power minimization in full scan sequential circuits," in *Proc. IEEE/ACM Design Automation and Test in Europe (DATE) Conference*, March 2000, pp. 715–722.

M. Nourani and M. H. Tehranipoor, "RL-Huffman encoding for test compression and power reduction in scan applications," *ACM Transactions on Design Automation of Electronic Systems*, vol. 10, pp. 91–115, January 2005.

I. Pomeranz, S. M. Reddy, and R. Guo, "Static test compaction for synchronous sequential circuits based on vector restoration," *IEEE Transactions on Computer-Aided Design of Integrated Circuits and Systems*, pp. 1040–1049, July 1999.

I. Pomeranz and S. M. Reddy, "3-weight pseudo-random test generation based on a deterministic test set for combinational and sequential circuits," *IEEE Transactions on Computer-Aided Design of Integrated Circuits and Systems*, vol. 12, pp. 1050–1058, 1993.

P. M. Rosinger, P. T. Gonciari, B. M. Al-Hashimi, and N. Nicolici, "Simultaneous reduction in volume of test data and power dissipation for systems-on-chip," *Electronic Letters*, vol. 37, no. 24, pp. 1434–1436, November 2001.

P. M. Rosinger, P. T. Gonciari, B. M. Al-Hashimi, and N. Nicolici, "Analysing trade-offs in scan power and test data compression for systems-on-a-chip," *IEE Proc.-Computers and Digital Techniques*, vol. 149, no. 4, pp. 188–196, July 2002a.

P. M. Rosinger, B. M. Al-Hashimi, and N. Nicolici, "Low power mixed mode BIST based on mask pattern generation using dual LFSR reseeding," in *Proc. International Conference on Computer Design (ICCD)*, pp. 474–479, 2002b.

P. Rosinger, B. M. Al-Hashimi, and N. Nicolici, "Scan architecture with mutually exclusive scan segment activation for shift- and capture-power reduction," *IEEE Transactions on Computer-Aided Design of Integrated Circuits and Systems*, vol. 23, pp. 1142–1153, 2004.

R. Sankaralingam, R. R. Oruganti, and N. A. Touba, "Static compaction techniques to control scan vector power dissipation," in *Proc. IEEE VLSI Test Symposium*, April 2000, pp. 35–40.

J. Saxena, K. Butler, and L. Whetsel, "An analysis of power reduction techniques in scan testing," in *Proc. International Test Conference*, October 2001, pp. 670–677.

M. Sugihara, H. Date, and H. Yasuura, "A novel test methodology for core-based system LSI's and a testing time minimization problem," in *Proc. International Test Conference*, October 1998, pp. 465–472.

N. Tamarapalli and J. Rajski, "Constructive multi-phase test point insertion for scan-based BIST," in *Proc. International Test Conference*, October 1996, pp. 649–658.

N. Touba and E. J. McCluskey, "Altering a pseudo-random bit sequence for scan-based BIST," in *Proc. International Test Conference*, October 1996, pp. 167–175.

R. Turakhia, W. R. Daasch, M. Ward, and J. van Slyke, "Silicon evaluation of longest path avoidance testing for small delay defects," in *Proc. International Test Conference*, pp. 1–10, October 2007.

B. Vermeulen, C. Hora, B. Kruseman, E. J. Marinissen, and R. van Rijsinge, "Trends in testing integrated circuits," in *Proc. International Test Conference*, October 2004, pp. 688–697.

J. Waicukauski, E. Lindbloom, E. Eichelberger, and O. Forlenza, "A method for generating weighted random test patterns," *IEEE Transactions on Computers*, vol. 33, no. 2, 1989.

S. Wang and S. K. Gupta, "ATPG for heat dissipation minimization during scan testing," in *Proc. International Test Conference*, October 1997a, pp. 250–258.

S. Wang and S. K. Gupta, "DS-LFSR: A new BIST TPG for low heat dissipation," in *Proc. International Test Conference*, November 1997b, pp. 848–857.

S. Wang and S. K. Gupta, "LT-RTPG: A new test-per-scan BIST TPG for low switching activity," *IEEE Transactions on Computer-Aided Design of Integrated Circuits and Systems*, vol. 25, no. 8, August 2006.

S. Wang and S. K. Gupta, "LT-RTPG: A new test-per-scan BIST TPG for low heat dissipation," in *Proc. International Test Conference*, September 1999, pp. 85–94.

S. Wang, "Minimizing heat dissipation during test application," *Ph.D. Dissertation*, University of Southern California, 1998.

S. Wang, "Low hardware overhead scan based 3-weight weighted random BIST," in *Proc. International Test Conference*, October 2001, pp. 868–877.

S. Wang, "Generation of low power dissipation and high fault coverage patterns for scan-based BIST," in *Proc. International Test Conference*, October 2002, pp. 834–843.

N. H. E Weste and K. Eshraghian, "Principles of CMOS VLSI design: A systems perspective." 2nd Edition, *Addison-Wesley*, MA, 1992.

Z. Wang and K. Chakrabarty, "Test data compression for IP embedded cores using selective encoding of scan slices," in *Proc. IEEE International Test Conference*, November 2005.

Z. Wang and K. Chakrabarty, "Test data compression using selective encoding of scan slices," *IEEE Transactions on VLSI Systems*, vol. 16, pp. 1429–1440, November 2008.

L. Whetsel, "Adapting scan architectures for low power operation," in *Proc. International Test Conference*, October 2000, pp. 863–872.

L. Xu, Y. Sun, and H. Chen, "Scan array solution for testing power and testing time," in *Proc. International Test Conference*, October 2001, pp. 652–659.

M. Yilmaz, K. Chakrabarty, and M. Tehranipoor, "Test pattern grading and pattern selection for small delay defects," in *Proc. VLSI Test Symposium*, April 2008, pp. 233–239.

K. Yang, K.-T. Cheng, and L.-C. Wang, "TranGen: A SAT-based ATPG for path-oriented transition faults," in *Proc. Asia South Pacific Design Automation Conference*, January 2004, pp. 92–97.

X. Zhang and K. Roy, "Peak power reduction in low power BIST," in *Proc. IEEE International Symposium on Quality of Electronic Design*, March 2000, pp. 425–432.

Y. Zorian, "A distributed BIST control scheme for complex VLSI devices," in *Proc. VLSI Test Symposium*, April 1993, pp. 4–9.

Chapter 6
Power-Aware System-Level Test Planning

Erik Larsson and C.P. Ravikumar

Abstract The high test power consumption, which can be several factors higher than the functional power consumption for which an integrated circuit (IC) is designed, may result in higher overall cost due to yield loss and potentially damaged ICs. As system-on-chips (SOCs) designed in modular fashion are becoming increasingly common, the testing can, in contrast to nonmodular SOCs, be performed in a modular manner. The key advantage is that modular test offers the possibility to plan the testing such that power consumption is controlled; modules are only activated when they are tested. This chapter contains an introduction to core-based testing, which is followed by a discussion on test power consumption and its modeling, and then the chapter discusses power-aware test planning for modular SOCs.

6.1 Introduction

The test power consumption can be several factors higher than the power consumption for which an integrated circuit (IC) is designed to handle during functional operation. High power consumption may lead to damaged ICs and power droops. Power droops make bits in the IC switch unintendedly. Due to high test power consumption, power droops may also occur and impact the testing. Power droops during testing may lead to correct ICs being rejected for the following two reasons. First, bits in the produced test responses can be changed due to power droops, which leads to that the produced test responses do not match the expected test responses; hence, a defect is indicated and the IC is classified as defective. Second, power droops may force bits to switch such that the *intended* test stimuli are no longer applied, and as the intended stimuli are not applied, the produced test responses will not match the expected test responses. The consequence is that the IC is classified as defective. Damaging ICs due to too high power consumption

E. Larsson (✉)
Linköping University, Linköping, Sweden
e-mail: erila@ida.liu.se

C.P. Ravikumar
Texas Instruments Inc., Bangalore, India

P. Girard et al. (eds.), *Power-Aware Testing and Test Strategies for Low Power Devices*,
DOI 10.1007/978-1-4419-0928-2_6, © Springer Science+Business Media, LLC 2010

during test and failing good ICs due to power droops are not desirable as the result is yield loss, which increases test cost as well as overall cost.

The high test power consumption can be addressed by design to *handle* test power consumption, design with *power-aware design-for-test* (DfT), and making use of *power-aware test planning*. We have seen in the previous chapters of this book the detailed sources of test power consumption (Chap. 2) and discussed test generation for low power (Chap. 3), design-for-test techniques to make ICs test power aware (Chap. 4), as well as test data compression and built-in self-test (BIST) (Chap. 5). In this chapter, we focus on power-aware test planning for modular ICs or modular system-on-chips (SOCs).

It is becoming increasingly common to design ICs in a modular fashion. The semiconductor technology development makes it possible to fabricate ICs with billions of transistor that are placed on a few square centimeters. In order to design and manufacture such advanced and complex ICs in a timely manner, it is increasingly common to make use of predesigned and preverified blocks of logic, so called cores. These predesigned and preverified cores, for example CPU cores, are used as building blocks in order to shorten the design time.

An IC designed in a modular fashion is tested as such by making the cores testable units. The advantage is the possibility to reduce the test application time and control the test power consumption. Assume, for illustration, the example in Fig. 6.1. The same SOC can be tested in a nonmodular and a modular way. Assume that core A has a scan chain of 100 flip-flops and is tested by 20 patterns while core B has a scan chain of 200 flip-flops and is tested by 10 patterns. The test application time for the SOC in the nonmodular alternative is in the range of $300 \times 20 = 6{,}000$ clock cycles, which is given by the length of the scan chains $(100 + 200)$ times the number of test patterns (max(20, 10)). In the modular alternative, where each core can be tested as a standalone unit, the test application time is the sum of the test time from the two cores, which is $100 \times 20 + 200 \times 10 = 4{,}000$ clock cycles. The example illustrates the potential savings in test time when applying modular testing. However, modular testing also allows the reduction and control of test power

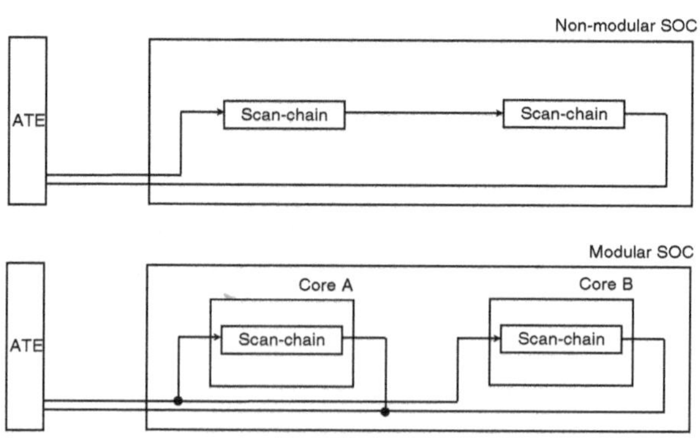

Fig. 6.1 Nonmodular SOC vs. modular SOC

consumption. Reduction of power consumption is achieved as follows. In the non-modular alternative, all scan chains are active during the test application, while in the modular alternative, only the scan chains that are related to the active test are in operation. The test power consumption can be controlled in the modular case as each module is a testable unit, it is possible to define when the modules are to be tested while test power consumption is under control.

Power-aware test planning can be used to guide the:

- Test-plan exploration, which is to find a test order with a minimal test application time while meeting power constraints
- Design exploration to find where to:
  - Insert power-aware DfT and/or
  - Over-design to handle high power

  while the test plan results in a minimal test application time and minimal additional cost while meeting power constraints.

This chapter is organized as follows. Section 6.2 outlines the key components to enable modular testing. We describe *core test wrapper*, *test access mechanism* (TAM), and *test scheduling*. The objective of a core test wrapper is to isolate a block of logic from the rest of the system such that the block becomes a testable unit (a wrapped core). In order to enable the transportation of test stimuli from IC pins to a wrapped core, and produced test responses from a wrapped core to IC pins, a test infrastructure is required. The test infrastructure, TAM, can utilize functional buses or dedicated test buses. For modular SOCs with TAM and a set of wrapped cores, test scheduling is the process of defining in which order to test each core. The overall objective of core test wrapper design, test infrastructure design, and test scheduling is to minimize the test application time, which is highly related to the test cost, while ensuring that all constraints, including test power consumption, are met.

In Sect. 6.3, power consumption and its modeling are discussed. We define the model to compute power consumption, and we discuss how to model test power: single-value model and multiple-value model. We discuss when power is consumed, at shift and capture, as well as where power is consumed in combinational logic and sequential logic. We also discuss power profile manipulations.

In Sect. 6.4, we discuss power-constrained test planning. In particular, the section contains discussion on power-constrained test scheduling, power-constrained test planning under single-power-value and multiple-power-value model, and power-aware test planning that utilizes power-aware DfT for shift-power as well as capture-power reduction.

In Sect. 6.5, we present results on low-power test planning for multiple clock domains (Sect. 6.5.1) and IDDQ test planning for core-based system chips (Sect. 6.5.2).

The chapter is summarized in Sect. 6.6.

6.2 Core-Based Test Architecture Design and Test Planning

Designing ICs in a modular fashion is becoming increasingly common as the possibility to design *advanced* ICs in a *timely* manner increases. The basic idea when designing a modular IC is to make use of predesigned and preverified blocks of logic, so-called cores. These cores can, for example, be CPU cores and DSP cores, and the cores are used as building blocks to compose the system.

A major obstacle with the fabrication of ICs, designed in a modular or nonmodular fashion, is that manufacturing is far from perfect, and therefore, each individual IC must be tested in order to separate defective ICs from correct ICs. Testing ICs is costly and difficult; especially for advanced ICs developed in later semiconductor technologies. However, ICs designed in a modular fashion can be tested in a modular manner. The advantage of modular test is the possibility to control and plan the test application such that, for example, test application time is minimized while constraints, such as test power, are controlled.

In this section, we will discuss *core test wrappers*, *test access mechanism* (TAM), and *test scheduling*. Core test wrappers and TAMs are fundamental requirements to enable modular test, and test scheduling is the process of planning the test application. The core test wrapper, discussed in Sect. 6.2.1, serves two purposes, namely the enabling of core *isolation* and core *access*. The TAM, discussed in Sect. 6.2.2, is the *infrastructure* to allow transportation of test data between the tester (test source) to the cores, and from the cores to the tester (test sink). Test scheduling, discussed in Sect. 6.2.3, is, given an IC with testable units, to define in which order the units (cores) are to be tested.

Figure 6.2 shows a modular SOC with a set of cores and some glue logic. The test source drives the TAM connected to the core test wrapper of the core-under-test (CUT) in order to provide test stimuli to the core. And the test responses from the CUT are transported on the TAM to the test sink. In this example, both the test source and the test sink are external, such as an automatic test equipment (ATE). However, the concept of test source and test sink is general, which means a test source and a test sink may either be off-chip, for example as an ATE, or on-chip as built-in self-test (BIST). Further, several test sources and test sinks may exist in the system.

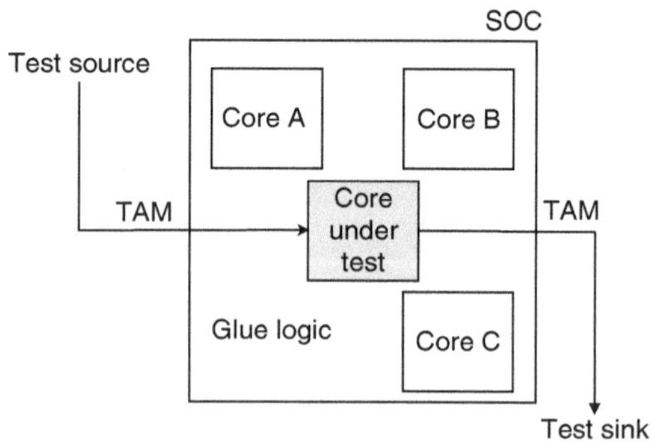

Fig. 6.2 Illustration of a modular SOC with modular test features

Significant amount of research has been devoted to outline the concepts of modular testing. Beenker et al. (1986) discussed how to join board test and IC test. And Bouwman et al. (1992) outlined core-based test planning. Later, a number of papers discussed modular test planning. For example, Bhatia et al. (1996) described the testing of custom logic blocks. Whetsel (1997), Zorian (1997, 1998), Gupta and Zorian (1997), and Zorian et al. (1998) discussed requirements for modular test. Xu and Nicolici (2005) made a survey of the work performed in modular test planning.

6.2.1 Core Test Wrapper

An IC designed in a modular fashion consists of a set of blocks-of-logic, so-called cores, and glue logic. In order to enable modular test, each logic block must be made testable units. A core test wrapper (or wrapper) makes a core a testable unit by enabling isolation and access for a given block-of-logic. Core isolation makes each wrapped core a stand-alone test unit, and core wrappers ease the test access by defining the interface between the core and the infrastructure for test data transportation, the TAM. The IEEE 1500 Standard for Embedded Core Test (SECT) (IEEE std 1500 2005) is developed for core access and core isolation.

For a given core with a number of scan chains, the core test wrapper is to be designed such that the scan chains interface with the TAM. The scan chains at a core are to be formed into a number of so-called wrapper chains. Each wrapper chain is connected to a TAM wire.

One problem is to form the scan chains at each core into a number of wrapper chains such that they can be connected to the TAM. Larsson and Peng assumed the test time to be $\tau = \tau_w / w$, where τ is the test time when all the scan chains in a core are connected into a single wrapper chain and w is the number of wrapper chains. A high number of wrapper chains gives a lower test application time as less shifting is required for stimuli load and responses unload. The cost of a high number of wrapper chains is the need of TAM interfacing the wrapper chains and the pins of the IC (Larsson and Peng 2001a, 2002a). The test time model by Larsson and Peng is linear and confirms well in the cases when the number of scan chains is relatively higher than the number of wrapper chains. To also address the general case, when there can be few scan chains and the scan chains can be of unequal length while the number of wrapper chains is relatively high compared to the number of scan chains, Iyengar et al. (2001) and Pouget et al. (2003b, 2005) proposed wrapper-design algorithms. Figure 6.3 shows the longest scan-in/scan-out, which is highly related to test application time, for each wrapper design at various number of wrapper chains for a core. At few number of wrapper chains, the test application time is higher compared to when the scan chains are formed into a higher number of wrapper chains. As the test time decreases in a stair-case function for each core where some stairs are longer than others, it is difficult to find the best wrapper configuration for each core when several cores are to share a TAM.

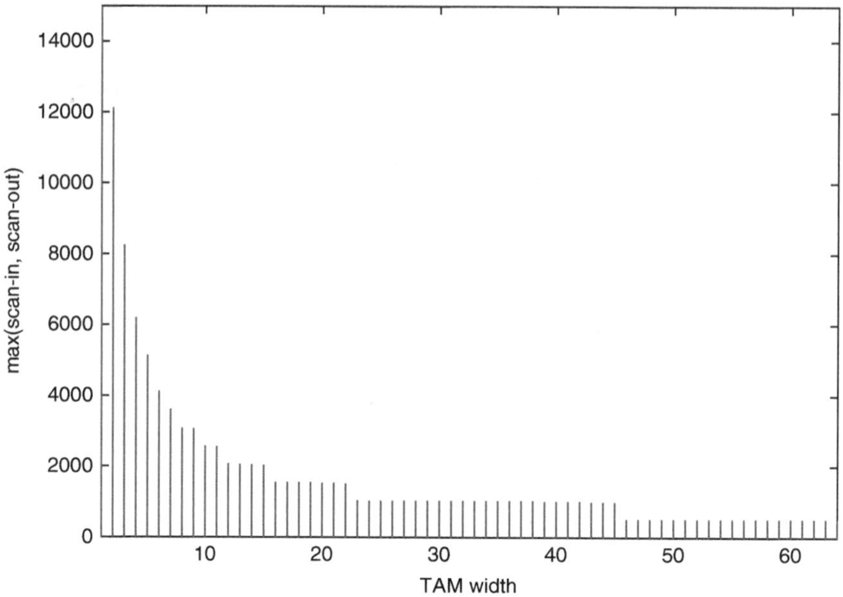

Fig. 6.3 The scan-in/scan-out at various number of wrapper chains (TAM width)

In contrast to the IEEE 1500 core wrapper (IEEE std 1500 2005), which allows the scan chains at a core to be grouped into *one* fixed set of wrapper chains, Koranne (2002) proposed a core test wrapper that allowed dynamic configuration of wrapper chains during test application. The scan chains may be configured into a number of wrapper chain designs, and the configurations are changed during the application of test.

6.2.2 Test Access Mechanism Design

For a modular IC, the cores (the testable units) do not have direct access to IC pins as the cores are embedded deep in the IC. In order to transport test data, that is test stimuli and test responses, to and from embedded cores, an infrastructure is needed (see Fig. 6.2). The TAM is such infrastructure and the TAM enables access of test data for each wrapped core.

Significant amount of work has been proposed on TAM design (Immaneni and Raman 1990; Harrod 1999; Aerts and Marinissen 1998; Varma and Bhatia 1998; Touba and Pouya 1997). Immaneni and Raman (1990) proposed the usage of direct access. In their approach, each core input is given direct access from a primary input and each core output is given direct access to a primary output. Harrod (1999) proposed the usage of the existing functional bus as the test access mechanism. A number of approaches have been proposed on dedicated TAMs. For example,

Fig. 6.4 TAM architecture for a modular IC

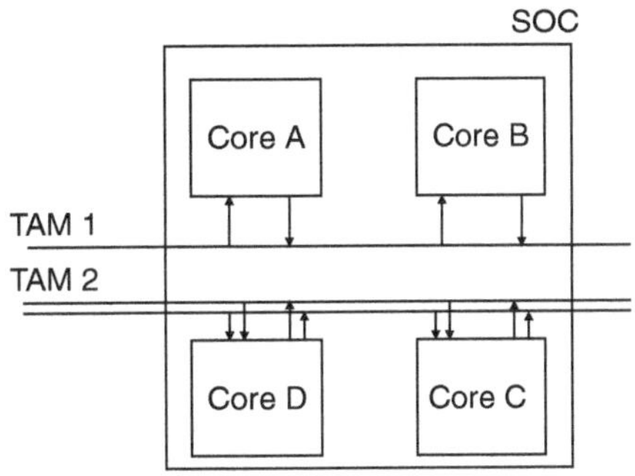

Aerts and Marinissen (1998) proposed multiplexing architecture, daisy-chain architecture, and distribution architecture, and Varma and Bhatia (1998) proposed bus-based architecture.

Figure 6.4 shows an example of a test bus design for a modular IC. The three TAM wires are partitioned into two test buses where TAM 1 is of width 1 and TAM 2 is of width 2. Core A and core B are assigned to TAM 1 while core C and core D are assigned to TAM 2. The scan chains at core A are formed into one wrapper chain as TAM 1 is of width 1.

6.2.3 *Test Scheduling*

For a given modular SOCs where cores are wrapped such that each core is a testable unit, and an infrastructure for test data transportation, a so-called TAM, exists, the test scheduling is the process of planning the order in which the cores are to be tested. The scan chains at each testable unit are formed into wrapper chains, and given the test patterns, each testable unit is associated with a test time. The objective of test planning is to plan the test application such that the overall test cost, which often is related to test application time, is minimal.

A number of approaches have been proposed for test scheduling. For IC with blocks-of-logic, testable units, that can be scheduled independently, Abadir and Breuer (1986) and Craig et al. (1988) have proposed techniques. Early work on test scheduling for modular designs was performed by Larsson and Peng (1999, 2000) and Chakrabarty (2000). And substantial work has been performed on integrated wrapper design, TAM design, and test scheduling. For example, by Iyengar et al. (2001), Yoneda and Fujiwara (2002), Yoneda et al. (2006), Goel and Marinissen (2003), Su and Wu (2004), and Larsson and Peng (2002a). Koranne (2002) proposed a test scheduling algorithm assuming reconfigurable core wrappers, and Larsson and Fujiwara (2006) showed it is possible to define an optimal test scheduling approach based on re-configurable core wrappers when making use of preemptive scheduling.

Figure 6.5 shows a possible test schedule for an IC with TAM as shown in Fig. 6.4. Each core is assigned to a TAM to enable testing. The available TAM wires

Fig. 6.5 A test schedule for the IC with TAM as in Fig. 6.4

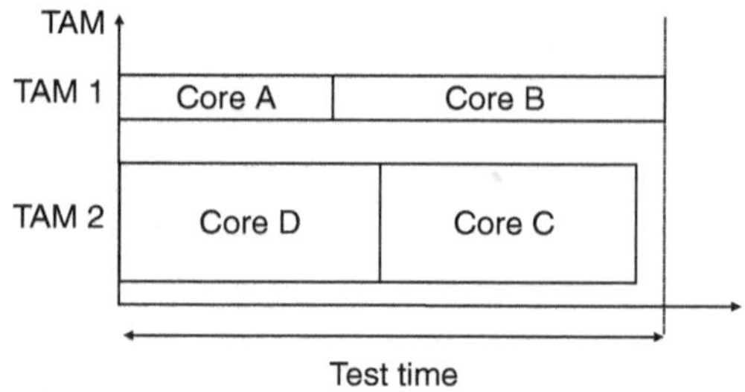

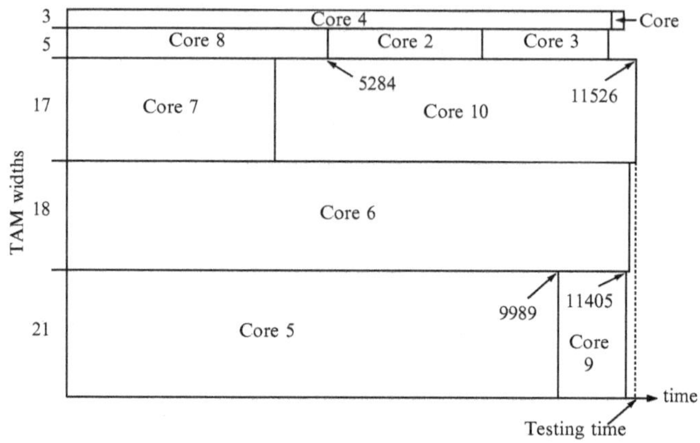

Fig. 6.6 An example of a test architecture and a test schedule (Samii et al. 2006)

connected to ATE channels for the feeding of test stimuli and receiving test responses are partitioned into TAM 1 and TAM 2; or two test buses. Core C and core D are assigned to TAM 2.

Figure 6.6 shows a slightly larger example of TAMs and the tests associated with each TAM for the ITC'02 design d695 [d695 is an ITC'02 benchmark circuit (Marinissen et al. 2002)]. The given TAM width of 64 is partitioned into five TAMs of width 3, 5, 17, 18, and 21. And, for example, core 5 and core 9 are associated with the largest TAM, and these cores are tested in sequence.

Table 6.1 shows the test application time from a number of approaches on the ITC'02 circuit p93791. The best results are collected in Fig. 6.7 and it is clear that the approaches, in general, produce good results since all results are in the range of 6% from the lower bound.

The work on test architecture design and test planning often takes a given test architecture and optimizes the test schedule. A major drawback is that the actual placement of cores in the system is not taken into account. In practice, it means that modifying the circuit slightly, and replanning the test, leads to potentially costly rerouting of TAMs; however, that is often not taken into account. Larsson et al. (2002, 2004), on the other hand, assume a given floor-plan where each core is given x, y coordinates. The optimization function optimizes both the test application time

Table 6.1 Test time comparison on P93791

Approach	Test application time						
	$N_{TAM}=16$	$N_{TAM}=24$	$N_{TAM}=32$	$N_{TAM}=40$	$N_{TAM}=48$	$N_{TAM}=56$	$N_{TAM}=64$
Lower bound (Goel and Marinissen 2002b)	1,746,657	1,164,442	873,334	698,670	582,227	499,053	436,673
Enumerate (Iyengar et al. 2002c)	1,883,150	1,288,380	944,881	929,848	835,526	537,891	551,111
ILP (Iyengar et al. 2002c)	1,771,720	1,187,990	887,751	(698,583)	599,373	514,688	460,328
Par-eval (Iyengar et al. 2002a)	1,786,200	1,209,420	894,342	741,965	599,373	514,688	473,997
GRP (Iyengar et al. 2002b)	1,932,331	1,310,841	988,039	794,027	669,196	568,436	517,958
Cluster (Goel and Marinissen 2002a)	–	–	947,111	816,972	677,707	542,445	467,680
Binpack (Huang et al. 2001)	1,791,860	1,200,157	900,798	719,880	607,955	521,168	459,233
CPLEX (Koranne 2002)	1,818,466	(1,164,023)	919,354	707,812	645,540	517,707	453,868
ECTSP (Koranne 2002)	1,755,886	(1,164,023)	919,354	707,812	**585,771**	517,707	453,868
ECTSP1 (Koranne 2002)	1,807,200	1,228,766	967,274	890,768	631,115	562,376	498,763
TB-serial (TRA) (Goel and Marinissen 2002b)	1,791,638	1,185,434	912,233	718,005	601,450	528,925	455,738
TR-serial (Goel and Marinissen 2002b)	1,853,402	1,240,305	940,745	786,608	628,977	530,059	461,128
TR-parallel (Goel and Marinissen 2002b)	1,975,485	1,264,236	962,856	800,513	646,610	540,693	477,648
K-tuple (Koranne and Iyengar 2002)	2,404,341	1,598,829	1,179,795	1,060,369	717,602	625,506	491,496
Larsson and Fujiwara (2006)	**1,752,336**	**1,174,252**	**877,977**	**703,219**	592,214	**511,925**	**442,478**

and the cost of additional TAM routing. Figure 6.8 shows the modeling of cores where each core includes one or several blocks and test sources and test sinks. In the example, an ATE is used as test source and test sink where the coordinates refer to the connection of the IC.

6.3 Power Modeling, Estimation, and Manipulation

In order to make use of power-aware test planning, there is a need of test power models that accurately model the test power consumption. Power consumption has been detailed in Chap. 2. The total power consumption (P_{total}) can be expressed as the sum of the three components:

$$P_{total} = P_{stat} + P_d + P_{sc}, \tag{6.1}$$

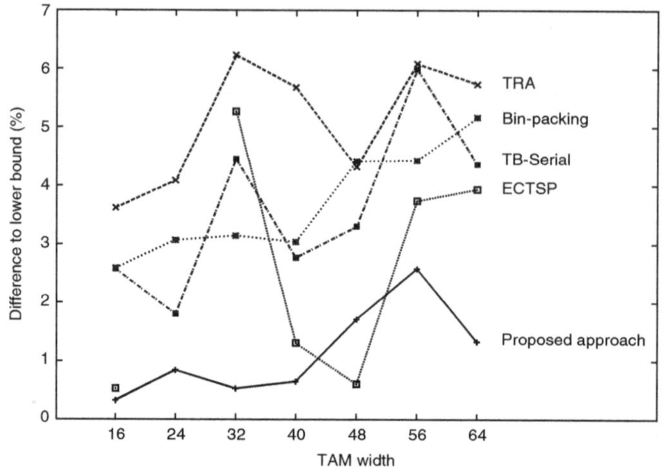

Fig. 6.7 The best test scheduling results in relation to lower bound where the label proposed approach refers to Larsson and Fujiwara (2006)

```
[Test source] #name     x       y
              ATE       0       10

[Test sink]   #name     x       y
              ATE       20      10

[Cores]       #name     x       y       {blocks}
              CoreA     10      20      {blockA1 blockA2}
              CoreB     20      20      {blockB1}
              CoreC     20      10      {blockC1 blockC2 blockC3}
              CoreD     10      10      {blockD1}
```

Fig. 6.8 Modeling a floor-plan for the example in Fig. 6.4

where P_{stat} is the static part, P_{d} the dynamic part, and P_{sc} is the short-circuit power (see Chap. 2 for details).

It is difficult to estimate the absolute power consumption, as it is technology dependent and computationally intensive. However, it is possible to make a technology-independent estimate. The dynamic part depends on the actual activity, that is the switching in the combinational logic and in the sequential elements, of the circuit. The dynamic power consumption is given by (6.2):

$$P_{\mathrm{d}} = C_{\mathrm{L}} \times V_{\mathrm{dd}}^2 \times f_{0\rightarrow 1}, \tag{6.2}$$

where C_{L} is the capacitance, V_{dd} is the supply voltage, and $f_{0\rightarrow 1}$ is the number of rising transitions (see Chap. 2 for details). During normal (functional) operation, the switching activity (rising transitions) depends on the inputs. While, during test mode, the switching activity depends on the test data.

In this section we will discuss:

- Modeling of test power consumption and constraints (Sect. 6.3.1)
- Estimation of test power consumption (Sect. 6.3.2)
- Test data manipulation (Sect. 6.3.3)

6.3.1 Modeling Power Consumption and Constraints

6.3.1.1 Power Modeling

For a given SOC with a number of testable units, Chou et al. (1997) approximated the test power consumption for each block (core) to a single fixed value. The single value is selected to be the peak power consumption over the test time of the block. Rosinger et al. (2002) refer to this as the *global peak power (approximation) model.* Figure 6.9 shows the actual power consumption and the modeled power consumption based on a single value. The false power is the mismatch between the actual power consumption and the modeled power consumption. The single-value power model is pessimistic, but it guarantees that the maximum power consumption will not be violated, and is simple to be handled by a test scheduling algorithm as it needs little computational effort. For the modeling, attached to each core, $\{c_1, c_2, \ldots, c_n\}$, is a test time, $\{\tau_1, \tau_2, \ldots, \tau_n\}$, and a test power consumption, $\{p_1, p_2, \ldots, p_n\}$.

Rosinger et al. (2002) proposed a two-value model in order to better model the test power consumption. The modeling becomes a bit more complicated as attached to each core, $\{c_i\}$, is two test times, $\{\tau_{i1}, \tau_{i2}\}$, and a value of test power

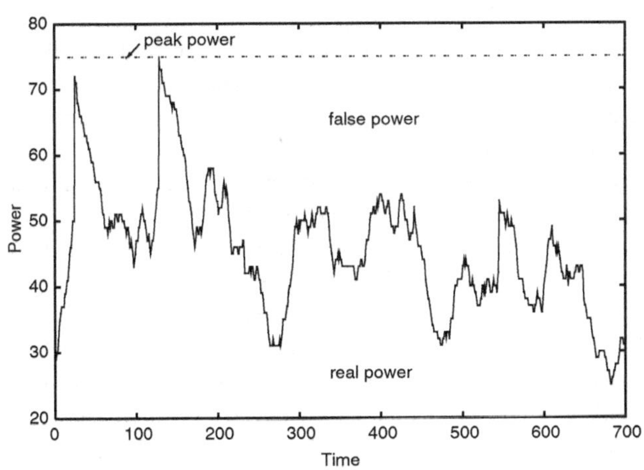

Fig. 6.9 Global peak power model

consumption for each part of the test, $\{p_{i1}, p_{i2}\}$. Samii et al. (2006) took a further step and proposed the usage of a cycle-accurate power model. The model keeps track of the power consumption in every clock cycle, hence, each clock cycle is attached to a test power value. The cycle-accurate power model obviously eliminates the false power and models the real power consumption accurately. The computational cost of making use of a cycle-accurate model during test planning is higher compared to making use of a single-value model; however, Samii et al. (2006) showed that a cycle-accurate model is applicable at little computational cost.

The dynamic power consumption, formulated by (6.2), is highly related to the input data. The input data causes transitions in the sequential logic (flip-flops) and in the combinational logic, and due to the switching activity, power is consumed. The input data, during the test application, is the test data. For scan-tested circuits, the flip-flops are turned into scan flip-flops (scan elements) such that virtual primary inputs and outputs are added. And during the testing, the test data is applied not only at the primary inputs but also to the added virtual inputs. The application of test data is, due to the nature of scan, applied during the following cycles:

- *Shift-in cycle*. The test stimuli are shifted into the scan elements
- *Launch-and-capture cycle*. The test stimuli are applied to the circuit and the test responses are the captured data
- *Shift-out cycle*. The captured test responses are shifted out

During the shift-in cycle, test stimuli are shifted through the scan elements, and as a result, there are switches in every clock cycle in the sequential elements as well as in the combinational logic. At the launch-and-capture cycle, the shifted-in (loaded) test stimuli are applied to the circuit, and as a result there are switches in both the sequential elements and the combinational logic. Finally, during the shift-out cycle, the captured responses are shifted through the scan elements, and hence there are switches in both the sequential elements and the combinational logic. Hence, there are power consumed in every cycle.

In order to reduce test time, it is common practice to pipeline the application of test data such that while the *current* test response is shifted out, the *following* test stimulus is shifted in. Such a pipelining scheme reduces test time; however, in all cycles, there are switches in the sequential elements and the combinational logic due to the application of test data.

While all scan cycles result in power consumption and shift-in and shift-out are overlapped, the origin of the switches can be related to test stimuli switches at shift-in and test responses switches at shift-out. Figure 6.10 shows at each clock cycle the contribution of transitions due to shift-in/launch-and-capture/shift-out for a 8-bit long scan chain (Samii et al. 2008). Figure 6.10 shows that initially, in terms of time (clock cycles), most transitions are due to shift-out of current test response while few transitions are due to shift-in of next test stimulus. While the shift process proceeds over time, Fig. 6.10 shows that the transitions originating from test stimulus increases and transitions due to test response shift-out decreases.

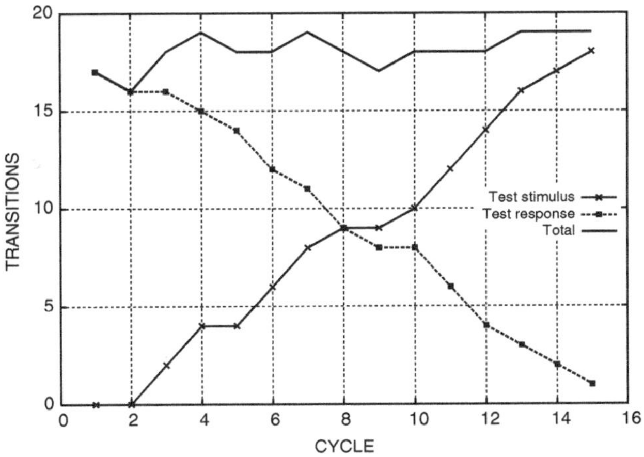

Fig. 6.10 Switches due to shift-in, launch-and-capture, and shift-out (Samii et al. 2008)

6.3.1.2 Power Constraint Modeling

Power-aware test planning requires a constraint to optimize against. The most straight forward is a maximal power constraint. At any time, the sum of the power consumption of the activated must be kept under a given constraint ($P_{\max}$).

A single/global power constraint model does not consider where the power is consumed. Activity, leading to power consumption at any place in the IC, is added together, and matched against the global power constraint. However, as discussed in Sect. 2.4 in Chap. 2, it is common practice that ICs contain a power distribution network. Hence, there are a number of power islands where each power island has its local power constraint. Instead of a single power constraint value that never is to be exceeded, each power island or power grid has its limit on deliverable power. In such a scenario, it is needed to also keep track on which power grid a testable unit belongs to. Larsson (2004) proposed the modeling of power grids as well as global power constraint. Each core is not only associated with a test time and a power constraint, but also associated with a power grid, and for each power grid there is a local power constraint. Figure 6.11 shows specification for the example in Fig. 6.4 where each core consists of a set of blocks (testable units), and each block is denoted by an idle test power consumption, a power grid to which it belongs, and a number of tests. A number of options for the specification is omitted, for details check the paper (Larsson 2004). The power grids are specified with their maximal allowed power. And for each test, the required test power consumption and test time are given.

```
[Cores]        #name        x      y        {blocks}
               CoreA        10     20       {blockA1 blockA2}
               CoreB        20     20       {blockB1}
               CoreC        20     10       {blockC1 blockC2 blockC3}
               CoreD        10     10       {blockD1}

[Blocks]       #name        idle_power     power_grid {tests}
               blockA1       5               grid_1    {testA1}
               blockA2       10              grid_1    {testA3}
               blockB1       7               grid_3    {testB1}
               blockC1       2               grid_2    {testC1}
               blockC2       3               grid_1    {testC2}
               blockC3       5               grid_3    {testC4}
               blockD1       8               grid_1    {testD1}

[Tests]        #name        power test time //rest of parameters omitted
               blockA1       5          10
               blockA2       10         20
               // specification for rest of tests omitted

[Power Grid] #name          limit
               grid_1        50
               grid_2        55
               grid_3        60
```

Fig. 6.11 Modeling power grids for the example in Figs. 6.4 and 6.8

6.3.2 Power Estimation

A power model needs data that corresponds to the power consumption. On the one hand, different semiconductor technologies result in different power consumption, but on the other hand, (6.2) states that power consumption is highly related to the input data, which during testing is the test data. We will discuss technology-independent estimation of power consumption.

In Sect. 6.3.1.1, we discussed transitions during shift-in, launch-and-capture, and shift-out. Given the cycle-accurate power modeling framework by Samii et al. (2008), we can model test power more accurately. There are two obvious problems (Samii et al. 2008). First, all gates in a circuit do not dissipate the same amount of power during switching. And, second, gates with the same applied input stimuli, switches at different probabilities. Table 6.2 lists some power properties capturing these issues, where $P_{0\rightarrow 1}$ is the probability that a transition from 0 to 1 occurs at the output with random input.

Given the problems to estimate power consumption, three ISCAS circuits, namely s1423, s3271, and s5378, were used in the simulations (Samii et al. 2008). The characteristics of the circuits are given in Table 6.3.

Two types of simulations were performed. The first simulations were aiming at extracting information about the switching activity in the three circuits. During the

Table 6.2 AMS c35 core cells (the output load is 20 fF for each core cell)

Gate	Power (μW/MHz)	$P_{0\to 1}$
CLKIN1	0.32	25%
NAND21	0.35	19%
NOR21	0.43	19%
XNOR21	0.50	50%
XOR21	0.61	50%
DFS1	1.27	–

Table 6.3 Characteristics of the three ISCAS'89 benchmarks used

	s1423	s3271	s5378
Gates	341	726	729
Flip-flops	74	116	160
Scan chains	1	4	4
Inputs	17	26	35
Outputs	5	14	49

Table 6.4 Transition count vs. real test power for s1423

Scan-in	Scan-out	Total	Power (mW)
2,774	2,774	5,548	22.72
2,774	1,406	4,180	18.90
1,406	1,406	2,812	14.02
722	1,406	2,128	10.80
380	1,406	1,786	8.99
722	722	1,444	7.89
722	380	1,102	6.91
380	380	760	5.34

Table 6.5 Transition count vs. real test power for s3271

Scan-in	Scan-out	Total	Power (mW)
1,736	1,736	3,472	41.50
896	1,736	2,632	33.20
896	896	1,792	24.96
476	896	1,372	19.00
476	476	952	15.21
272	476	748	12.53
272	272	544	10.01

simulation, the number of transitions were counted, and the transitions due to scan-in and scan-out were counted separately. For the second experiment, the real test power consumption for the circuits was simulated using a commercial tool. In both simulations, both the test stimuli and the test responses were considered.

The results are presented in Tables 6.4–6.6. The first two columns in each table show the transition count for scan-in and scan-out separately, while the third column shows the total transition count, which is the sum of transitions due to scan-in and scan-out. The total transition count is then compared to the real power dissipation in the fourth column. It is obvious from the tables that a power model that only

Table 6.6 Transition count vs. real test power for s5378

Scan-in	Scan-out	Total	Power (mW)
3,276	3,276	6,552	38.92
1,680	3,276	4,956	31.95
1,680	1,680	3,360	24.71
720	1,680	2,400	20.26
720	720	1,440	16.63
320	720	1,040	13.99
320	320	640	11.34

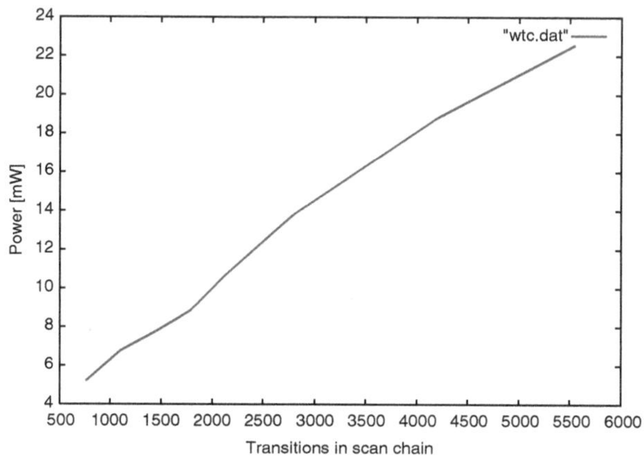

Fig. 6.12 The total transition count vs. power dissipation for s1423

considers the scan-in transitions does not correlate well with the test power dissipation. For example, the first two rows in Table 6.4 show that the scan-in transitions are equal, but the test power dissipation values for the two cases are different. Similarly, considering only the scan-out transitions leads to a power model that does not correlate with the test power simulations. This can, for example, be seen on the second to fifth line in Table 6.4 where the same number of transitions in scan-out results in different actual power consumption.

However, when transitions in both test stimuli and test responses are taken into account, there is a close correlation between transition count and power consumption. The last two columns in the Tables 6.4, 6.5, and 6.6 are plotted and illustrated in Figs. 6.12, 6.13, and 6.14, respectively. The figures show an almost linear correlation between the test power model that takes transitions in both test stimuli and test responses into account and the real power simulations.

Finally, the Pearson coefficient (Runyon et al. 1996) is used to quantify the correlation between the test power model proposed by Samii et al. and the real power. The obtained values for the three circuits considered in the experiments are listed in Table 6.7. The coefficients are very close to 1, indicating a good correlation between the test power model and the real test power dissipation.

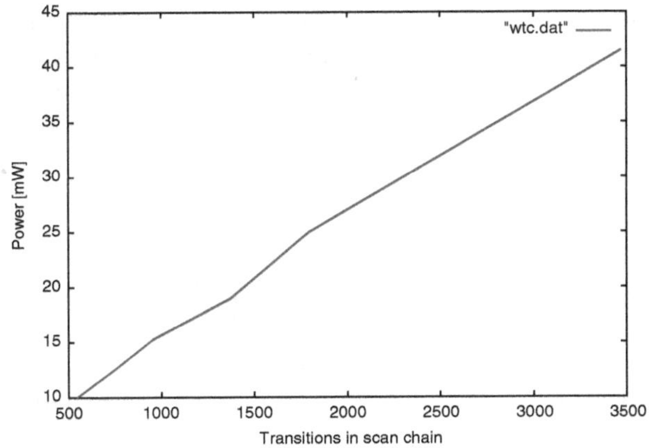

Fig. 6.13 The total transition count vs. power dissipation for s3271

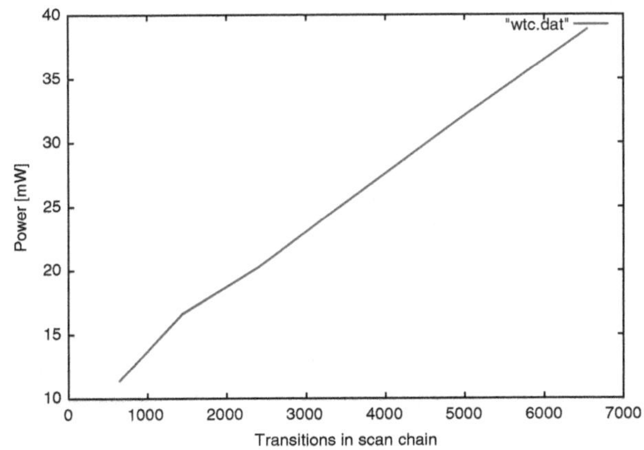

Fig. 6.14 The total transition count vs. power dissipation for s5378

Table 6.7 Pearson coefficients for the three ISCAS'89 circuits

Circuit	Pearson coefficient
s1423	0.997
s3271	0.999
s5378	0.999

6.3.3 Power Manipulation

The dynamic power consumption given by (6.2) depends on the switching activity $f_{0\rightarrow 1}$, the capacitance C_L, and the supply voltage V_{dd}. For a given IC, C_L and V_{dd} are fixed, while $f_{0\rightarrow 1}$ depends on the input, which during testing is the test data.

In this section, we will discuss the manipulation of test data in order to control and handle test power consumption. We will first discuss power-aware wrapper design (Sect. 6.3.3.1), and then the ordering and manipulation of test data in order to control test power consumption (Sect. 6.3.3.2).

6.3.3.1 Power-Aware Wrapper Design

A modular SOC can be tested in a modular fashion (discussed in Sect. 6.2). The cores are made testable units by the use of core test wrappers (Sect. 6.2.1) and test data is transported using a TAM (Sect. 6.2.2). The focus in Sect. 6.2 was to describe the design of the test architecture such that test planning can be applied, and demonstrate how test application time can be minimized. In this section, we discuss core wrapper design and its impact on test power consumption.

Core wrapper design imposes that the scan elements, that is scan chains, functional inputs, and functional outputs, for a given core are formed into a number of wrapper chains, which are to be connected to the TAM.

A power model is required for each core (testable unit) (power modeling is detailed in Sect. 6.3.1.1). Each core can be associated with a single power value. A number of such approaches assuming a single power value per testable unit have been proposed (Chou et al. 1997; Chakrabarty 2000; Iyengar and Chakrabarty 2001; Yoneda et al. 2006; Zhao and Upadhyaya 2003; Su and Wu 2004).

Figure 6.15 shows a core with two scan chains (functional inputs and functional outputs are omitted). Each scan chain contains four flip-flops. The core can be designed to have two wrapper chains, each scan chain forms a wrapper chain, or one wrapper chain, the scan chains are connected into one long chain. The test time is impacted by the wrapper design (discussed in Sect. 6.2.1). A single wrapper chain requires eight clock cycles for a shift-in/shift-out while two wrapper chains reduces the shift-in/shift-out time to four clock cycles.

However, the test data is defined depending on the core wrapper design. For the example in Fig. 6.15, in the case when a single wrapper chain is used, the test stimuli is shifted-in a single bit stream, while if two wrapper chains are used, two bit streams are used. Samii et al. (2006) analyzed the test power consumption per wrapper configuration. For the example core in Figs. 6.15 and 6.16 shows the corresponding power consumptions. Interestingly, the power profiles from the two wrapper design configurations are very different and they do not resemble each other. If a single power value is to be used when making use of core wrapper design, an analysis of power consumption for all possible wrapper configurations is needed in order to find the peak power consumption, which will define the single value. If a power

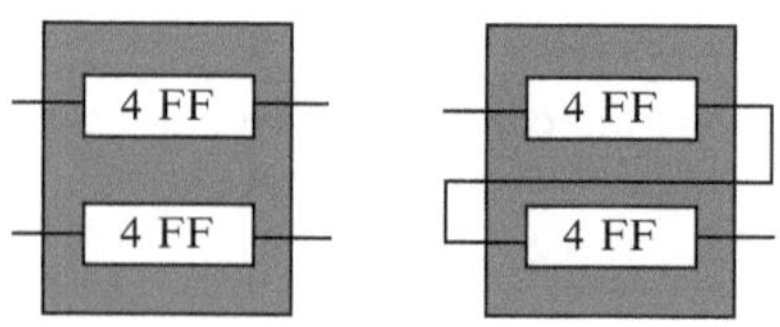

Fig. 6.15 A core with two scan chains connected as (1) two wrapper chains and (2) one wrapper chain (Samii et al. 2006)

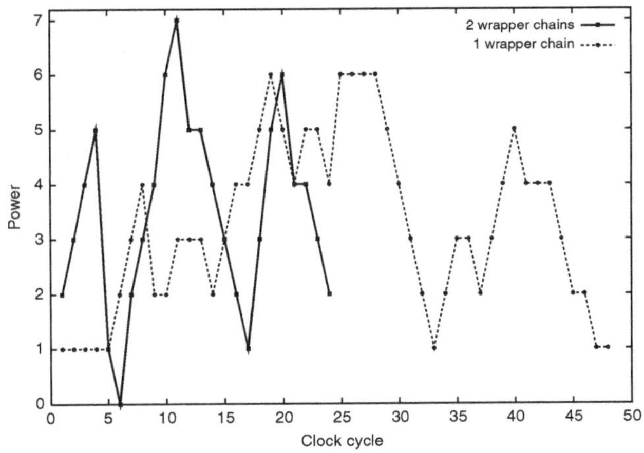

Fig. 6.16 The transition count profiles (scan-in/scan-out) for the two wrapper chain configurations illustrated in Fig. 6.15 when tested with five test patterns (Samii et al. 2006)

value per wrapper configuration is to be used, an analysis of all wrapper configurations has to be done, and the power value per wrapper configuration is given. Samii et al. (2006) used a clock-cycle accurate power model and from an analysis of the transitions (Fig. 6.16) the power profile per wrapper design configuration is given.

6.3.3.2 Ordering of Test Data

Application of test implies that test stimuli have to be applied. The quality of the test is not impacted by the order in which the test stimuli are applied. The importance is to apply all patterns. A number of approaches have been proposed to modify the order in which the test patterns are applied or the order in which the scan elements are connected to scan chains (Flores et al. 1999; Girard et al. 1997; Ghosh et al. 2003; Dabholkar et al. 1998; Bonhomme et al. 2002; Tudu et al. 2009).

Dabholkar et al. (1998) used test vector reordering to achieve average power reduction, while Bonhomme et al. (2002) used a scan-chain reordering technique to minimize the total number of transitions to reduce average power. However, these approaches aim at minimizing the average test power. The peak power, which is more crucial, is only minimized as a by-product. Tudu et al. (2009) addressed test vector ordering with the aim of reducing peak test power. Samii et al. (2006) showed that the contribution from scan-in and scan-out can be separated out. Hence, there is the possibility to assign a power value to each transition from pattern i to pattern j by taking the scan-out contribution from pattern i and the scan-in contribution from pattern j. Tudu et al. explored that fact and presented a graph-based approach to traverse the pattern set. Further, additional patterns were added in order to find better test pattern sequences. The problem, detailed below, is for a given test set to find a test vector ordering such that the peak power is minimal (Tudu et al. 2009).

T_1 :111111101010111	R_1 :111010101111000
T_2 :111111111010101	R_2 :101010111111111
T_3 :111111111110101	R_3 :101011111111001
T_4 :111111111111101	R_4 :110111111110110

Fig. 6.17 Example of test set size 5 (test stimuli and expected test responses) for a scan chain of length 15

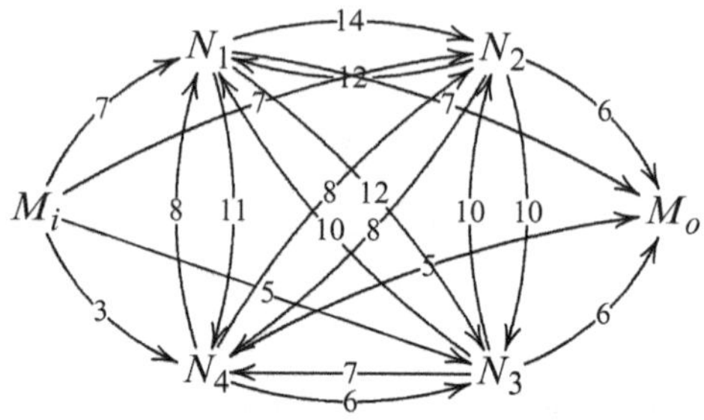

Fig. 6.18 A weighted digraph for the test data in Fig. 6.17

The example in Fig. 6.17 shows test data (test stimuli and test responses) for a circuit with a scan chain of 15 flip-flops. For each test stimulus T_i, the corresponding test response is R_i. Let each test stimulus (T_i) form a node N_i. A directed edge $E(i, j)$ between two nodes (N_i, N_j) exists when T_i and T_j can be applied consecutively. The weight of edge E_{ij}, EW_{ij} is the maximum of transitions that occur in the scan chain per clock cycle for the complete scan operation including load/unload and launch and capture, while applying test responses R_i followed by T_j [as discussed in Sect. 6.3.1.1 where Samii et al. (2006) showed how to separate transitions due to shift-out of current test response and shift-in of next test stimulus]. Figure 6.18 shows the weighted diagraph for the test set in Fig. 6.17. Two dummy nodes (M_i, M_o) are added. Dummy node M_i is added to scan-in the first test stimulus assuming the scan chain is initially in reset state and dummy node M_o is added to scan-out the last response.

Tudu et al. (2009) formulated three problems on obtaining the minimum peak power for a given test vector set and the order of test vectors. The first problem defines the order in which the test stimuli are to be applied without time penalty and the other two problems give the test data orders with marginal increase in time. Tudu et al. (2009) also defined a lower bound on the minimum achievable peak power.

6.4 Power-Constrained Test Planning

The objective of power-constrained test planning is to define a test plan, the order in which the cores are to be tested, such that a cost function, often related to test application time, is minimized while ensuring that the test plan has no problems related to test power consumption. Given is a modular SOC enabled for modular test (described in Sect. 6.2) such that each core is a testable unit. For each testable unit, there is a model of the test power consumption and for the SOC there is a model of the test power constraint (discussed in Sect. 6.3).

Much research have been devoted to power-constrained test planning. Zorian (1993), Chou et al. (1997), Larsson and Peng (1999), Ravikumar et al. (2000), and Iyengar and Chakrabarty (2001) proposed test scheduling approaches under power constraints. And, Chakrabarty (2000), Larsson and Peng (2001a, 2002a,b, 2006), Pouget et al. (2003a,b, 2005), Larsson et al. (2002, 2004), Su and Wu (2004), Huang et al. (2002), Xia et al. (2003), Zhao and Upadhyaya (2003), and Samii et al. (2006, 2008) proposed test architecture and test scheduling techniques. Approaches that integrate power-aware DfT with test planning have been proposed by Larsson and Peng (2001b, 2006), Singh and Larsson (2008), and Larsson (2004). A number of other approaches have been proposed. For example, He et al. (2005) proposed power-constrained test scheduling for BIST, Yoneda et al. (2006) and Xu et al. (2005) tackled multiple-clock domains, Sehgal et al. (2008) addressed test power in port-scalable testers, Rosinger et al. (2005) proposed thermal-aware scheduling, and Nicolici and Al-Hashimi (2003) discussed power-constrained test synthesis.

In this section, we will discuss power-constrained test planning. In Sect. 6.4.1, general power-constrained test planning is discussed, in Sect. 6.4.2 co-optimization of power-aware test architecture design and test scheduling are discussed, and in Sect. 6.4.3 power-constrained test planning that makes use of power-aware DfT is outlined.

6.4.1 Power-Constrained Test Scheduling

An SOC tested using BIST is an SOC where each testable unit has its dedicated test source and test sink (Zorian 1993). Zorian addressed test planning for BIST systems. Attached to each core is a *fixed* test time and a *single-fixed power value*, and the system is allowed to tolerate a power dissipation corresponding to a *single global peak power constraint* (discussed in Sect. 6.3.1) (Zorian 1993). The system can be modeled as a set of cores $\{c_1, c_2, \ldots, c_n\}$ with test time $\{\tau_1, \tau_2, \ldots, \tau_n\}$ and power consumption $\{p_1, p_2, \ldots, p_n\}$ such that core c_i is associated with test time τ_i and power consumption p_i. When core c_i is tested, it consumes power corresponding to p_i during a period of time τ_i and while c_i is *not* tested, the power dissipation is zero. The cores are grouped into sessions $\{S_1, S_2, \ldots, S_n\}$. Cores assigned to the same session S_i are tested concurrently, and no new test can be started until all tests in current session are completed. The optimization objective is twofold. First, the defined test plan, assigning cores to sessions, should minimize the test application time. Second, in order to minimize the routing overhead of added controller lines from the BIST controller to the cores (required to start the testing), cores that can share control lines, i.e., they are physically close, are to be grouped in the same test session. The defined test plan is at any time not allowed to consume more power than $P_{\max}$.

Zorian makes use of ASIC Z (details in Table 6.8 where each core is associated with a test time and a power dissipation) and presents a test plan as in Figs. 6.19 and 6.20, where the power constraint $P_{max} = 900$ mW. In the first session S_1, the cores RAM1, RAM4, and RF are tested. The total power consumption in session S_1 is the sum of the power consumed by each core, which comes out as $282 + 96 + 95 = 473$. The power consumed in the session is well below the given power constraint. The length of the session is $\max\{\tau(\text{RAM1}), \tau(\text{RAM4}), \tau(\text{RF})\} = \max\{69, 23, 19\} = 69$. The length of the test plan, the test application time, is the sum of the test time for each session.

Table 6.8 ASIC Z test length and test power dissipation (Zorian 1993)

Core i	Test time τ_i	Power consumption p_i
RL1	134	295
RL2	160	352
RF	10	95
RAM1	69	282
RAM2	61	241
RAM3	38	213
RAM4	23	96
ROM1	102	279
ROM2	102	279

```
{RAM1, RAM4, RF}      Length of test session = 69
{RL1, RL2}            Length of test session = 160
{RAM2, RAM3}          Length of test session = 61
{ROM1, ROM2}          Length of test session = 102
                      Total test length = 392
```

Fig. 6.19 Test sessions for ASIC Z using the approach by Zorian (1993)

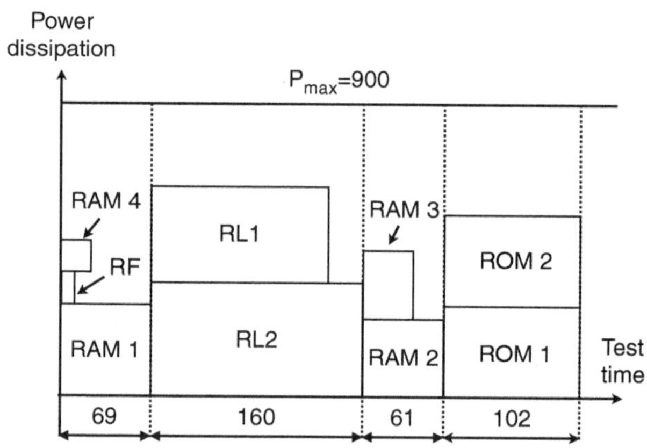

Fig. 6.20 Test schedule for ASIC Z using the approach by Zorian (1993)

Chou et al. assume, as Zorian, that each testable unit is associated with a fixed test time and a single fixed test power. Different from Zorian, Chou et al. assume that there may be conflicts among the testable units. In order to handle test conflicts such that conflicts are captured, the problem is formulated as a graph problem (Chou et al. 1997). A test compatibility graph TCG(V, E) is used, and cores are vertices (nodes) and compatibility is modeled through the edges where an edge between two tests means that the corresponding cores can be tested at the same time. A power compatibility graph (PCG) is used to derive power compatible alternatives. An example of a PCG is shown in Fig. 6.21 where each node is a test and attached to each node is a test time and a test power consumption.

Chou et al. also made experiments using ASIC Z (results are presented in Fig. 6.22). The test plan defined by Chou et al. includes only three sessions, and the length of the sessions are such that the total test application time is 331, which is an improvement over Zorian's result of 392. Larsson and Peng (1999) used the same assumptions as Chou et al. and formulated a fast heuristic to schedule the tests. While the heuristic is rather simple, it manages to define a test plan as in Fig. 6.23, which has a test application time of only 300.

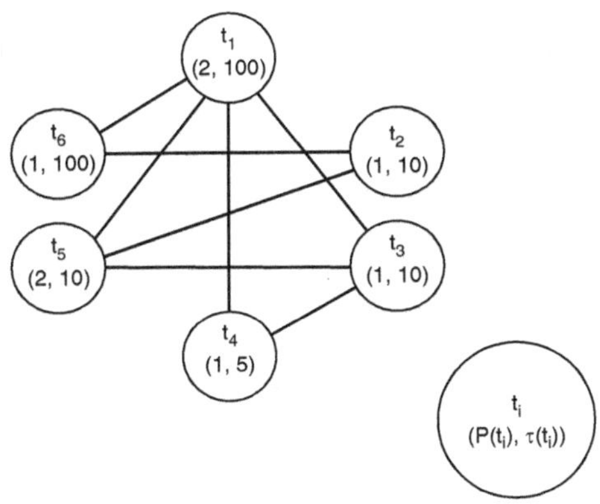

Fig. 6.21 Test conflict graph and power constraints where each node is attached with a test time and a power dissipation

```
{RAM1, RAM3, RAM4, RF}      Length of test session = 69
{RL1, RL2}                  Length of test session = 160
{ROM1, ROM2, RAM2}          Length of test session = 102
                            Total test length = 331
```

Fig. 6.22 Test sessions for ASIC Z using the approach by Chou et al. (1997)

```
{RL1, RL2, RAM2}           Length of test session = 160
{RAM1, ROM1, ROM2}         Length of test session = 102
{RAM3, RAM4, RF}           Length of test session = 38
                           Total test length = 300
```

Fig. 6.23 Test sessions for ASIC Z using the approach by Larsson and Peng (1999)

Muresan et al. (2000) and Mureşan et al. (2004) explored a number of test scheduling algorithms and Ravikumar et al. (2000) proposed a technique to define the test resources in the system.

6.4.2 Power-Aware Test Architecture Design and Test Scheduling

Test planning without considering test power consumption is discussed in Sect. 6.2. In that section, each core (testable unit) is associated with a test time and a requirement on TAM wires/wrapper chains, and the objective is to define a test plan where the test application time is minimized while constraint on TAM width is not violated. In Sect. 6.4.1, power-constrained test planning is discussed. In the section, each core c_i is associated with a test time τ_i and a test power consumption p_i. The optimization objective is to define a test plan such that test application time is minimized while the constraint on power dissipation is not violated at any time. In this section, we will integrate the two approaches. It means, we have for each core i, a model of test time, power consumption, and TAM wire/wrapper-chain requirement. The most straight-forward approach is to associate each core i with a fixed value on test time τ_i, a single fixed value on test power consumption p_i, and a fixed value on TAM (wrapper-chain) requirement w_i (illustrated in Fig. 6.24). The objective is to define a test plan where the box for each core is assigned a start time such that constraints on power dissipation ($P_{\max}$) and TAM width ($\text{TAM}_{\max}$) are not violated at any time (shown in Fig. 6.25).

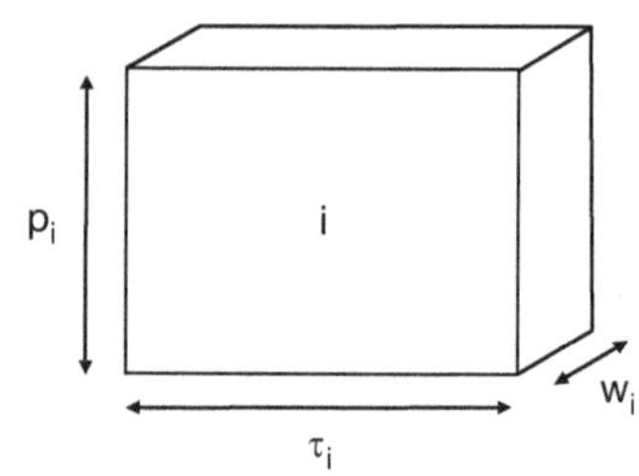

Fig. 6.24 A model of test time (τ_i), test power consumption (p_i), and wrapper-chain requirement (w_i) for a core i

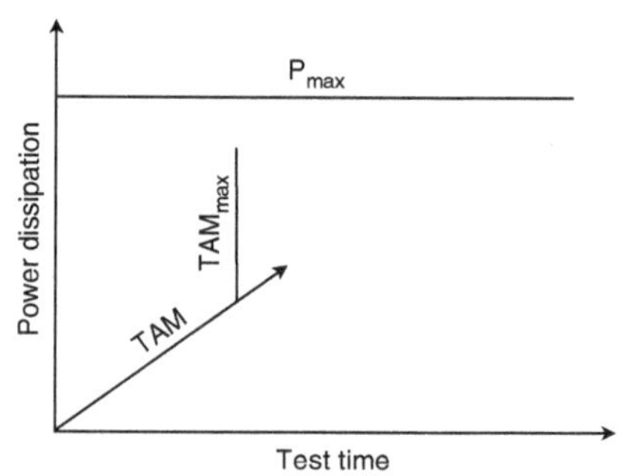

Fig. 6.25 Scheduling tests under constraint from test power consumption ($P_{\max}$) and TAM width ($\text{TAM}_{\max}$)

The following two observations can be made. First, the optimization is not trivial as packing three-dimensional boxes is difficult. Second, the simple modeling with only one value per parameter leads to a test plan that meets the constraints but making use of more accurate models would lead to more optimized test plans.

Instead of a fixed number of wrapper chains at each core, a number of approaches have been proposed for defining test plans under the assumption that each core i is associated with a test power consumption p_i, the test stimuli and produced test responses are stored at the ATE (the ATE serves as test source and test sink), while the test time $\tau_i(w)$ depends on the number of assigned wrapper chains. The optimization objective remains the same as above; design the TAM to connect the ATE with the cores and assign cores to the TAM such that test application time is minimal while the constraint on power consumption is met at any time during the test application. An example of a test infrastructure for an SOC is shown in Fig. 6.4. The ATE channels are connected to the TAM, and the TAM is partitioned into a number of test buses. The width of the test buses determines the maximal number of wrapper chains that can be used at each core (discussed in Sect. 6.2.1). The developed plan is to be made such that not only TAM width constraints are met but also test power constraint ($P_{\max}$). And, examples of approaches for the above problem are the ones by Huang et al. (2002), Pouget et al. (2003a,b), Su and Wu (2004), and Zhao and Upadhyaya (2003).

Core wrapper design determines the organization of scan elements in wrapper chains, which impact the way test data is stored in the ATE, and hence it impacts the number of transitions at shift-in and shift-out. Based on this fact, Samii et al. proposed the usage of a power model per wrapper-chain configuration (see Fig. 6.26). The proposed model is a cycle-accurate model where a power value is assigned to each clock cycle, and showed its feasibility to test architecture design and test scheduling (Samii et al. 2006, 2008).

A number of approaches have been proposed to address test infrastructure while considering test power consumption. Chakrabarty (2000) discussed the design of test architectures under place-and-route and power constraints. And, Larsson and Peng (2001a, 2002a) proposed assumed x, y coordinates for each core and added a minimal test infrastructure.

In contrast to previous power constraint models that made use of a single power constraint value, Larsson (2004) and Larsson and Peng (2006) proposed the usage

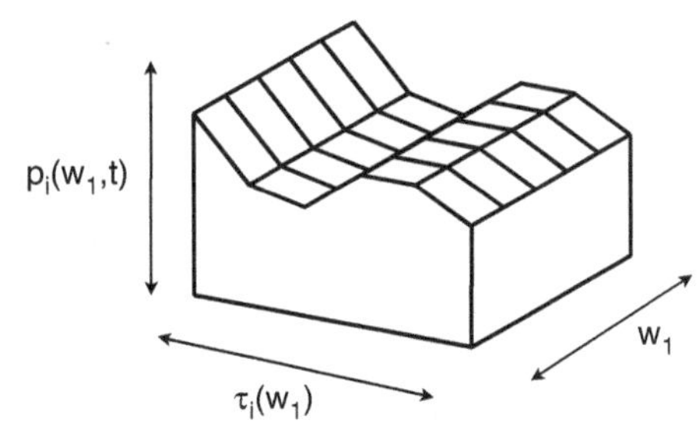

Fig. 6.26 The model by Samii et al. for test time ($\tau_i(w_1)$), test power consumption ($p_i(t, w_1)$), and wrapper-chain requirement (w_1) for a core i (Samii et al. 2006, 2008)

of power grids (see discussion on power grids in Sect. 6.3.1.2). Each core is assigned to a power grid, and, for each power grid there is a power constraint. The result is local power constraints for each power grid as well as global power constraints.

6.4.3 Power-Constrained Test Planning Utilizing Power-Aware DfT

In this section, we discuss the utilization of power-aware DfT in the test planning process. A number of power-aware DfT techniques have been proposed (detailed in Chap. 4). For scan-tested designs, there are techniques to address shift-power consumption and techniques to address capture power consumption. As these techniques do come at a cost, it is appropriate to include them in the test planning process in order to find where the techniques are most efficient. A power-aware DfT technique may not be needed at all places in the circuit.

6.4.3.1 DfT for Shift-Power Reduction

The shift process is a necessity to enable loading of next test stimulus and unloading of current produced test responses; however, the shift process does not contribute to improve the test quality. And the transitions during the shift process causes power to be consumed. This power is useless. A number of techniques, such as gating scan chains, have been proposed to reduce shift-power consumption (detailed in Sect. 4.3.2). Saxena et al. (2001) and Bonhomme et al. (2001) proposed clock gating for scan chains.

Figure 6.27 shows an example without clock-gating (top) and with clock-gating (bottom). At the same wire cost and test time, less sequential elements are activated with clock-gating, and consequently less combinational logic is switched. During the shift process with clock-gating, at any moment only one scan chain in Fig. 6.27 is active.

A number of approaches have been proposed to include power-aware DfT during the test planning. Larsson and Peng proposed a test scheduling approach where the power consumption for each test is not a fixed value but a variable that depends on the number of associated wrapper chains (Larsson and Peng 2001a; Larsson and

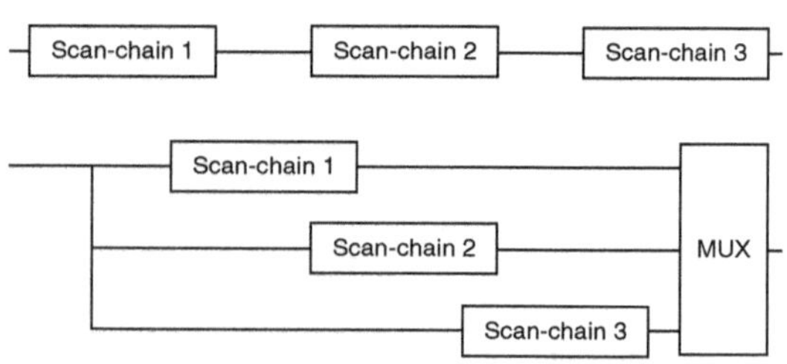

Fig. 6.27 Scan chains without clock-gating (*top*) and with clock-gating (*bottom*)

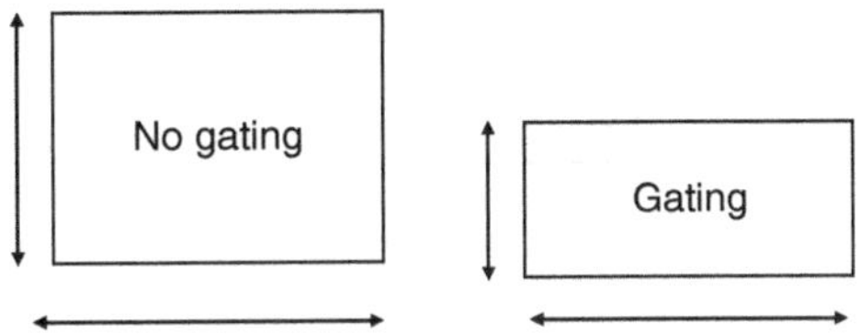

Fig. 6.28 Power model when applying and not applying clock-gating for the example in Fig. 6.27

Peng (2001b); Larsson and Peng (2002a)). For each core i, the power consumption p_i depends on the *degree* of scan-chain clock-gating (see Fig. 6.28). Let w be the number of wrapper chains at core i, the power consumption is then $p_i(w) = p_i \times w$. The penalty of assigning a high number of wrapper chains is not only that a high number of TAM wires are to be used, but also that the possibility of making use of clock-gating is reduced. The effect of clock-gating is highest for a core where the scan elements are formed into a few number of wrapper chains, as that allows more clock-gating, and therefore more savings in power.

6.4.3.2 DfT for Capture-Power Reduction

Scan-tested ICs consume power during the shift process and the capture cycle. While the shift process itself does not improve on the test quality, the power consumed during the shift process is useless (discussed above). Techniques have been proposed to address useless power consumption. The capture cycle, on the other hand, does contribute to test quality. Due to the increasing use of at-speed scan where the capture cycle is applied at normal clock speed while the shift process is in low speed, the capture power is far higher than that during the shift process. It is therefore important to address the capture power consumption.

Addressing capture power consumption is difficult for nonmodular ICs. The scan elements are connected into scan chains and all scan chains are active during testing. For modular ICs that consist of testable blocks of logic, test planing can be used to reduce test time and control test power consumption. For modular IC, it is also possible to address capture power consumption.

The following example illustrates the problem with capture power for modular SOCs (assume that the capture power is much higher than shift power due to at-speed testing or that shift-power consumption is reduced significantly using shift-power DfT).

Example: Assume two cores being tested simultaneously. Core 1 has a scan chain of length 4 and core 2 has a scan chain of length 5. Therefore, core 1 needs 4 cycles to shift in test stimulus and the test response is captured in cycle 5; and the capture cycle repeats after every 5 cycle (Fig. 6.29). Similarly, core 2 shifts in a test stimulus in 5 cycles and captures the response in cycle number 6; and the capture cycle repeats every 6 cycles (Fig. 6.29). Both cores, when scheduled concurrently will have simultaneous capture test responses at cycles 30, 60, 90, …(which are multiples of least common multiplier of 5 and 6). It may result into power droop problem that causes chips to falsely fail the test, if the sum of capture power of

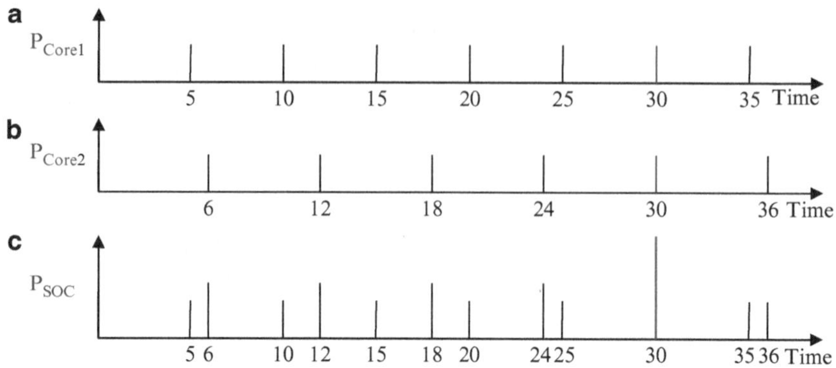

Fig. 6.29 Capture power profile

core 1 and core 2 exceeds a threshold for the following vectors: (1) 6th vector of core 1 and 5th vector of core 2, (2) 12th vector of core 1 and 10th vector of core 2, and so on as illustrated in Fig. 6.29. Therefore, it is important to care about capture power when capture cycle of various cores scheduled together coincides.

Larsson and Singh proposed strategies for a given test schedule reordering test vectors for capture power reduction, and insertion of idle cycle to prevent capture cycle to coincide (Singh and Larsson 2008).

6.5 Hierarchical Test Planning Strategies for SOCs

In this section we discuss a couple of practical solutions based on hierarchical test planning. The focus is on multiple clock domains and IDDQ testing.

6.5.1 Low-Power Test Planning for Multiple Clock Domains

SOC designs with multiple clock domains, as many as 20, are increasingly common. Multiclock domain scan capture can lead to incorrect data capture since the arrival of the capture clocks is not synchronized. A solution is to capture in one domain at a time. Note that the time and energy spent in shifting data through the flip-flops of a domain in which the test response is not captured is wasted. In this section we summarize two approaches: the divide-and-conquer (DNC) scan which enables testing of individual blocks in an SOC and its improvement by using a clock-domain-based partitioning called virtual divide-and-conquer (VDNC), so as to reduce test application time and test power (Ravikumar et al. 2005).

Consider an SOC named XDSL with two subchips and four blocks: ARM, EMIF, CPU, and DDR. Assume that there are four clock domains, one corresponding to each block. The ATE has a limit on the number of high-speed clocks that it can supply and the number of scan chains, k, it can support. A total of k scan chains are enforced in each of the four blocks and the chains are concatenated at the top level. For any particular fault type (stuck-at, transition-delay, IDDQ, etc.) scan test involves a separate *test mode*, where the scan chains are loaded through the k scan inputs, and the responses are unloaded using the k scan outputs. Assuming that the lengths of the longest chains in the blocks are l_{ARM}, l_{EMIF}, l_{DDR}, and l_{CPU}, the scan test application time will be proportional to $l_{\text{ARM}} + l_{\text{EMIF}} + l_{\text{DDR}} + l_{\text{CPU}}$. Since the pattern responses are captured in only one domain at a time, there is a high number of wasted test cycles.

The essential idea in DNC is to provide a scan access mechanism to allow scan testing of individual portions of the SOC. If there are n subchips in the SOC, DNC scan will use the available bandwidth of k scan pins to route k scan chains through each of the subchips. A scan multiplexer logic (also known as scan router) is used to permit testing of one subchip at a time. Since subchips may interact through glue logic, it becomes necessary to also permit a daisy-chain mode. In the daisy-chain mode, the target fault list includes all faults that are not already caught in the n individual scan-test modes. Since only portions of the SOC are tested at a time, the sequential elements in the remaining parts of the chip can be initialized to constant values to reduce test power (Ravikumar and Hetherington 2004; Butler et al. 2004).

DNC can be applied to XDSL as follows. The chip is partitioned into two subchips, namely, ARM + EMIF and CPU + DDR. If the chip has k scan-in and k scan-out ports, balanced scan chains are inserted in the two subchips and the scan chains are connected to a scan router. In test mode 0, the ARM + EMIF subchip will be scan tested through the scan path scanin–ARM–EMIF–scanout and the flip-flops in the DDR and CPU subblocks will be initialized to constants. In test mode 1, the CPU + DDR subchip will be scan tested through the scan path scanin–CPU–DDR–scanout and the flip-flops in the ARM and EMIF subblocks will be initialized to constants. In mode 2, the daisy-chain mode, the scan path would be scanin–ARM–EMIF–CPU–DDR–scanout.

DNC scan fits well into a physical design hierarchy; as it is natural to partition the chip into logical partitions such as ARM + EMIF and CPU + DDR, so as to balance the gate counts across partitions. Another consideration during physical partitioning is the connectivity between the blocks, so that an effective floorplan can be derived. This partitioning strategy also works well from the view point of DNC scan, since balancing the gate counts would tend to balance the number of faults across the partitions, leading to a balance in ATPG run-times on the individual partitions. Similarly, keeping physically related modules together will lead to a smaller target fault set for the daisy-chain mode. DNC scan architecture allows ATPG to run concurrently for each partition and the only dependence in the ATPG flow is that the daisy-chain mode ATPG cannot be started without completing the ATPG runs for the partitions (Ravikumar and Hetherington 2004). The daisy-chain mode ATPG has a dependency on the test group fault list. Since the daisy-chain mode targets faults that are not detected during the test group ATPG. Therefore, the speedup of a distributed implementation of the ATPG is impacted adversely by a long run of the daisy chain.

In the VDNC scan scheme (Senthil et al. 2007), the design is partitioned into test groups based on clock domain information. Since the partition may not preserve hierarchical boundaries, it is referred as virtual partitioning. A test group in VDNC consists of scan chains that are clocked by a single clock or domains of same frequency but independent of each other. Two clock domains are considered independent if there exists no path between them or all the paths between them are false paths. Test patterns are generated for each test group separately. Since there is only one clock per test group, the shift and capture are completely safe on all flops in the scan chains. Hence, all flops scanned with test data are also used to capture new data. In the VDNC architecture, since test partitioning is based on clock domains, it is possible to not only reduce scan shift power, but also the clock tree power. The instantaneous power will also be smaller in VDNC in comparison with DNC scan, since the number of flops that toggle at a point of time will be smaller in the case of VDNC.

Analyzing the interaction among cores and using partial residual test modes can be used to improve the test coverage for transition delay faults; the partial residual test modes include a much smaller number of flops than the full residual test mode, thereby resulting in a lower peak power, lower test volume, lower test application time, and lower test cost.

6.5.2 IDDQ Test Planning for Core-Based System Chips

IDDQ testing has been used to supplement voltage testing of CMOS chips (Chakravarty and Thadikaran 1997; Gattiker et al. 1996; Rajsuman 1995; Sachdev 1997). The idea is to declare a chip as faulty if the steady-state current drawn from the power supply after the application of a test vector exceeds a threshold value. A CMOS circuit only consumes leakage power after the switching transients settle down, and a large quiescent power-line current indicates a defective chip. With device counts in system chips crossing into millions, the leakage power is no more insignificant, making IDDQ tests unsafe. Yet, IDDQ tests are invaluable since they can catch faults that are not testable using voltage testing.

The quiescent power-line current after the application of a test vector to a CMOS circuit is referred to as IDDQ and consists mainly of subthreshold leakage current and PN-junction leakage current. If there is a defect in the circuit, such as a short circuit between two nodes, a direct resistive path may be formed between VDD and VSS, causing an increase in IDDQ.

IDDQ testing becomes hard to practice for SOCs implemented in nanometer technologies, since system chips have a larger number of transistors, each of which draws a larger subthreshold current. *Test planning* can be an alternate to the problem. There is yet another practical reason that provides motivation for DNC IDDQ testing. For a large SOC, generating IDDQ patterns offers a number of challenges. Due to the growing size of the chips, the run-times for pattern generation can be high when high fault coverage is targeted. Generating IDDQ patterns for subchips and integrating them at the top level, e.g., generating IDDQ patterns for memories and IDDQ tests for logic blocks and integrating these together – is a tough task. Fault simulation of IDDQ patterns is also a major challenge since it involves device simulation. Run-times of fault simulation will be tolerable when a DNC scheme is followed. Furthermore, concurrent execution of multiple subchip level fault simulations can be performed in a distributed environment, reducing the turn-around-time.

A simple modification to IEEE 1500 scheme to permit current testing has been proposed by Ravikumar and Kumar (2002). A high-threshold-voltage switch similar to the one described by Rajsuman (1998) can be used with the core wrapper for isolating a core from the power supply. The gate voltage of the switch can be controlled to turn off the switch, cutting off the power supply to the core to which the switch is connected. A 1500-compliant test architecture with an isolation control register selects which cores are powered off during testing. The outputs of the register control the gating of the high-threshold switches. The high-threshold switch can be regarded as part of the core wrapper. The wrapper also consists of scan flip-flops, which, depending on whether they are placed on the input side or the output side, are useful for scanning in test data and scanning out test responses (for voltage testing). A bypass register is useful for isolating the core from the TAM, so that test data can be forwarded to another core. Power switches are common in SOCs of today for the purpose of power management. This architecture can reuse the power switches for implementing the hierarchical IDDQ scheme.

Let there be n cores in the system. When current testing is applied individually to each core, let the IDDQ for a fault-free core j be given by IDDQ_j. The total IDDQ for a fault-free chip is given by $\mathrm{IDDQ} = \mathrm{IDDQ}_1 + \mathrm{IDDQ}_2 + \cdots + \mathrm{IDDQ}_n$. Note that IDDQ_j are random variables, since the current depends on the operating conditions, input pattern, and the variations in the manufacturing process, temperature, and voltage. Usually, IDDQ_j are taken to be Gaussian random variates. Let μ_j and σ_j be the mean and standard deviation of current IDDQ_j. Then the mean μ and standard deviation σ of the cumulative current IDDQ are given by $\mu = \mu_1 + \mu_2 + \cdots + \mu_n$ and $\sigma^2 = \sigma_1^2 + \sigma_2^2 + \cdots + \sigma_n^2$.

The faulty chip IDDQ can be written as $\mathrm{IDDQ_f} = \mathrm{IDDQ} + I_\mathrm{f}$ where I_f corresponds to the extra current that the SOC will sink due to a resistive path from VDD to VSS. The mean and standard deviations of $\mathrm{IDDQ_f}$ are given by $\mu_{\mathrm{IDDQ_f}} = \mu_{\mathrm{IDDQ}} + I_\mathrm{f}$ and $\sigma^2_{\mathrm{IDDQ_f}} = \sigma^2_{\mathrm{IDDQ}} + \sigma I_\mathrm{f}^2$.

It is common to set the IDDQ threshold limit to $\mu_{\mathrm{IDDQ}} + 3 \times \sigma_{\mathrm{IDDQ}}$. Due to the intrinsic leakage of the system, the distribution of the fault-free IDDQ may overlap with the distribution of the faulty IDDQ. As a result, the confidence in a tested product suffers, and there is an increased chance of *aliasing*. Suppose we partition the set of cores into k groups and test each group separately. The threshold limit of $\mu_{\mathrm{IDDQ}} + 3 \times \sigma_{\mathrm{IDDQ}}$ will be applicable to each group, and since the associated mean and standard deviations are small, the chances of *aliasing* are smaller for a subset of cores. Therefore, the confidence in the tested product improves.

Let $C = \{C_1, C_2, \ldots, C_n\}$ be the set of cores in the SOC. Let $P = P_1 \oplus P_2 \oplus \cdots \oplus P_k$ be a partition of C, where P_j are subsets of C such that $P_i \cap P_j = \varphi$ if $i \neq j$, and $P_1 \cup P_2 \cdots \cup P_k = C$. Two extreme cases of partitioning occur when $k = 1$ and $k = n$. In the former case, all the cores are in the same partition, and the resulting IDDQ may be too large for ensuring reliable testing. In the latter case, the cores are tested one at a time, increasing the total test application time. An optimal solution is one that minimizes the test execution time while ensuring the reliability of the IDDQ test procedure. The inputs to the partitioning problem include the description of the system with details such as the number of cores, the descriptions of the cores, and the upper threshold on the mean value of IDDQ that is acceptable from the viewpoint of reliability. Because finding the optimal solution is computationally intractable, partitioning heuristics have been developed in practice for this purpose.

6.6 Summary

The power consumption during test is significantly higher than the power consumption during functional operation. The problem is that ICs are designed for functional operation, and that the high power consumption during test may lead to low yield, which is costly. The power consumption during test must therefore be carefully considered. In this chapter, we have discussed test planning as a course to cope with high test power consumption. Modular testing is enabled by the fact that ICs are increasingly designed in a modular fashion. The advantage with modular testing is that it allows test planning, which is a low-cost alternative to control the test power consumption.

In Sect. 6.2 we discussed the requirements to employ modular testing. We outlined core test wrappers, which are used to *isolate* and *interface* an embedded core. The isolation makes the core a standalone testable unit and the interface ensures that test stimuli can be received from the test access mechanism and that produced test responses can be sent to the test access mechanism. The test access mechanism connects the test sources with the cores, and the cores with the test sinks. In Sect. 6.2, we also discussed the design and optimization of the core test wrappers and the test access mechanism, as well as the test scheduling, which is to define when each core is to be tested. The overall objective is to define a test plan such that test application time and test architecture are minimized.

In Sect. 6.3 we discussed test power modeling and constraints. We discussed *when* and *where* power is consumed. We assumed scan-based circuitry, and focused on the shift cycles and the launch-and-capture cycles. We discussed accuracy of power modeling: single value vs. multiple values. And, we discussed the estimation of power consumption, which is needed for the models. And we found that test power consumption can be accurately estimated at clock-cycle granularity when taking transitions during shift-in and shift-out into account. We also discussed power constraints, which define the upper limit on the allowed activity. We discussed single global constraint as well as multiple local constraints.

Given test architecture design and test scheduling from Sect. 6.2 and modeling of power consumption and constraints from Sect. 6.3, we discussed in Sect. 6.4 the combination of the two. We described techniques that modeled test time and power consumption in different ways. We also discussed test planning when making use of power-aware DfT. As power-aware DfT comes at a cost, the objective is to minimize the usage of power-aware DfT, and apply it only when necessary. We discussed power-aware DfT for shift-power reduction as well as techniques to avoid capture power violations.

In Sect. 6.5, we discussed test planning strategies for multiple clock domains and IDDQ test planning for core-based system chips.

References

Abadir MS, Breuer MA (1986) Test schedules for VLSI circuits having built-in test hardware. IEEE Trans Comput 35(4):361–367. DOI http://dx.doi.org/10.1109/TC.1986.1676771

Aerts J, Marinissen EJ (1998) Scan chain design for test time reduction in core-based ICs. In: Proceedings of IEEE international test conference (ITC), pp 448–457

Beenker FPM, Eerdewijk KJE, Gerritsen RBW, Peacock FN, Star MD (1986) Macro testing: unifying ic and board test. IEEE Design Test Comput 3(6):26–32. DOI http://dx.doi.org/10.1109/MDT.1986.295048

Bhatia S, Gheewala T, Varma P (1996) A unifying methodology for intellectual property and custom logic testing. In: Proceedings of IEEE international test conference (ITC), pp 639–648

Bonhomme Y, Girard P, Guiller L, Landrault C, Pravossoudovitch S (2001) A gated clock scheme for low power scan testing of logic ICs or embedded cores. In: Proceedings of IEEE Asian test symposium (ATS), pp 253–258

Bonhomme Y, Girard P, Landrault C, Pravossoudovitch S (2002) Power driven chaining of flip-flops in scan architectures. In: Proceedings of IEEE international test conference (ITC), pp 796–803

Bouwman F, Oostdijk S, Stans R, Bennetts B, Beenker FPM (1992) Macro testability: the results of production device applications. In: Proceedings of IEEE international test conference (ITC), pp 232–241

Butler KM, Saxena J, Fryars T, Hetherington G, Jain A, Lewis J (2004) Minimizing power consumption in scan testing: pattern generation and DFT techniques. In: Proceedings of IEEE international test conference (ITC), pp 355–364

Chakrabarty K (2000) Design of system-on-a-chip test access architectures under place-and-route and power constraints. In: Proceedings of ACM/IEEE design automation conference (DAC), pp 432–437. DOI http://doi.acm.org/10.1145/337292.337531

Chakravarty S, Thadikaran PJ (1997) Introduction to IDDQ testing. Kluwer, Dordrecht

Chou R, Saluja K, Agrawal V (1997) Scheduling tests for VLSI systems under power constraints. IEEE Trans VLSI Syst 5(2):175–185

Craig GL, Kime CR, Saluja KK (1988) Test scheduling and control for VLSI built-in self-test. IEEE Trans Comput 37(9):1099–1109. DOI http://dx.doi.org/10.1109/12.2260

Dabholkar V, Chakravarty S, Pomeranz I, Reddy S (1998) Techniques for minimizing power dissipation in scan and combinatorial circuits during test application. IEEE Trans Comput Aided Des 17(12):1325–1333

Flores P, Costa J, Neto H, Monteiro J, Marques-Silva J (1999) Assignment and reordering of incompletely specified pattern sequences targetting minimum power dissipation. In: Proceedings of IEEE international conference on VLSI design (ICVD), pp 37–41

Gattiker A, Nigh P, Grosch D, Maly W (1996) Current signatures for production testing. In: Proceedings of IEEE international workshop on IDDQ testing (IDDQ)

Ghosh S, Basu S, Touba NA (2003) Joint minimization of power and area in scan testing by scan cell reordering. In: Proceedings of IEEE computer society annual symposium on VLSI, pp 246–249

Girard P, Landrault C, Pravossoudovitch S, Severac D (1997) Reduction of power consumption during test application by test vector ordering. Electron Lett 33(21):1752–1754

Goel SK, Marinissen EJ (2002a) Cluster-based test architecture design for system-on-chip. In: Proceedings of IEEE VLSI test symposium (VTS), pp 259–264

Goel SK, Marinissen EJ (2002b) Effective and efficient test architecture design for SOCs. In: Proceedings of IEEE international test conference (ITC), pp 529–538

Goel SK, Marinissen EJ (2003) SOC test architecture design for efficient utilization of test bandwidth. ACM Trans Des Automat Electron Syst 8(4):399–429. DOI http://doi.acm.org/10.1145/944027.944029

Gupta RK, Zorian Y (1997) Introducing core-based system design. IEEE Des Test Comput 14(4):15–25. DOI http://dx.doi.org/10.1109/54.632877

Harrod P (1999) Testing reusable IP – a case study. In: Proceedings of IEEE international test conference (ITC), p 493

He Z, Jervan G, Peng Z, Eles P (2005) Power-constrained hybrid bist test scheduling in an abort-on-first-fail test environment. In: Proceedings of Euromicro conference on digital system design (DSD), pp 83–87. DOI http://dx.doi.org/10.1109/DSD.2005.63

Huang Y, Cheng WT, Tsai CC, Mukherjee N, Samman O, Zaidan Y, Reddy SM (2001) Resource allocation and test scheduling for concurrent test of core-based SOC design. In: Proceedings of IEEE Asian test symposium (ATS), pp 265–270

Huang Y, Reddy S, Cheng W, Reuter P, Mukherjee N, Tsai C, Samman O, Zaidan Y (2002) Optimal core wrapper width selection and SOC test scheduling based on 3-d bin packing algorithm. In: Proceedings of IEEE international test conference (ITC), pp 74–82

IEEE std 1500 – Standard for embedded core test (2005). DOI http://http://grouper.ieee.org/groups/1500/

Immaneni V, Raman S (1990) Direct access test scheme-design of block and core cells for embedded asics. In: Proceedings of IEEE international test conference (ITC), pp 488–492. DOI 10.1109/TEST.1990.114058

Iyengar V, Chakrabarty K (2001) Precedence-based, preemptive, and power-constrained test scheduling for system-on-a-chip. In: Proceedings of IEEE VLSI test symposium (VTS), pp 368–374

Iyengar V, Chakrabarty K, Marinissen EJ (2001) Test wrapper and test access mechanism co-optimization for system-on-chip. In: Proceedings of IEEE international test conference (ITC), pp 1023–1032

Iyengar V, Chakrabarty K, Marinissen E (2002a) Efficient wrapper/TAM co-optimization for large SOCs. In: Proceedings of design, automation, and test in Europe (DATE). IEEE Computer Society, Washington, DC, pp 491–498

Iyengar V, Chakrabarty K, Marinissen EJ (2002b) On using rectangle packing for SOC wrapper/TAM co-optimization. In: VTS '02: Proceedings of the 20th IEEE VLSI test symposium. IEEE Computer Society, Washington, DC, pp 253–258

Iyengar V, Chakrabarty K, Marinissen EJ (2002c) Test wrapper and test access mechanism co-optimization for system-on-chip. J Electron Test Theory Appl 18(2):213–230. DOI http://dx.doi.org/10.1023/A:1014916913577

Koranne S (2002) A novel reconfigurable wrapper for testing of embedded core-based SOCs and its associated scheduling algorithm. J Electron Test Theory Appl 18(4–5):415–434

Koranne S, Iyengar V (2002) On the use of k-tuples for SOC test schedule representation. In: Proceedings of IEEE international test conference (ITC). IEEE Computer Society, Washington, DC, p 539

Larsson E (2004) Integrating core selection in the SOC test solution design-flow. In: Proceedings of IEEE international test conference (ITC), pp 1349–1358

Larsson E, Fujiwara H (2006) System-on-chip test scheduling with reconfigurable core wrappers. IEEE Trans VLSI Syst 14(3):305–309. DOI http://dx.doi.org/10.1109/TVLSI.2006.871757
Larsson E, Peng Z (1999) An estimation-based technique for test scheduling. In: Proceedings of electronic circuits and systems conference
Larsson E, Peng Z (2000) A technique for test infrastructure design and test scheduling. In: Proceedings of IEEE design and diagnostics of electronic circuits and systems workshop (DDECS)
Larsson E, Peng Z (2001a) An integrated system-on-chip test framework. In: DATE '01: Proceedings of the conference on design, automation and test in Europe. IEEE, Piscataway, NJ, pp 138–144
Larsson E, Peng Z (2001b) Test scheduling and scan-chain division under power constraints. In: Proceedings of IEEE Asian test symposium (ATS), pp 259–264
Larsson E, Peng Z (2002a) An integrated framework for the design and optimization of SOC test solutions. J Electron Test Theory Appl 18(4–5):385–400
Larsson E, Peng Z (2002b) An integrated framework for the design and optimization of SOC test solutions. In: Chakrabarty K (ed) SOC (system-on-a-chip) testing for plug and play test automation, frontiers in electronics testing, vol 21. Kluwer, Dordrecht, pp 21–36
Larsson E, Peng Z (2006) Power-aware test planning in the early system-on-chip design exploration process. IEEE Trans Comput 6(2):227–239
Larsson E, Arvidsson K, Fujiwara H, Peng Z (2002) Integrated test scheduling, test parallelization and TAM design. In: Proceedings of IEEE Asian test symposium (ATS), p 397
Larsson E, Arvidsson K, Fujiwara H, Peng Z (2004) Efficient test solutions for core-based designs. IEEE Trans Comput Aided Des 23(5):758–775. DOI 10.1109/TCAD.2004.826560
Marinissen EJ, Iyengar V, Chakrabarty K (2002) A set of benchmarks for modular testing of SOCs. In: Proceedings of IEEE international test conference, pp 519–528
Muresan V, Wang X, Muresan V, Vladutiu M (2000) A comparison of classical scheduling approaches in power-constrained block-test scheduling. In: ITC '00: Proceedings of the 2000 IEEE international test conference. IEEE Computer Society, Washington, DC, p 882
Mureşan V, Wang X, Mureşan V, Vlăduţiu M (2004) Greedy tree growing heuristics on block-test scheduling under power constraints. J Electron Test Theory Appl 20(1):61–78. DOI http://dx.doi.org/10.1023/B:JETT.0000009314.39022.78
Nicolici N, Al-Hashimi BM (2003) Power-conscious test synthesis and scheduling. IEEE Design Test Comput 20(4):48–55. DOI http://dx.doi.org/10.1109/MDT.2003.1214352
Pouget J, Larsson E, Peng Z (2003a) SOC test time minimization under multiple constraints. In: Proceedings of Asian test symposium (ATS), pp 312–317
Pouget J, Larsson E, Peng Z, Flottes M, Rouzeyre B (2003b) An efficient approach to SOC wrapper design, TAM configuration and test scheduling. In: Proceedings of IEEE European test symposium (ETS), pp 51–56
Pouget J, Larsson E, Peng Z (2005) Multiple-constraint driven system-on-chip test time optimization. J Electron Test Theory Appl 21(6):599–611. DOI http://dx.doi.org/10.1007/s10836-005-2911-4
Rajsuman R (1995) IDDQ testing for CMOS VLSI. Artech Publishing, Italy
Rajsuman R (1998) Design for IDDQ testing for embedded cores based sysem-on-chip. In: Proceedings of IEEE international workshop on IDDQ testing (IDDQ), pp 69–73
Ravikumar CP, Hetherington G (2004) A holistic parallel and hierarchical approach towards design-for-test. In: Proceedings of IEEE international test conference (ITC), pp 345–354
Ravikumar CP, Kumar R (2002) Divide-and-conquer IDDQ testing for core-based system chips. In: Proceedings of international conference on VLSI design (VLSID), pp 761–766
Ravikumar CP, Chandra G, Verma A (2000) Simultaneous module selection and scheduling for power-constrained testing of core based systems. In: Proceedings of international conference on VLSI design (VLSID), p 462
Ravikumar CP, Dandamudi R, Devanathan VR, Haldar N, Kiran K, Kumar PS (2005) A framework for distributed and hierarchical design-for-test. In: Proceedings of international conference on VLSI design (VLSID), pp 497–503

Rosinger PM, Al-Hashimi BM, Nicolici N (2002) Power profile manipulation: a new approach for reducing test application time under power constraints. IEEE Trans Comput Aided Des 21(10):1217–1225

Rosinger P, Al-Hashimi B, Chakrabarty K (2005) Rapid generation of thermal-safe test schedules. In: Proceedings of the design, automation and test in Europe conference, pp 840–845

Runyon RP, Haber A, Pittenger DJ, Coleman KA (1996) Fundamentals of behavioral statistics, 2nd edn. McGraw-Hill, New York

Sachdev M (1997) Deep submicron IDDQ testing: issues and solutions. In: Proceedings of European design and test conference (ED&TC), pp 271–278

Samii S, Larsson E, Chakrabarty K, Peng Z (2006) Cycle-accurate test power modeling and its application to SOC test scheduling. In: Proceedings of IEEE international test conference (ITC), pp 1–10. DOI 10.1109/TEST.2006.297693

Samii S, Selkälä M, Larsson E, Chakrabarty K, Peng Z (2008) Cycle-accurate test power modeling and its application to SOC test architecture design and scheduling. IEEE Trans Comput Aided Des 27(5):973–977

Saxena J, Butler KM, Whetsel L (2001) An analysis of power reduction techniques in scan testing. In: Proceedings of the IEEE international test conference 2001. IEEE Computer Society, Washington, DC, pp 670–677

Sehgal A, Bahukudumbi S, Chakrabarty K (2008) Power-aware SOC test planning for effective utilization of port-scalable testers. ACM Trans Des Automat Electron Syst 13(3):1–19. DOI http://doi.acm.org/10.1145/1367045.1367062

Senthil AT, Ravikumar CP, Nandy SK (2007) Low-power hierarchical scan test for multiple clock domains. J Low Power Electron 3(1):106–118

Singh V, Larsson E (2008) On reduction of capture power for modular system-on-chip test. In: Digest of papers of IEEE workshop on RTL and high level testing (WRTLT)

Su CP, Wu CW (2004) A graph-based approach to power-constrained SOC test scheduling. J Electron Test Theory Appl 20(1):45–60. DOI http://dx.doi.org/10.1023/B:JETT.0000009313.23362.fd

Touba NA, Pouya B (1997) Using partial isolation rings to test core-based designs. IEEE Des Test Comput 14(4):52–59. DOI http://dx.doi.org/10.1109/54.632881

Tudu JT, Larsson E, Singh V, Agrawal V (2009) On capture power reduction for modular system-on-chip test. In: Proceedings of IEEE European test symposium (ETS)

Varma P, Bhatia S (1998) A structured test re-use methodology for core-based system chips. In: Proceedings of IEEE international test conference (ITC), pp 294–302

Whetsel L (1997) An IEEE 1149.1-based test access architecture for ICs with embedded cores. In: Proceedings of IEEE international test conference (ITC), pp 69–78

Xia Y, Chrzanowska-Jeske M, Wang B, Jeske M (2003) Using a distributed rectangle bin-packing approach for core-based SOC test scheduling with power constraints. In: Proceedings of international conference on computer-aided design (ICCAD), pp 100–105. DOI http://dx.doi.org/10.1109/ICCAD.2003.148

Xu Q, Nicolici N (2005) Resource-constrained system-on-a-chip test: a survey. Comput Digital Tech, IEE Proc 152(1):67–81

Xu Q, Nicolici N, Chakrabarty K (2005) Multi-frequency wrapper design and optimization for embedded cores under average power constraints. In: Proceedings of ACM/IEEE design automation conference (DAC). ACM, New York, NY, pp 123–128. DOI http://doi.acm.org/10.1145/1065579.1065615

Yoneda T, Fujiwara H (2002) Design for consecutive testability of system-on-a-chip with built-in self testable cores. J Electron Test Theory Appl 18(4–5):487–501

Yoneda T, Masuda K, Fujiwara H (2006) Power-constrained test scheduling for multi-clock domain SOCs. In: Proceedings of design, automation, and test in Europe (DATE), pp 297–302

Zhao D, Upadhyaya S (2003) Power constrained test scheduling with dynamically varied TAM. In: Proceedings of IEEE VLSI test symposium (VTS). IEEE Computer Society, Washington, DC, p 273

Zorian Y (1993) A distributed BIST control scheme for complex VLSI devices. In: Proceedings of VLSI Test Symposium, pp 4–9

Zorian Y (1997) Test requirements for embedded core-based systems and IEEE p1500. In: Proceedings of IEEE international test conference (ITC), p 191

Zorian Y (1998) Challenges in testing core-based system chips. IEEE Commun Mag 37(6):104–109

Zorian Y, Marinissen EJ, Dey S (1998) Testing embedded-core based system chips. In: Proceedings of IEEE international test conference (ITC), pp 130–143

Chapter 7
Low-Power Design Techniques and Test Implications

Kaushik Roy and Swarup Bhunia

Abstract This chapter provides a brief overview of the prevalent design techniques for dynamic and leakage power reduction in both logic and memory circuits. It also provides an introduction to power specification format, which allows specification of circuit properties with respect to power dissipation in a consistent manner. Next, it discusses the impact of existing low-power design techniques on test. Finally, it covers the test implications of the post-silicon adaptation approaches for power reduction.

7.1 Introduction

In the nanometer technology regime, power dissipation has emerged as a major design consideration (Rabaey and Pedram 1995; Roy and Prasad 2000; Yeo and Roy 2005). On the other hand, variations in the device parameters, both systematic and random, manifest as variations in circuit parameters such as delay and leakage, leading to loss in *parametric yield* (Borkar et al. 2003). Numerous design techniques have been investigated for both logic and memory circuits to address the growing issues with power and variations. Low-power and *process-tolerant designs*, however, impose new test challenges and may even have conflicting requirements for test – affecting delay fault coverage, *quiescent current* (I_{DDQ}) testability, parametric yield, and even stuck-at tests. Hence, there is a need to consider test and yield, while designing for low-power and robustness under variations.

Although dynamic power traditionally has been the significant form of power consumption in submicron process nodes, aggressive technology scaling has exposed the secondary problem of leakage power (Roy et al. 2003), which contributes to nearly 20–50% of total power in deep submicron modern microprocessors. Increased power dissipation also manifests as increase in junction temperature due

K. Roy (✉)
Purdue University, West Lafayette, IN, USA
e-mail: kaushik@purdue.edu

S. Bhunia
Case Western Reserve University, Cleveland, OH, USA

P. Girard et al. (eds.), *Power-Aware Testing and Test Strategies for Low Power Devices*,
DOI 10.1007/978-1-4419-0928-2_7, © Springer Science+Business Media, LLC 2010

to limited cooling capacity of the package. To improve battery-life in portable devices and to reduce temperature-induced reliability concerns, numerous power saving techniques have been investigated at circuit and architecture level that target reduction of leakage and/or dynamic power. Due to quadratic dependence of dynamic power on supply voltage, voltage scaling has emerged as a popular choice for dynamic power reduction. Besides scaling of supply voltage, other important low-power design techniques that target dynamic power reduction are gate sizing that reduces effective switching capacitance, clock gating, supply gating, and frequency scaling. On the other hand, dominant leakage saving techniques for logic and memory circuits include transistor *stacking*, dual or multiple threshold voltage CMOS, and *body biasing*. Although these techniques provide effective power saving solutions, many of these techniques cause undesirable consequences on test and parametric yield of the design.

Another major design challenge in the nanometer regime is increased process parameter variations (Borkar et al. 2003; Jacobs and Berkelaar 2000; Yuan and Qu 2006). Process imperfections due to subwavelength lithography lead to device level variations in small-geometry devices. Variations in device parameters such as length, width, oxide thickness, and flat-band voltage of devices along with *random dopant fluctuations* (RDF) and *line edge roughness* (LER) are making the devices exhibit large variations in circuit-level parameters, particularly in the *threshold voltage* (V_{th}). Threshold voltage is a strong determinant of circuit speed: low-V_{th} chips are typically faster than high-V_{th} ones (since low-V_{th} corresponds to higher drive current). Statistical variations in device parameters lead to a statistical distribution of V_{th}. Consequently, delay of a circuit (and thus the maximum allowable frequency of operation) also follows a statistical distribution (Chang and Sapatnekar 2003; Kang et al. 2005). Hence, parametric yield of a circuit (probability to meet the desired performance or power specification) is expected to suffer considerably, unless an overly pessimistic worst-case design approach is followed. Since leakage power of a circuit has exponential dependence on device threshold voltage (V_{th}), parameter variations result in large variability in leakage power (Rao et al. 2004; Srivastava and Sylvester 2004) along with variation in circuit delay. Moreover, threshold voltage variation poses concern in robustness of operation, particularly in *Static Random Access Memory* (SRAM) and dynamic logic circuits (such as domino).

Since worst-case design approach may incur prohibitive design overhead in terms of power dissipation, multitude of research efforts has been devoted to explore alternative design methodologies under variations. Broadly, three classes of techniques are proposed to ensure/enhance yield under variations while incurring minimal design overhead. (1) *Statistical design approach*, where a circuit parameter (e.g., delay or leakage) is modeled as a statistical distribution (e.g., Gaussian) and the circuit is designed to meet a constraint on yield (or to maximize it) with respect to a target value of the parameter (Agarwal et al. 2005; Jacobs and Berkelaar 2000; Mani et al. 2005; Srivastava and Sylvester 2004). *Gate sizing* or *dual-V_{th}* CMOS are examples of techniques that can be used to vary circuit delay or leakage distribution. (2) *Variation avoidance*, where a given circuit is synthesized using nominal parameter values; however, any possible failures due to delay variations are identified at run time and avoided by adaptively switching to two-cycle operations (Ghosh et al. 2007). (3) *Post-silicon compensation and correction*, where parameter shift is detected (using delay or leakage sensor) and adjusted after manufacturing by changing operating parameters such as supply voltage, frequency, or body bias.

Variations in process parameters (in particular, threshold voltage) can also lead to failures in a SRAM array, degrading memory yield (Bhavnagarwala et al. 2001; Mukhopadhyay et al. 2004a). Intra-die process variation is a major concern for memory design since it introduces mismatch in strength between two identical transistors in a memory cell. Similar to logic circuit, different circuit and architecture-level design techniques have been investigated (Kim et al. 2006; Mukhopadhyay et al. 2005) to improve yield of nanoscaled SRAM.

Parameter variations can have large negative impact on test affecting both test-quality and cost (Cheng et al. 2000; Krstic et al. 2003; Liou et al. 2002). In particular, delay testing under probabilistic path delay model can be challenging in terms of path selection and pattern generation for path sensitization (Mak et al. 2004). Parameter variations also affect noise margin of dynamic circuits, which in turn puts burden on test to check robustness of these circuits after manufacturing. The combined impact of advanced power management techniques [such as dynamic voltage scaling (DVS) or clock gating] and process-induced uncertainty in device parameters bring new challenges to conventional ATE-based testing. One of the difficulties is to create the worst-case operating condition during test. Considering the large number of operating points in today's high-performance chips (defined by supply voltage, frequency and temperature), ensuring correct operation under all possible conditions has become a major test challenge. Low-power and process-tolerant design techniques may also have conflicting requirements for test. Hence, there is a need to consider test and yield, while designing for low-power and variation tolerance.

In this chapter, we highlight the major test challenges associated with nanoscale CMOS designs. In particular, we discuss test challenges related to low-power and variation-tolerant designs and discuss a new class of design techniques based on self-calibration and self-repair that can potentially reduce the burden on test and help achieve increased test confidence and higher yield.

The rest of the chapter is organized as follows. Section 7.3 presents the current trend towards a power specification format while Sect. 7.2 presents major techniques for low-power logic and memory design. Section 7.4 discusses test considerations associated with the low-power design approaches. In Sect. 7.5, we present the effectiveness of low-power design approaches on improving test power and test coverage. We note that effective use of low-power design techniques (and their incorporation/utilization in CMOS testing) can lead to large improvements in test power and test cost. In Sect. 7.6, we analyze self-calibration and self-repair techniques for improving yield and reliability of low-power designs under variations. Section 7.7 concludes the chapter with a summary of observations and future trend.

7.2 Low-Power Design Trends

Power reduction has been addressed at different levels of design abstraction: from system to architecture to circuit. Existing power reduction approaches can be broadly classified into two categories (1) dynamic power reduction techniques and (2) static or leakage power reduction techniques. In this section, we will cover some of the major dynamic and leakage power reduction techniques in details.

7.2.1 Dynamic Power Reduction Techniques

With technology scaling, active power per switching reduces due to scaling of V_{DD} and switching capacitance. However, faster clock and increasing device integration cause significant rise in overall dynamic power. The increase in dynamic power manifests as increase in average power as well as in power density of the chip. Higher power density translates to higher junction temperature in the device layer, giving rise to localized "hotspots" due to limited cooling capacity of the packages. The power density for high performance microprocessors has been reported to be over 50 W cm^{-2} for 100-nm technology and is increasing further with scaling (Xu 2006). Interestingly, *localized hotspots* are also a leakage concern, since the static power, in particular, the subthreshold leakage component increases exponentially with temperature (Meterelliyoz et al. 2005), potentially causing thermal runaway condition. It has been almost mandatory to incorporate power reduction techniques in nanoscale CMOS designs to reduce average power dissipation and avoid temperature-induced reliability concerns as well. Next, we will discuss some major dynamic power reduction techniques at circuit and architecture level. Note that large volume of literature exits on different power reduction techniques and it is difficult to cover all such design methodologies in this section. However, interested readers should refer to Rabaey and Pedram (1995), Roy and Prasad (2000), and Yeo and Roy (2005).

7.2.1.1 Circuit Optimization for Low Power

Circuit-level design techniques for dynamic power reduction typically include delay-constrained sizing of logic gates (in order to reduce effective switching capacitance) (Jacobs and Berkelaar 2000) and static assignment of multiple threshold voltages (Wei et al. 1999) or multiple supply voltages (Srivastava and Sylvester 2004). These techniques essentially exploit the timing slack available in the shorter paths and make them slower, effectively equalizing the timing paths. The sizing or multi-V_{th}/multi-VDD assignment or combination of them can be formulated as an optimization problem typically with power as the optimization objective and critical path delay as primary constraint. Such a formulation can then be solved using one of the multitude of solution approaches including integer linear programming and Lagrangian Relaxation (LR) method.

7.2.1.2 Clock Gating

Clock is a major source of dynamic power in high-performance circuit. In a digital circuit such as microprocessor, the clock line drives a large capacitive load since the clock is connected to large number sequential elements as well as dynamic logic circuits. Besides, to facilitate routing of clock from one end of a die to another with minimal jitter and skew, the clock network is typically associated with many buffers and de-skewing elements, which also add to the clock power. *Clock gating* is an effective low-overhead technique for reducing power in the clock line by shutting off clock switching in the idle logic blocks. Typically, the clock line is "gated" by AND-ing a *clock gating control* with the clock line. The clock gating prevents charging and discharging of the capacitive load (primarily contributed by the gate capacitance of the clock fanout nodes) as well as switching of clock buffers in the gated clock network, thereby saving dynamic power. An important part of incorporating clock gating into a design is to determine the clock gating control logic, which decides when and how long the clock can be gated. We need to ensure that the output of the gated logic is not used during the time the clock line is shut off. In Chap. 9, we will discuss the clock gating technique in more details.

7.2.1.3 Operand Isolation

Present day circuit designs contain many datapath modules which occasionally perform useful computations but spend a large amount of time in idle states. However, the switching activity at the inputs of these modules causes redundant computations which are not useful for downstream circuit computations. This unnecessary circuit activity causes significant increase in power consumption. *Operand isolation* is an effective technique that prevents unnecessary switching in a module by utilizing isolation circuitry at the input of the module. Enabling the isolation circuitry forces the modules in their idle states to prevent redundant computation. Leakage power, however, becomes an important issue in this idle state of the modules. It should be ensured that the isolation circuitry be designed in a way that the isolated module consumes minimal leakage power as well.

The concept of the operand isolation is illustrated with the following example. Figure 7.1a shows a small part of a computational module of a circuit consisting of a multiplier and an adder as computational blocks. For certain configurations of the control signals S0 and S1 of the multiplexers and signals G1 and G2 of the registers, the outputs of multiplier mu1 and adder ad1 are not used to compute the final stored values in the registers r1 and r2. For instance, the multiplier is selected for computing the final output values only when (S0 = "0" and G1 = "1"). On the other hand, the adder is selected for computation of final outputs only when (S0 = "1" and G1 = "1") and/or (S1 = "0" and G2 = "1"). However, whenever there is a switching activity to the input signals A, B, C, or D, there are redundant computations in mu1 and ad1 even when their outputs are not being used. For example, let us assume the initial values of signals A and B were set to "0" and S0 = "1."

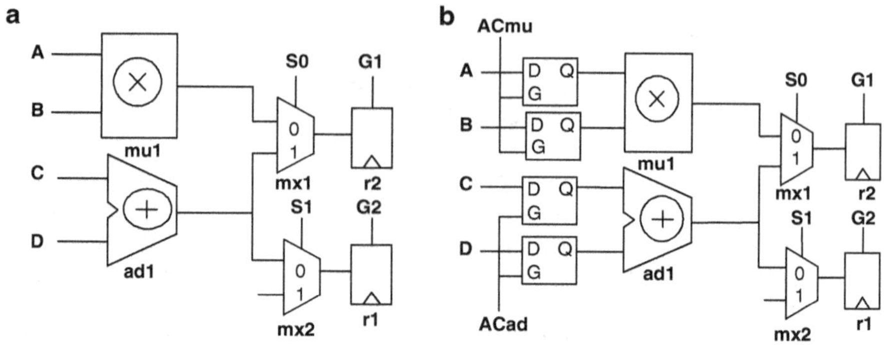

Fig. 7.1 (**a**) Design consisting of multiplier and adder; (**b**) The same design after operand isolation is incorporated

The multiplier is not selected for computation of the final outputs of this module in this case. Let us suppose that the circuit generating signals A and B sets their values both to "1." The computation performed by the multiplier with the changed inputs is redundant, since S0 value is still at "1." Especially, when the outputs of a module are not useful for a considerable period of time, the power dissipation due to such redundant computations can be significant.

Now suppose there are activation signals AC_{mu}, and AC_{ad} which indicate when the outputs of the multiplier and/or the adder are useful for downstream computations. These signals can be effectively utilized to freeze the inputs of these modules (e.g., by insertion of transparent latches as shown in Fig. 7.1b) and prevent input switching activity from propagation into the modules during redundant computations, thereby enabling the modules to perform useful operations only. Figure 7.1b shows the same circuit which has been operand isolated with transparent latches. In the context of our example circuit, the signals AC_{mu} and AC_{ad} can evaluate to "0" when the redundant computation is to occur and let inputs A, B, C, and D to retain their previous values and prevent redundant switching. Effectiveness of an operand isolation approach largely depends on (a) finding low-overhead isolation circuitry and (b) generation of proper activation signals to indicate a redundant computation. Research efforts have been directed toward both improving the overhead of the isolation circuitry (Banerjee et al. 2006) as well as automatic determination of activation signals (Tiwari et al. 1998).

7.2.1.4 Advanced Power and Thermal Management

Due to quadratic dependence of dynamic power of a circuit on its operating voltage, supply voltage scaling (along with commensurate scaling of operating frequency – Dynamic Voltage and Frequency Scaling or DVFS) has been extremely effective in reducing the power dissipation. Such schemes can be effectively utilized to reduce power dissipation at the system level – under low-load condition; the supply voltage is scaled down along with operating frequency, while under normal condition the supply voltage and frequency are maintained. Chapter 8 provides detailed description about the voltage scaling approach for power reduction.

As we have noted earlier, high performance systems, such as processors or system-on-chips, also suffer from high power density issue that results in high junction temperature. The temperature issue is typically addressed by monitoring the temperature of processing units (using distributed temperature sensors) and throttling clock frequency or reducing voltage (similar to DVFS) when the temperature goes beyond a threshold (McGowen et al. 2006). Since the best way to reduce power dissipation is to scale the supply voltage, some of the recent techniques for logic design include smart approaches to scale down supply voltages with no frequency scaling. Such DVS techniques avoid frequency scaling at scaled supply voltage by isolating the critical timing paths and taking special action when they are activated. In one solution, the critical paths are made rare by design using gate sizing or logic restructuring and multicycle operation is enabled on activation of a critical path [CRISTA approach (Ghosh et al. 2007)]. Allowing single-cycle operation in the noncritical path and multicycle in critical ones, delay failures in all paths are prevented at lower voltage. In another solution, flip-flops in critical timing paths are associated with shadow latches, which are triggered by a delayed clock [RAZOR approach (Ernst et al. 2003)]. Timing failure in a critical path at scaled voltage is detected by comparing the latched value in the original functional flip-flop with that in the shadow latch. Once detected, a failure is corrected by recomputing at a higher voltage.

7.2.2 Leakage Power Reduction Techniques

Increasing leakage power with technology scaling poses both design and test concerns (Roy et al. 2003). The total contribution of all leakage components constitutes a major source of power dissipation in sub-100 nm logic and memory circuits. Next, we discuss some major leakage control techniques in logic and memory circuits.

7.2.2.1 Input Vector Control

For each logic gate, the quiescent current depends on its input combinations. Consider a three-input CMOS NAND gate as an example. For the "111" input combination, the three NMOS transistors are turned on and act as a short circuit; the gate's leakage current is the sum of the leakage current through the three PMOS transistors. For the "001," "010," "100," and "000" combinations, there are at least two NMOS transistors that are turned off in the pull-down network. In these cases, the "off" transistor on top of the stack has a positive source voltage, V_S. In the quiescent state, the leakage currents through all the transistors are equal. So, we can consider only the first "off" transistor on top in the pull-down tree as pertinent to

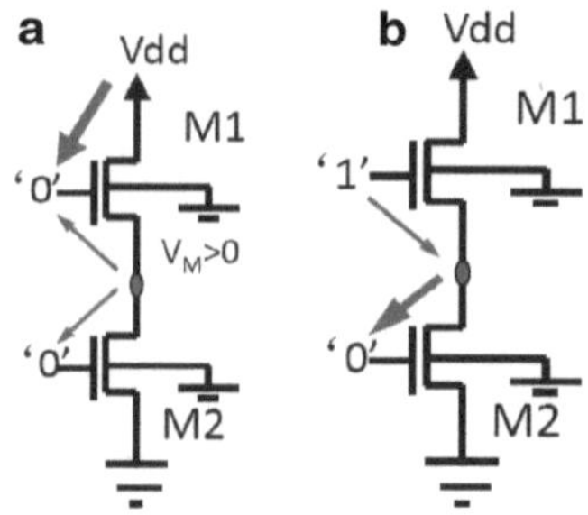

Fig. 7.2 (**a**) Gate current with "00" and (**b**) with "10." Since stacking effect reduces subthreshold leakage, "00" is the best input vector for subthreshold leakage reduction. Gate tunneling current increases with increased gate to source/drain/body voltage. Gate current with "10" is lower than gate current with "00"

our analysis. A positive V_S means a negative V_{GS}, which greatly reduces the leakage. A positive V_S also indicates the existence of body effect and a reduction in V_{DS}. Both effects increase the threshold voltage leading to exponential reduction in subthreshold leakage. This is called "*stacking effect*." Since a circuit's total leakage current depends on its primary inputs, applying the best input vectors to some circuits can cause the leakage current to decrease significantly (Roy et al. 2003). Because of the exponential complexity with respect to the number of primary inputs, efficient algorithms to determine near-optimal solution based on random search or genetic algorithm have been developed (Johnson et al. 1999). Investigation shows that for a reasonably complex circuit, input vector control (IVC) can result in about 30–35% saving in standby leakage using proper selection of input vector.

Stacking effect has been shown to be very effective for subthreshold leakage current reduction. Since gate leakage and band-to-band junction tunneling leakage (BTBT) are becoming increasingly dominant in nanoscale CMOS technologies, one should also consider the impact of stacking on these leakage components. It is observed that BTBT is not very sensitive to stacking. However, gate leakage is a strong function of stacking and interestingly, the input vector for minimum gate tunneling current is different from that for minimum subthreshold current. As shown in Fig. 7.2, input "00" provides the minimum subthreshold current, while "10" provides minimum gate current. Hence, total leakage reduction with stacking requires considering the relative magnitude of different leakage components.

7.2.2.2 Dual-V_{th} Design

For a logic circuit, we can assign a higher threshold voltage to some transistors in noncritical paths to reduce leakage current, while maintaining performance by using low-threshold transistors in critical paths. Therefore, no additional leakage-control transistors are necessary and we can achieve both high-performance and low-power dissipation simultaneously. Figure 7.3a illustrates the idea of a dual-V_{th} circuit.

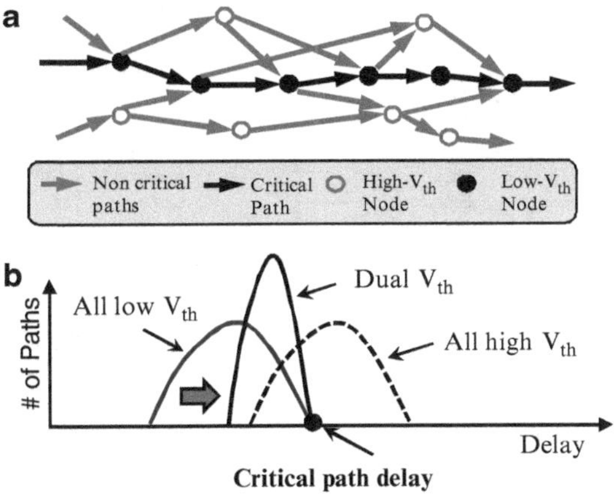

Fig. 7.3 (**a**) Dual-threshold-voltage CMOS circuit; (**b**) Path distribution of dual- and single-V_{th} CMOS

Figure 7.3b shows the path distribution of dual- and single-V_{th} CMOS for a 32-bit adder. Dual-V_{th} CMOS has the same critical delay as a single-low-V_{th} CMOS circuit, but we can assign the transistors in noncritical paths a high V_{th} to reduce leakage power. Hence, this dual-threshold technique can effectively reduce leakage power during both standby and active modes without incurring delay or area overhead. Because it can reduce background leakage, it can be beneficial for I_{DDQ} testing.

Let us investigate the benefits of combining the dual-threshold CMOS design technique and a vector-control technique for I_{DDQ} testing. For simplicity, we map the benchmark circuits to a library containing NAND gates, NOR gates, and inverters. The supply voltage is 1 V and the low threshold voltage is 0.2 V. Using the algorithm described in Wei et al. (1999), we can transform the single-low-V_{th} circuit to a dual-V_{th} circuit with the optimal value for a high threshold voltage. We can then use a random search to choose the best vector from 1,000 randomly generated vectors. Thus, we capture the benefit of the vector-control technique on I_{DDQ} testing of a dual-threshold circuit. Results indicate that, for some shorts, combining the dual-threshold voltage design and vector-control techniques can increase the fault current ratio by a factor of more than 10.

7.2.2.3 Supply Gating

A more promising technique is to force "stacking," supplying V_{SS} or V_{DD} through another control transistor (Roy et al. 2003) as shown in Fig. 7.4a. The additional transistor in the stack effectively "gates" the V_{SS} or V_{DD} line during idle mode of the circuit to save leakage power. A variant of this gating technique, called *Multi-Threshold CMOS (MTCMOS)* uses high-V_t gating transistor along with low-V_t core (Tschanz et al. 2002) to maximize the leakage saving (Fig. 7.4b). This fits particularly well with regular structures such as data paths, where the gating transistor can be easily shared. The additional gating transistor in the charging/discharging path

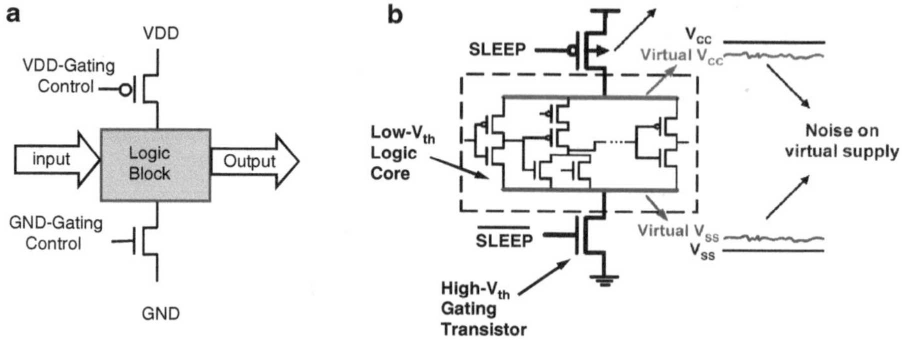

Fig. 7.4 (**a**) Supply gating for leakage reduction. (**b**) Multithreshold CMOS (MTCMOS) design approach for leakage reduction. In active mode, sleep transistors introduce noise on the virtual supply lines (Tschanz et al. 2002)

is a performance issue. A shared gating transistor requires careful sizing such that it is wide enough to sustain worst-case switching condition with acceptable performance loss. The virtual supply lines experience noise in active mode of operation which can affect the reliability of operation. Moreover, since in the sleep mode, some output nodes are floated (using the small leakage current to hold their states), noise immunity becomes a robustness concern. The circuits in the sleep mode become susceptible to coupling noise or other power-transient events. Test engineers must face the challenge of deciding how to test the noise margin as well as the worst-case delay overhead due to the gating transistor.

7.2.2.4 Shannon Cofactoring-Based Dynamic Supply Gating

Low-leakage circuit design technique can directly help in improving I_{DDQ} *testability*. However, leakage control techniques based on transistor stacking that target active leakage reduction in logic circuits can also improve test power and test time. In Ghosh et al. (2005), a circuit synthesis approach is proposed that can result in low active power dissipation, while enhancing test cost and test confidence. The synthesis technique is based on structural transformation of a design using *Shannon's decomposition* and supply gating. Using a control variable x_i, a circuit is decomposed into two cofactors, only one of which will functionally contribute at any time depending on the state of the control variable. As shown in Fig. 7.5, CF_1 will be active when $x_i = 1$. Similarly, CF_2 will be active when $x_i = 0$. Therefore, one of the cofactors can be supply gated at any time using x_i as gating control. The procedure can be applied recursively using multilevel Shannon decomposition to increase the power saving. The advantage of power saving, however, comes at the cost of area and delay overhead in addition to the robustness concern associated with supply gating.

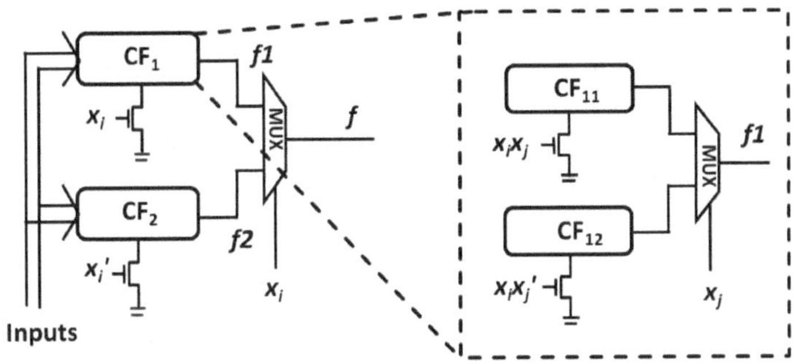

Fig. 7.5 Single or multilevel Shannon decomposition based supply gating reduces leakage current which improves the IDDQ testability

7.2.2.5 Leakage Control in Memory

Leakage from embedded memory cells constitute a major part of system static power, typically in high-performance computing systems such as processor, SoC, etc., which requires large on-chip memory. The de facto standard of embedded memory design is six-transistor SRAM. Leakage saving techniques in memory are primarily based on variants of supply gating technique. A common scheme is *source biasing* (Roy et al. 2003) that applies "gating" at the source terminal of the NMOS transistors and applies a fixed bias at the virtual ground node to ensure data retention.

7.3 Power Specification Format

Different circuit blocks in a design can be treated differently with respect to power analysis and optimization. During design, verification, and implementation of complex electronic system, we need to specify the properties of a circuit block with respect to power dissipation in a consistent manner so that they can be correctly interpreted by the designers and the design automation tools. For instance, we require identification of inactive blocks and their various power modes of operation, identification of always-on blocks, proper isolation of different blocks, state retention by using shadow latches, proper insertion of power supply switches, proper layout of multiple voltage lines, insertion of level shifters for proper voltage level compatibility between the interfaces of different blocks operating in different power modes, etc. In particular, the existing design flow needs enhanced capabilities for managing the following low-power chip design requirements (DasGupta 2007): (1) specification of low-power design intent, (2) architectural trade-offs, (3) library design, (4) logic implementation, (5) physical implementation, (6) design verification, and (7) testability.

Different EDA tools independently chose to include features for addressing many of these issues as electronic design advanced into nanoscale technologies. But these highly fragmented solutions create more problems, especially for multivendor design flows, as different tools interpret different formats differently. The specifications may need to be repeated and re-entered several times in various formats, leading to excessive redundancy. To achieve a unified and efficient design flow, it is necessary for the various tools to communicate among each other using a common language used by all the tools. Most of the time, these power-aware design information cannot be specified in conventional *Hardware Description Language* (HDL) code and sometimes, even though they can be described, it may be undesirable to do so, since that would tie the logic specification directly to a constrained power implementation (UPF 2007). It is highly desirable that they should be present as simple add-on features, which does not require major modifications in existing design flows.

Currently, unification efforts have led to the emergence of two well-recognized standards for power specification, which are widely supported by most of the EDA industry. Cadence Design Systems designed the earlier versions of *Common Power Format* (CPF) (Hsu 2006) and it is being standardized by the *Silicon Industry Initiative* (Si2)'s *Low Power Coalition* (LPC). An alternative effort, supported by Synopsys, Mentor Graphics, and Magma, led to the development of the *Unified Power Format* (UPF) (http://en.wikipedia.org/wiki/Unified_Power_Format), which is being standardized by IEEE as part of the IEEE P1801 standards working group (http://en.wikipedia.org/wiki/IEEE_P1801). Both formats have 90% of the same concepts using completely different syntaxes (Goering 2007). Both formats are based upon the *Tool Control Language* (TCL), embedded in most EDA tools. Figure 7.6a provides a list of UPF commands. Figure. 7.6b shows an example power specification for a top-level module using CPF.

These power specification formats basically provide a way to describe the following specifications, unique to low-power design techniques:

- Voltage domains or blocks operating at different voltage levels with level shifters inserted at all domain crossings
- Power domains or blocks with a separate power supply that can be turned off
- Multiple supply nets with different names and connections
- Isolation logic, placed at the outputs of power domains, which may remain powered on
- Retention registers, which are flip-flops within an always-on power domain, to retain state when the domain supply is shut off
- Always-on cells and paths for logic, which must remain powered on even when the domain supply is switched off
- Power switches, which are large on-chip switching transistors to shut off the power to a power domain

The various power modes of operation also need to be specified along with the control logic (always-on) which provides the necessary signals for turning on or off various blocks. Using the power specifications, one can reuse the same logic in different power domains, without having to rewrite the entire block or cell. The format also needs to specify the different timing library data for timing analysis tools, so that the same cell can be used in different power domains.

a

UPF Commands
create_power_domain
add_domain_element
connect_supply_net
create_supply_net
create_supply_port
get_supply_net
merge_power_domains
set_domain_supply_net

b ***CPF file of a top level design***

```
# Define top design
set_design top
# Set up logic structure for all power domains
include IPB.cpf
create_power_domain -name PD1 -default
create_power_domain -name PD2 -instances inst_A -shutoff_condition {!pinst.penable} -
base_domains PD1
# Define static behavior of all power domains and specify timing constraints
set_instance inst_B -domain_mapping { PDX PD2 }
create_nominal_condition -name high -voltage 1.2
create_nominal_condition -name low -voltage 1.0
create_power_mode -name PM1 -domain_conditions {PD1@high PD2@low}
update_power_mode -name PM1 -sdc_files ./cm.sdc -activity_file act.tcf -activity_file_weight 1
# Set up required isolation and state retention rules for all domains
create_state_retention_rule -name sr1 -domain PD2 -restore_edge {!pinst.pgenable}
create_isolation_rule -name ir1 -from PD2 -isolation_condition {pinst.ienable} -isolation_output
high
create_level_shifter_rule -name lsr1 -to {PD1 PD2}
end_design
```

Fig. 7.6 (**a**) A list of commands used in UPF; (**b**) An example power specification using CPF

An operating condition is determined by the voltages of all power supplies applied to a power domain, including the power voltage, ground voltage, and the body bias voltage for PMOS and NMOS transistors (si2 Sep 2008). Depending on the technology used, this set of voltages determines whether the state of a power domain is on, off, or in standby mode. It can even support partially on domains where a threshold voltage defines full on/off state (si2 Jan 2008). The power format files are part of the design source and together with the Register Transfer Level (RTL) description, they convey the designer's intent to the various EDA tools for simulation, synthesis, formal verification, ATPG, place and route, etc. Each tool has a TCL parser which can understand the format to read in the files and create new files or update the existing ones as necessary.

The primary difference between the two formats is that UPF does not contain any commands to define library elements such as level shifter or retention register (Allen 2008). It is assumed that some other library format exists such as *Synopsys Liberty format* (.lib) to capture this information. Other minor differences include the presence of specific commands for handling multiprocess corner timing analysis in CPF, presence of commands and options to provide the right simulation semantics for data corruption and voltage resolution in UPF, presence of CPF power switch library command to specify a "partly on" state for the power switch with two enable inputs, along with specification of current limits and other parameters.

Both formats need to improve their handling of embedded IP. A design team or third-party IP vendor will often provide RTL for a block, which is integrated into a System-on-Chip (SoC) design. In this case, there may be a number of options for low-power design but the IP vendor will want to limit the options to ensure correct functionality, but neither CPF nor UPF provides a straight-forward way to do so.

7.4 Implications to Test Requirement and Test Cost

7.4.1 Impact of Dynamic Power Reduction Techniques on Test

7.4.1.1 Static Design-Time Techniques

Circuit level dynamic power optimization approaches such as gate sizing or multi-voltage assignment typically exploit the available timing margin. The undesirable effect of such optimization on test is large increase in critical timing paths, which complicates the path selection process for delay testing and speed binning. This also becomes a major reason for yield loss due to parameter variations. Under parameter variations, low-power designs become more susceptible to variation-induced delay failures, which degrade the parametric yield.

On the other hand, power gating causes conditional switching in the clock line (in case of clock gating) or in datapath block (in case of operand isolation). Such conditional shut-off and wakeup occurs in a localized manner. It causes test concern with respect to both test generation and application. Clock gating increases temporal variations in supply current drawn from the power grid (which can be modeled as a big RLC network) causing inductive voltage droop ($L \times \mathrm{d}i/\mathrm{d}t$, where L is the inductance on the power line and $\mathrm{d}i/\mathrm{d}t$ is the rate of change in supply current). Such local transient fluctuations in power grid affect signal propagation through logic gates resulting in timing failure unless sufficient margin is maintained during design time. Delay test generation and application require mimicking the worst-case droop in power grid to realistically capture the delay variation.

Similar to clock gating, turning on/off large datapath blocks using operand isolation results in temporal variation in supply current. It can also cause inductive voltage drop in the power grid, which needs to be considered during delay testing. Besides, incorporation of operand isolation adds to the test complexity since the isolation circuitry and the activation signal generation logic need to be tested for proper functionality. The isolation logic may add to the delay of the critical timing path and hence delay test generation and application require accounting for the extra logic.

7.4.1.2 Dynamic Power Reduction Techniques

Dynamic power reduction and thermal management techniques are attractive since they can achieve maximum performance under a power-temperature envelope. DVS has emerged as an effective approach for both power and temperature control due to the quadratic dependence of dynamic power on supply voltage. However, voltage scaling increases the path delay and hence cause delay failures, as illustrated in Fig. 7.7a. The situation becomes more complex with multiple voltage domains (Fig. 7.7b) since they can have different delay margins, thus requiring careful selection of scaled voltage levels.

They can have undesirable consequences on test. Circuit delay changes in a nonlinear fashion with voltage and temperature. Moreover, temperature-induced variations are often local due to the presence of localized thermal gradient. This makes a static design-time delay calibration at different operating conditions very unrealistic. However, an important test challenge is to define the worst-case timing condition during test. Different activity levels in different parts of a die cause variations in junction temperature. The worst-case condition may correspond to a nonuniform power level, which may be difficult to emulate in test mode using an ATE. Testing all the processing units for the worst-case condition may cause overtesting leading to yield loss. On the other hand, leaving some paths untested under worst-case temperature distribution may result test escape. Finally, during functional testing, an ATE needs to correctly predict the thermal trigger point in order to avoid false alarm. The problem aggravates for emerging multicore platforms that distribute workload (with the help of operating system) among multiple cores to

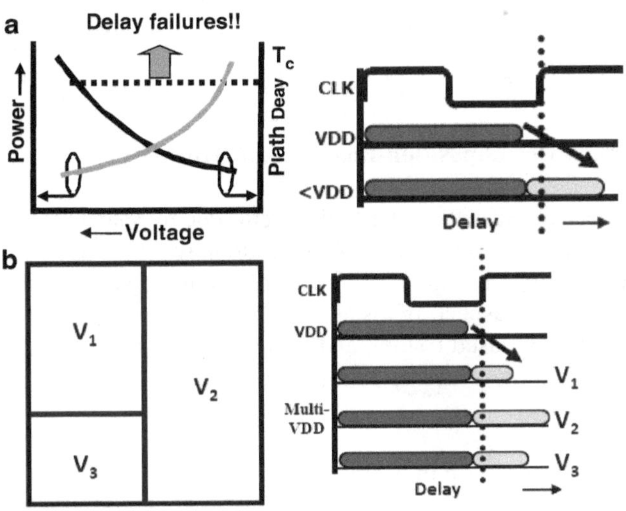

Fig. 7.7 (**a**) Scaling down supply voltage increases the probability of delay failure due to reduced delay margin. (**b**) For a multivoltage design, different regions experience different delay margins

achieve power efficiency. Since the thermal conditions on different cores are functions of applications and operating system, it is difficult to structure delay test for the worst-case thermal distribution.

7.4.2 Impact of Leakage Power Reduction Techniques on Test

While increasing leakage power has triggered circuit and architecture-level leakage control techniques, it has also affected design testability significantly. Two major impacts of increasing leakage on testability are (1) I_{DDQ} Testability: Technology scaling challenges the effectiveness of current-based test techniques such as I_{DDQ} testing. Sensitivity of I_{DDQ} testing reduces drastically due to high intrinsic leakage. (2) Impact on Burn-In: The exponential dependence of subthreshold leakage on temperature leads to positive feedback that can result in thermal runaway condition and yield loss during burn-in test (when stressed voltage and temperature are applied).

7.4.2.1 Leakage Reduction Using IVC

This reduction in background leakage can improve the effectiveness of I_{DDQ} testing, particularly for testing complex circuits, e.g., SoC, where all modules on chip except the one being tested can be applied the best vector for leakage reduction. Note that IVC may require hard-wiring the best input vector in the first-level logic gates of a logic block or control point insertion (Yuan and Qu 2006). Proper functioning of this extra logic needs to be checked during test while ensuring that it does not affect normal functionality.

7.4.2.2 Shannon Decomposition-Based Logic Synthesis

It is observed that tree structure of a logic circuit due to Shannon's decomposition makes it intrinsically more testable than conventionally synthesized circuit, while at the same time entailing an improvement in active power. Significant improvement can be observed in three aspects of testability of a circuit (a) I_{DDQ} test sensitivity, (b) test power during scan-based testing, and (c) test length (for both ATPG-generated deterministic and random patterns) (Ghosh et al. 2005).

7.4.2.3 Leakage Reduction in Memory

Leakage reduction techniques in memory will have positive impact on static current testing as well as on burn-in. In Bhunia et al. (2002), improvement in I_{DDQ} testability for a GND-gating scheme applied to SRAM cells is proposed. During test mode, idle (not accessed) parts of the memory are "gated" using the most significant bits of the address line as gating control. Supply gating and source biasing techniques for memory, however, introduce new test challenges. A source-biased memory will have two distinct states (normal and supply-gated) and desired behavior in each state need to be checked during test. While read/write and access time failures need to be validated with the gating transistor "on" (normal mode of operation), the primary concern in the power saving mode (gating transistor "off") is data retention in memory cells. Test engineers need to ensure that the bias voltage is large enough to retain stored content in power-gated cells.

7.4.2.4 Thermal Stability During Burn-In

Leakage is a major issue during burn-in test, which is used to detect infant mortality types of defects. Leakage power is a dominating component of total power dissipation during burn-in test condition due to applied high supply voltage and temperature. In scaled technologies, during burn-in there is an exponential increase in junction temperature due to drastic increase in leakage power, higher transistor density and increase in die-to-package thermal resistance. An effective solution to the problem is to design a negative feedback system to stabilize the junction temperature by controlling the leakage power of a chip dynamically. In Meterelliyoz et al. (2005), such a system is proposed that continuously monitors the junction temperature and compares it with the target burn-in temperature. If the junction temperature is higher (lower) than the target temperature, the system decreases (increases) leakage current by decreasing (increasing) the reverse body bias of the chip.

7.5 Low-Power Design Techniques for Test Power and Coverage Improvement

Power dissipation during test mode can be significantly higher than that during functional mode, since the input vectors during functional mode are usually strongly correlated compared with the statistically independent consecutive input vectors during testing. Zorian showed that the test power could be twice as high as the power consumed during the normal mode (Zorian 1993). Test power is an important design concern to increase battery lifetime in hand-held electronic devices that incorporate *built-in self test* (BIST) circuitry for periodic self-test. It is also important to improve test cost, since reduced test power of a module allows parallel testing of multiple embedded cores in an IC (Whetsel 2000). Increased peak power is likely to create noise problems in a chip by causing a drop in the supply voltage (Bushnell and Agarwal 2000). Peak and average power reduction during test contributes to enhanced reliability of the test and improvement of yield (Rosinger et al. 2002). It is, therefore, important to ensure reduction in power dissipation during the test mode.

Scan architectures represent the prevalent *Design for Testability* (DFT) approach to test digital circuits (Bushnell and Agarwal 2000). During test application in a scan-based circuit, power is dissipated in both the sequential scan elements and in the combinational logic. While scan values are loaded into a scan chain, the effect of scan-ripple propagates to the combinational block and redundant switching occurs in the combinational gates during the entire scan-in/out period. It is observed that about 78% of total test energy is dissipated in the combinational block alone (Gerstendorfer and Wunderlich 1999). Hence, a low-power scan design should address techniques to reduce power dissipation in the combinational block.

There has been multitude of research exploring efficient techniques to reduce test power in scan-based circuits. Wang et al. proposed *automatic test pattern generation* (ATPG) technique to redesign test vectors for reducing power dissipation during scan testing (Wang and Gupta 1998). With their ATPG, redundant transitions in combinational logic can be reduced but not completely eliminated. Moreover, test application time may increase to trade off power. Scan-latch reordering (Dabholkar et al. 1998) or input vector reordering (Girard et al. 1998) techniques have been proposed for reduction in test power. However, these techniques target reduction of transitions at the output of scan flip-flops and cannot eliminate redundant switching in combinational block. In Whetsel (2000), the author provided a solution for reduction in average and peak power dissipation by transforming conventional scan architecture into desired number of selectable, separate scan paths. Each scan path is in turn filled with stimulus and emptied of response. The authors in Sankaralingam et al. (2001) proposed a solution to the peak power problem during external testing by selectively disabling the scan chain. In this scheme, the test-set is generated and ordered in such a way that only changing portions of consecutive tests are shifted into the scan chains. In Rosinger et al. (2002) and Basturkmen et al. (2002), the authors provide a solution to prevent peak power violation during both shift and capture cycle using scan-chain partitioning. However, the modification of the scan flip-flop in Basturkmen et al. (2002) results in a substantial increase in area and degradation in performance. Redundant power loss in combinational logic is not completely prevented in the above cases, since part of the scan chain is always active during shifting.

Inserting blocking logic into the stimulus path of the scan flip-flops (as shown in Fig. 7.8) to prevent propagation of scan ripple effect to logic gates offers a simple and effective solution to significantly reduce test power, independent of test set. Werstendorfer et al. have proposed NOR or NAND gate-based blocking method in Gerstendorfer and Wunderlich (1999). Blocking gates (of type NOR or NAND) are controlled by the test enable signal, and the stimulus paths remain fixed at either logic "0" or logic "1" during the entire scan-shift operation. Zhang et al. (Zhang and Roy 2000) have used multiplexers at the output of the scan cells, which hold the previous states of the scan register during shifting and, thus, prevent activity in combinational logic. Another method for reduction in combinational power using blocking is to use a scan-hold circuit as a sequential element. This technique is called *enhanced scan* (Bushnell and Agarwal 2000), which also helps in delay fault testing by allowing application of arbitrary two-pattern test. In a scan-hold design,

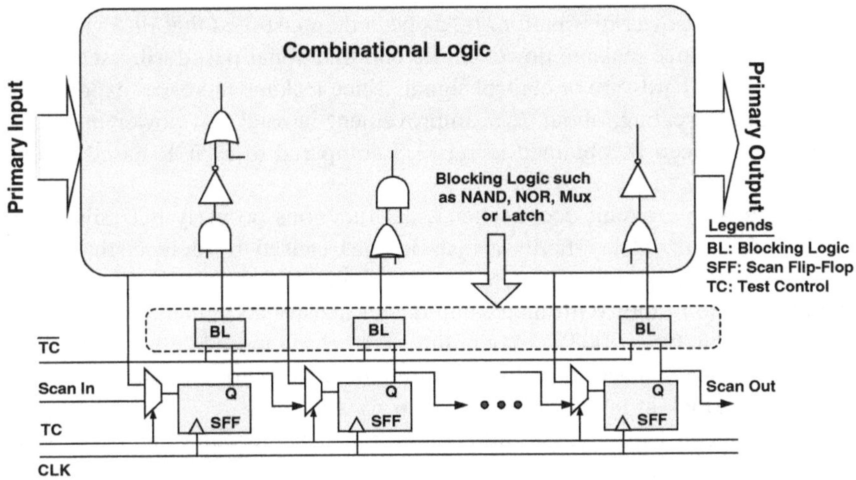

Fig. 7.8 Scan architecture with existing blocking circuitry to reduce power during scan operation (Bhunia et al. 2005)

each sequential element contains an additional storage cell called the hold latch, and the stimulus path for the combinational part is connected to the output of the hold latch, which is not used in scan shifting. Therefore, it also prevents redundant switching in combinational logic.

The problem with the blocking logic is that it adds significant delay in the signal propagation path from the scan flip-flop to logic (Gerstendorfer and Wunderlich 1999). Moreover, they have large overhead in terms of area and switching power in normal operation of the circuit. In Bhunia et al. (2005), the authors present a better signal blocking technique, which is referred as *First-Level Supply* (FLS) gating, to reduce power dissipation in the combinational logic during scan shifting. This is achieved by inserting a supply gating transistor in the first level of logic connected to the scan cell outputs, which essentially "gates" the VDD or GND line. This method is as effective as the other blocking methods in terms of reducing peak power and total energy dissipation during scan testing. However, since it introduces just one transistor in the charge/discharge path of the first-level logic, the delay penalty is significantly reduced compared to other blocking methods, which insert additional level of logic into signal propagation path. The overhead incurred in die-area and switching power in normal mode due to extra DFT logic is also significantly lower than the existing methods using NOR (Gerstendorfer and Wunderlich 1999), MUX (Zhang and Roy 2000), and Hold-latch (Bushnell and Agarwal 2000). The area overhead for FLS, however, depends on the number of unique first-level fanout gates. The authors have also presented a low-complexity algorithm to reduce fanouts of the scan flip-flops under delay constraint, which helps to further reduce the area overhead in FLS. Besides saving dynamic power in the combinational logic during test application, FLS can also be used to reduce leakage power by IVC mechanism (Johnson et al. 1999).With technology scaling, leakage power is becoming a notable source of power dissipation. It has been demonstrated that FLS can be easily adapted to reduce leakage power in the combinational part during scan testing without any extra hardware or control signal. Since leakage increases exponentially with technology scaling, about 25% improvement in total test power in a 45-nm technology node can be obtained using FLS compared to a NOR-based blocking scheme.

Delay faults in a circuit occur when a net functions properly but fails to meet timing requirement. Delay faults are sometimes caused by defects that are not large enough to cause a stuck-at failure by changing logic level, but affect the signal propagation time. With increasing defect density and unanticipated process variations (Borkar et al. 2003), delay failures are getting more likely to arise in sub-100 nm technologies. Therefore, it is becoming mandatory for manufacturing test to cover not only stuck-at faults, but delay faults as well.

Scan architectures provide an efficient way to test for delay faults with good fault coverage. Scan-based structural delay testing not only helps detection but also helps diagnosis of delay faults and, hence, is a popular choice for delay fault testing. However, testing for delay faults requires launching a transition at the input of the *Circuit Under Test* (CUT), and capturing the response of the circuit at rated clock. Although it is easier to apply a transition at the primary inputs of the CUT by the tester, it is not straight-forward to make a transition at the state inputs. Based on test application procedure, there are three prevalent techniques for scan-based delay testing. In the first one, called *broad-side* delay test, no transition is applied to the state inputs. State portion of the second pattern is derived as the combinational circuit's response to the first pattern. Although, the testing process is simple and it does not require any additional DFT logic, the *broad-side* case can suffer from poor fault coverage (Wang et al. 2004; Mao and Ciletti 1994). In the second method, referred as *skewed-load* delay testing, transition in the state inputs is induced by shifting the scan values by one-bit position. However, design requirement for *skewed-load* case can be costly because of fast switching scan enable signal (Wang et al. 2004). Moreover, since the second pattern (launching pattern) is highly correlated to the first one (initialization pattern), the test generation for high fault coverage can be difficult (Bushnell and Agarwal 2000).

The third approach, referred as *enhanced scan* method, allows easy application of a state transition and enables deterministic choice of any launching pattern in the scan flip-flops for best possible fault coverage (Bushnell and Agarwal 2000; Mao and Ciletti 1994). Enhanced scan method improves fault coverage for all delay fault models, however, it is particularly useful for path delay fault testing, where a set of critical timing paths need to be sensitized and tested for delay violations. Although enhanced scan has high combinational path testability, it involves high-DFT overhead due to addition of an extra latch, named as hold latch, at the output of each scan flip-flop to hold the initialization pattern (Bushnell and Agarwal 2000). The latch resides in the stimulus path between the scan flip-flops and the combinational logic and can considerably affect circuit performance during normal mode of operation. Adding to the overhead, the latch takes up significant amount of die-area and consumes power in normal mode. There have been a large number of investigations to devise alternative delay fault testing strategies with reduced DFT overhead and acceptable coverage (Cheng et al. 1991; Savir 1997; Tekumalla and Menon 1997; Wang et al. 2004). However, these techniques are either not as efficient as enhanced scan method with respect to fault coverage and required number of test patterns, or they complicate the test generation/application considerably.

The *Level Sensitive Scan Design* (LSSD) (DasGupta et al. 1978, 1981) is a test scheme for sequential designs, that can be used for enhanced-scan-like arbitrary two-pattern test application. Several alternative implementations of LSSD have been explored for enhanced-scan-like test application. Compared with muxed scan approach, LSSD has the advantage of isolating the functional flip-flop from the scan path that reduces the delay and power (in normal mode) overhead. However, LSSD has some major constraints and drawbacks. LSSD requires at least two clock signals, one for the scan chain and another for system flip-flops (DasGupta et al. 1978). Moreover, it uses extra latches per input flip-flop resulting in considerable area overhead. Extra latches increase leakage power as well. Although the extra latch is not in the signal propagation path, the extra loading on the first latch (due to the additional DFT hardware) increases power and delay of the system flip-flop. A recently proposed scan design by Intel (Kuppuswamy et al. 2004), based on hold-scan, uses a scan gadget along with the system latch. This technique can be referred to as *Hold Scan Using Scan Gadget* (HSSG) scheme (Bhunia et al. 2008). The scan chain is implemented by scan gadget element, which provides the basic scan test functions (shift, load, and capture). In this scheme, two extra latches added for the scan-chain implementation do not switch during the normal mode, however, they add to the leakage power and area overhead. More importantly, the system flip-flop is a more complicated flip-flop with two clock and two data inputs, one for system operation and the other for loading to/from the scan chain. The internal circuit for the flip-flop is provided in Kuppuswamy et al. (2004). The connection from the second input to the slave latch for load function is implemented by a transmission gate. This extra circuitry adds to the system flip-flop power in the normal mode of operation.

Muxed scan and LSSD both have some advantages and disadvantages and the choice of either of them depends on several design constraints (on die-area, delay, time to market, etc.). In Bhunia et al. (2008), the authors propose a circuit technique alternative to enhanced scan, which allows enhanced scan-like test application, but comes at a much lower hardware overhead. The technique is suitable for muxed scan implementation. This technique is referred to as *First Level Hold* (FLH) that employs the principle of "supply gating," in a novel way, to hold the state of combinational logic. Instead of holding the initialization pattern at the scan-hold latch as done in the case of enhanced scan (Bushnell and Agarwal 2000), it holds the state of the combinational circuit in response to the first pattern by gating the VDD and GND of the first-level gates. The scheme uses two extra transistors, one in the pull-up and the other in pull-down, to "gate" the supply lines for the first-level logic gates during scan shifting, thus, cutting off any charge/discharge path for the output logic level of the gates. Hence, the output state of the first-level logic gates at the fanout cone of the scan flip-flops hold their state, irrespective of the activity in the scan registers due to rippling of scan values. Once the first-level logic gates hold their states, the other levels also retain their states, since no signal activity propagates to them. Test application remains as in enhanced scan approach, except that the control for holding state is now moved from the hold latches to the gating control of the first-level logic. FLH does not require any extra control signals and does not change the test generation/application process. Moreover, unlike enhanced scan test, it does not introduce extra level of logic in the timing path of a circuit and hence, the delay overhead reduces greatly compared to the enhanced scan. It is worth noting that FLH also maintains the power-saving advantage of enhanced scan in the test mode, since it prevents redundant switching in the combinational block by isolating it from the activities in scan register.

Neither enhanced scan nor FLH, on the other hand, is effective in saving power dissipated in scan chain due to rippling of scan values. There are two primary components of power dissipated in scan chain: switching power in the scan flip-flop and power in the clock line due to transitions of clock. While clock power is independent of the load capacitance at the output of a scan flip-flop, switching power of a scan element is almost linearly dependent on the output load. In the case of enhanced scan, output load of a scan flip-flop is an optimally designed hold latch. However, in the case of FLH, the load varies depending on fanout of the scan flip-flop. Hence, FLH is likely to consume more power in the scan chain during test mode. However, power dissipation for the scan gadget scheme is expected to be higher than FLH due to additional load on the clock line during scan shifting.

7.6 Self-Calibrating and Self-Correcting Systems for Power-Related Failure Detection

Post-silicon strategies for self-calibration and self-repair constitute a promising class of solutions to address power- and variation-induced test challenges. Below, we discuss some important calibration and repair schemes for logic and memory circuits that can simplify the test procedure and reduce test cost with moderate design overhead.

7.6.1 Self-Calibration and Repair in Logic Circuits

As discussed earlier, process variation in logic circuit primarily manifest as variations in delay, leakage, and noise margin. The shift in circuit parameters can be detected using on-chip process sensor and deviation in parameters due to variations can be compensated by appropriate technique.

7.6.1.1 RAZOR

One such technique, called *RAZOR*, uses dynamic detection and correction of circuit timing errors to adjust the supply voltage (Ernst et al. 2003). It potentially eliminates the requirement of delay margin during design phase. RAZOR relies on a combination of architectural and circuit-level techniques for efficient detection and correction of delaypath failures by using a shadow latch controlled by a delayed clock corresponding to each critical flip-flop. In a given clock cycle, if the combinational logic meets the timing requirement for the main flip-flop, then it writes the correct data. On the other hand, if the combinational logic does not complete computation in time, the main flip-flop will latch an incorrect value, while the shadow latch will write the late-arriving correct value. A simple correction scheme restores the correct value from the shadow latch. Such an adaptive technique definitely helps to address the uncertainty in path delay due to variations reducing cost for delay test and speed binning.

7.6.1.2 Body Biasing and Effect on Delay Test

Body bias has strong impact on leakage and performance of a die and thus has been investigated as a potent process adjustment tool. While *forward body bias* (FBB) helps to improve performance in active mode (by lowering the V_{th}), *reverse body bias* (RBB) is effective to reduce leakage power (by increasing the V_{th}). A practical application of body bias to adjust process variations requires accurate detection of process shift at different parts of a circuit and application of an optimal body bias voltage, which maximizes the performance under leakage constraint. Typically, on-chip process sensors for delay or leakage monitoring are used to determine the process shift during test. In Tschanz et al. (2002), a bidirectional *adaptive body bias* (ABB) technique, shown in Fig. 7.9a is used to compensate for die-to-die parameter variations by applying optimum PMOS and NMOS body bias voltages to each die. To account for intra-die variations, an enhancement of this technique is proposed that requires a phase detector (PD) to determine frequency of a block from its critical path replica. The central bias generator considers output of all PDs to

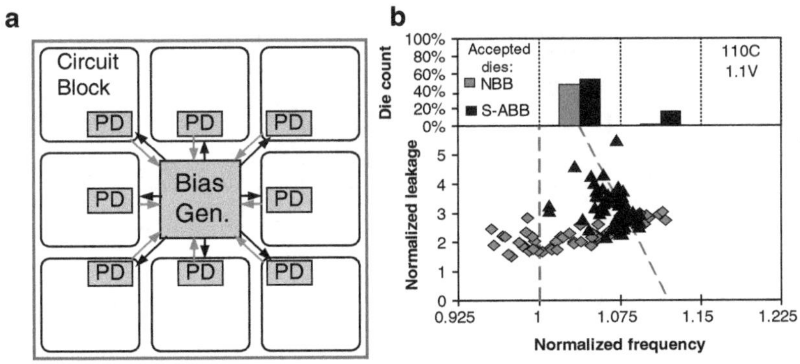

Fig. 7.9 (**a**) Adaptive body biasing scheme considering within-die delay variations; (**b**) Leakage vs. frequency distribution of an adaptive body biasing scheme that considers both inter- and intra-die variations (Tschanz et al. 2002)

determine the optimal bias. Measurement results show that the technique results in an increase in the number of acceptable dies as well as the number of high-frequency dies (Fig. 7.9b).

An ABB technique effectively reduces the delay spread in each chip, thereby improving path delay testability. An investigation was performed in Paul et al. (2004) to observe the impact of body biasing on delay fault test under both inter- and intra-die process variations. Simulation results show that with a fixed optimum forward body bias one can considerably reduce the delay fault test overhead due to process parameter variations. However, with the ABB technique one requires to test only a few paths for delay faults, while achieving very high test quality.

7.6.1.3 Process Compensation in Dynamic Circuits

Increasing I_{OFF} with process scaling has forced designers to upsize the keeper in dynamic circuits to obtain an acceptable robustness under worst-case leakage conditions. However, large (over 20×) variation in die-to-die NMOS I_{OFF} indicates that (1) a large number of low leakage dies suffer from the performance loss due to an unnecessarily strong keeper, while (2) the excess leakage dies still cannot meet the robustness requirements with a keeper sized for the fast corner leakage. A *process-compensating dynamic* (PCD) circuit technique that improves robustness and delay variation spread by restoring robustness of worst-case leakage dies and improving performance of low-leakage dies is presented in Kim et al. (2006). Figure 7.10 shows the PCD scheme with a digitally programmable 3-bit keeper applied on an eight-way register file *local bitline* (LBL). Such a keeper enables 10% faster performance, 35% reduction in delay variation, and 5× reduction in robustness failing dies over conventional static keeper design in 90nm dual-V_{th} CMOS process (Kim et al. 2006). As before, effectiveness of the compensation scheme largely depends on efficient process detection mechanism. Together, they can be very effective to improve test cost and yield for dynamic circuits.

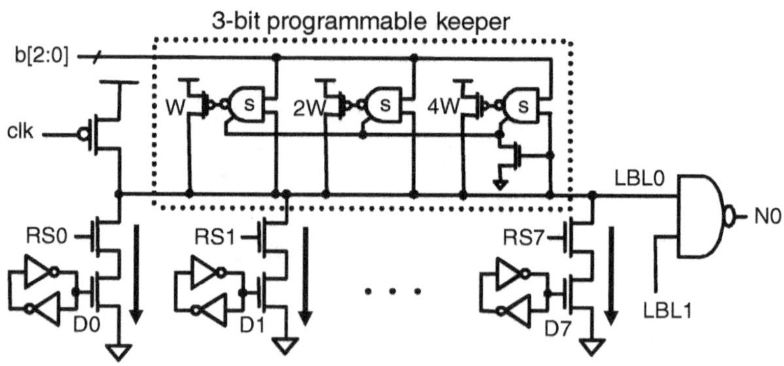

Fig. 7.10 Register file with process compensating dynamic circuit technique (digitally programmable keeper size can be configured to be: $0, W, 2W, \ldots, 7W$) (Kim et al. 2006)

7.6.1.4 Delay Calibration

The wide variation in operating frequency (e.g., ~30% in a processor) has introduced the concept of frequency or speed binning. Speed binning requires calibration of *maximum operating frequency* (F_{max}) at different operating conditions such as supply voltage, temperature, etc. In the simplest scenario, it is desired to determine F_{max} corresponding to a given operating voltage under worst-case temperature condition. The process is expensive in terms of both test application time and complexity of the test hardware since it requires testing at multiple frequencies for a given supply voltage. Consequently, test cost associated with speed binning is significant. The situation becomes worse when it is required to calibrate F_{max} at multiple operating voltages. Calibration of F_{max} at different operating voltages is required primarily for two reasons (a) in a *Dynamic Voltage and Frequency Scaling* (DVFS) system (Ernst et al. 2003), the adaptation hardware is required to apply correct operating frequency corresponding to a scaled supply; and (b) to sort chips in correct *voltage–frequency* ($V{-}F_{max}$) bins, so that chips in different bins can be used for different applications. It has been observed that frequency vs. voltage relationship not only changes from chip to chip but changes in an unpredictable manner at different voltage points for the same chip as well. Thus a static design-time calibration cannot provide a practical solution (Paul et al. 2007).

Given the complexity and cost of speed binning at just one voltage, it is important to develop design techniques to aid the binning process based on structural testing. Earlier it has been demonstrated that speed binning using structural delay testing correlates well with binning process based on functional tests. Conventional approach based on creating a critical-path replica cannot reliably represent the delay of the actual critical path due to local within-die variations. In order to measure the frequency shift accurately, it is better to consider the actual timing paths in the circuit. In Paul et al. (2007), a low-overhead design solution for characterizing F_{max} of a circuit at different operating voltages is presented. The basic idea is to choose a small set of representative paths in a circuit based on their voltage sensitivity and dynamically configure them into ring oscillator to compute F_{max}. The proposed calibration mechanism is all digital, robust with respect to parameter variations, reasonably accurate (with an average error of 2.8% for ISCAS89 benchmarks), and incorporates minimal hardware overhead.

7.6.2 Self-Repairing SRAM

With the limitations of the existing fault-tolerant techniques, SRAM that can repair itself and reduce the number of failures would be very effective for memory yield improvement.

Next, we will discuss a low-overhead circuit-level self-repair technique for SRAM array. A V_{th} shift toward low-V_{th} process corners, due to inter-die variation, increases the read and the hold failures of SRAMs. This is because of the fact that, lowering the V_{th} of the cell transistors increases V_{READ} and V_{TRIPRD}, thereby increasing read failures (Mukhopadhyay et al. 2004a). The negative V_{th} shift increases the leakage through the transistor N_L, thereby, increasing the hold failures. On the other hand, for SRAM arrays in the high-V_{th} process corners, the probabilities of access failures and write failures are high. This is principally due to the reduction in the current drive of the access transistors. The hold failure also increases at the high V_{th} corners, as the trip-point of the inverter PR-NR increases with positive V_{th} shift. Hence, the overall cell failure increases both at low- and high-V_{th} corners and is minimum for arrays in the nominal corner. Consequently, the probability of memory failure is high at both low-V_{th} and high-V_{th} process corners.

Let us now discuss the effect of the body-bias (applied only to NMOS) on different types of failures. Application of reverse body-bias increases the V_{th} of the transistors which reduces V_{READ} and increases V_{TRIPRD}, resulting in a reduction in the read failure (Mukhopadhyay et al. 2004a, b). The V_{th} increase due to RBB also reduces the leakage through the NMOS thereby reducing hold failures. However, an increase in the V_{th} of the access transistors due to RBB increases the access and the write failures. On the other hand, application of FBB reduces the V_{th} of the access transistors, which reduces both access and write failures. However, it increases the read (V_{READ} increases and V_{TRIPRD} reduces) and hold (leakage through NMOS increases) failures (Mukhopadhyay et al. 2004b).

To determine the correct body bias to apply to the SRAM chip for failure probability improvement, the process corner, in which the memory chip resides, needs to be determined. An effective way to perform V_{th} binning is to use leakage monitoring. The random intra-die variation in threshold voltage results in significant variation in cell leakage, particularly, the subthreshold leakage. In a self-repairing SRAM using "*Leakage Monitoring*," the measured leakage is compared with the reference currents to identify the inter-die process corner of the chip. Based on this measurement, the right body bias can be applied to the chip. The schematic of a self-repairing SRAM array with self-adjustable body-bias generator is shown in Fig. 7.11a (Mukhopadhyay et al. 2005). Experimental results on reduction in number of failures shown in Fig. 7.11b appear promising to contain process-induced failures in SRAM.

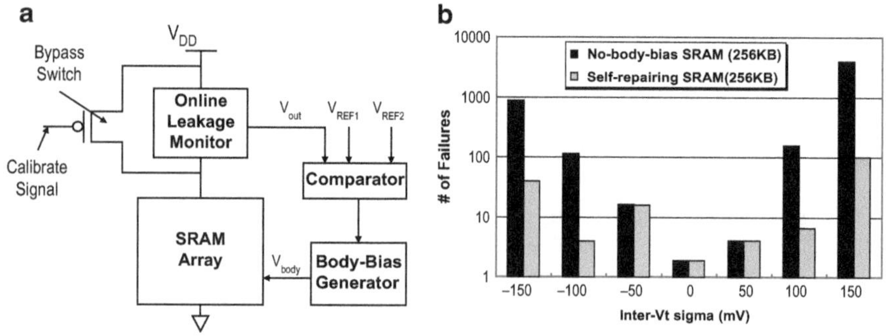

Fig. 7.11 (**a**) Self-repairing SRAM scheme; (**b**) Reduction in number of failures in 256 kB memory array (Mukhopadhyay et al. 2005)

7.7 Summary and Conclusions

Scaling of technology and higher levels of integration, while giving us unprecedented functionality with high speed of operation, has also introduced several adverse effects – high power dissipation and parameter variability. As a by-product, test cost has grown higher and yield has suffered. Today, test has to comprehend any design changes that may arise due to design techniques for low-power and variation tolerance. Also, applied test vectors should cause as low-power dissipation as possible. In this chapter, we discussed the growing impact of power and process variations in nanoscale design and their impact on manufacturing test and yield. New failure mechanisms in logic circuits and SRAM have emerged due to inter- and intra-die process parameter variations. Hence, new test methodologies are required. However, the test problem becomes more complicated with new design methodologies/paradigms being adopted to cope with power and variation problems. Existing techniques on testing logic and memory circuits, fault diagnosis, and fault tolerance may not work well under the new low-power statistical design environment. Besides, circuit and architectural techniques for low-power and variation-tolerant design often impose conflicting requirements on test resulting in increased test complexity and cost. We believe that designers need to consider testability and yield in the design optimization framework in order to limit the growing test complexity and test cost as well as to achieve higher test confidence. In fact, in some cases, test generation can effectively utilize the design concepts used for low-power design to reduce the number of test vectors and to reduce test cost at improved fault coverage. Self-calibration and self-repair techniques also appear promising to reduce test-cost, however, design overhead associated with these techniques should be minimized.

Acknowledgments We would like to express our appreciation to Dr. Swaroop Ghosh, Prof. Chris Kim, Mr. Seetharam Narasimhan, and Mr. Rajat Subhra Chakraborty for providing important help with the technical content and presentation of the chapter.

References

Agarwal A, Chopra K, Baauw D, Zolotov V (2005) Circuit optimization using statistical static timing analysis. In: Proceedings of the design automation conference, June 2005, pp 321–324

Allen D (2008) Power formats: you can have it your way. Electronic Design online, id # 18420, 27 March 2008

Banerjee N, Raychowdhury A, Roy K, Bhunia S, Mahmoodi H (2006) Novel low-overhead operand isolation techniques for low-power datapath synthesis. IEEE Trans VLSI Syst 14(9):1034–1039

Basturkmen NZ, Reddy SM, Pomeranz I (2002) A low power pseudo-random BIST technique. In: Proceedings of international on-line testing workshop, pp 140–144

Bhavnagarwala A, Tang X, Meindl JD (2001) The impact of intrinsic device fluctuations on CMOS SRAM cell stability. IEEE J Solid-State Circuits 36(4):658–665

Bhunia S, Hai L, Roy K (2002) A high performance IDDQ testable cache for scaled CMOS technologies. In: Proceedings of the Asian test symposium, pp 157–162

Bhunia S, Mahmoodi H, Ghosh D, Mukhopadhyay S, Roy K (2005) Low-power scan design using first-level supply gating. IEEE Trans VLSI Syst 13(3):384–395

Bhunia S, Mahmoodi H, Raychowdhury A, Roy K (2008) Arbitrary two-pattern delay testing using a low-overhead supply gating technique. J Electron Test Theory Appl 24(6):577–590

Borkar S, Karnik T, Narendra S, Tschanz J, Keshavarzi A, De V (2003) Parameter variations and impact on circuits and microarchitecture. In: Proceedings of the design automation conference, June 2003, pp 338–342

Bushnell ML, Agarwal VD (2000). Essentials of electronic testing for digital, memory, and mixed-signal VLSI circuits. Kluwer, Boston, MA

Chang H, Sapatnekar SS (2003) Statistical timing analysis considering spatial correlations using a single PERT-like traversal. In: Proceedings of the international conference on computer aided design, Nov. 2003, pp 621–625

Cheng K-T, Devadas S, Keutzer K (1991) A partial enhanced-scan approach to robust delay-fault test generation for sequential circuits. In: Proceedings of the international testing conference, Oct. 1991, pp 403–410

Cheng K-T, Dey S, Rodgers M, Roy K (2000) Test challenges for deep sub-micron technologies. In: Proceedings of the design automation conference, June 2000, pp 142–149

Dabholkar V, Chakravarty S, Pomeranz I, Reddy S (1998) Techniques for minimizing power dissipation in scan and combinational circuits during test application. IEEE Trans Comput Aided Des Integr. Circuits Syst 17(12):1325–1333

DasGupta S (2007) Low-power coalition, May 2007. [Online] http://www.si2.org/?page=729

DasGupta S, Eichelberger E, Williams TW (1978) LSI chip design for testability. In: Proceedings of the international solid-state circuits conference, Feb. 1978, pp 216–217

DasGupta S, Walther RG, Williams TW, Eichelberger EB (1981) An enhancement to LSSD and some applications of LSSD in reliability, availability, and serviceability. In: Proceedings of the international symposium on fault tolerant computing, June 1981, pp 32–34

Ernst D, Kim NS, Das S, Pant S, Rao R, Pham T, Ziesler C, Blaauw D, Austin T, Flautner K, Mudge T (2003) Razor: a low-power pipeline based on circuit-level timing speculation. In: Proceedings of the international symposium on microarchitecture, Dec. 2003, pp 7–18

Gerstendorfer S, Wunderlich H-J (1999) Minimized power consumption for scan-based BIST. In: Proceedings of the international test conference, Sep. 1999, pp 77–84

Ghosh S, Bhunia S, Roy K (2005) Shannon expansion based supply-gated logic for improved power and testability. In: Proceedings of the Asian test symposium, Dec. 2005, pp 404–409

Ghosh S, Bhunia S, Roy K (2007) CRISTA: a new paradigm for low-power, variation-tolerant, and adaptive circuit synthesis using critical path isolation. IEEE Trans Comput Aided Des Integr Circuits Syst 26(11):1947–1956

Girard P, Landrault C, Pravossoudovitch S, Severac D (1998) Reducing power consumption during test application by test vector ordering. In: Proceedings of the international symposium on circuits and systems, pp 296–299

Goering R (2007) IC power standards convergence falters. EETimes, 21 March 2007

Hsu C-P (2006) Pushing power forward with a common power format – The process of getting it right. EETimes, 5 Nov. 2006

Jacobs ETAF, Berkelaar MRCM (2000) Gate sizing using a statistical delay model. In: Proceedings of the design, automation and test in Europe conference, March 2000, pp 283–290

Johnson MC, Somasekhar D, Roy K (1999) Models and algorithms for bounds on leakage in CMOS circuits. IEEE Trans. Comput Aided Des Integr Circuits Syst 18(6):714–725

Kang K, Paul BC, Roy K (2005) Statistical timing analysis using levelized covariance propagation. In: Proceedings of the design, automation and test in Europe conference, March 2005, pp 764–769

Kim CH, Roy K, Hsu S, Krishnamurthy R, Borkar S (2006) A process variation compensating technique with an on-die leakage current sensor for nanometer scale dynamic circuits. IEEE Trans VLSI Syst 14(6):646–649

Krstic A, Wang L-C, Cheng K-T, Liou J-J, Mak TM (2003) Enhancing diagnosis resolution for delay defects based upon statistical timing and statistical fault models. In: Proceedings of the design automation conference, June 2003, pp 668–673

Kuppuswamy R, DesRosier P, Feltham D, Sheik R, Thadikaran P (2004) Full hold-scan systems in microprocessors: cost/benefit analysis. Intel Technol J 8(1):63–72

Liou J-J, Krstic A, Wang L-C, Cheng K-T (2002) False-path-aware statistical timing analysis and efficient path selection for delay testing and timing validation. In: Proceedings of the design automation conference, June 2002, pp 566–569

Mak TM, Krstic A, Cheng K-T, Wang L-C (2004) New challenges in delay testing of nanometer, multigigahertz designs. IEEE Des Test Comput 21(3):241–248

Mani M, Devgan A, Orshansky M (2005) An efficient algorithm for statistical minimization of total power under timing yield constraints. In: Proceedings of the design automation conference, June 2005, pp 309–314

Mao W, Ciletti MD (1994) Reducing correlation to improve coverage of delay faults in scan-path design. IEEE Trans Comput Aided Des Integr Circuits Syst 13(5):638–646

McGowen R, Poirier CA, Bostak C, Ignowski J, Millican M, Parks WH, Naffziger S (2006) Power and temperature control on a 90-nm itanium family processor. IEEE J Solid-state Circuits 41(1):229–237

Meterelliyoz M, Mahmoodi H, Roy K (2005) A leakage control system for thermal stability during burn-in test. In: Proceedings of the international test conference, Nov. 2005, pp 981–990

Mukhopadhyay S, Mahmoodi H, Roy K (2004a) Statistical design and optimization of SRAM for yield enhancement. In: Proceedings of the international conference of computer aided design, Nov. 2004, pp 10–13

Mukhopadhyay S, Mahmoodi-Meimand H, Roy K (2004b) Modeling and estimation of failure probability due to parameter variations in nano-scale SRAMs for yield enhancement. In: Proceedings of the Symposium on VLSI Circuits, June 2004, pp 64–67

Mukhopadhyay S, Kang K, Mahmoodi H, Roy K (2005) Reliable and self-repairing SRAM in nano-scale technologies using leakage and delay monitoring. In: Proceedings of the international test conference, Nov. 2005, pp 1135–1144

Paul BC, Neau C, Roy K (2004) Impact of body bias on delay fault testing of nanoscale CMOS circuits. In: Proceedings of the international test conference, Oct. 2004, pp 1269–1275

Paul S, Krishnamurthy S, Mahmoodi H, Bhunia S (2007) Low-overhead design technique for calibration of maximum frequency at multiple operating points. In: Proceedings of the international conference of computer aided design, Nov. 2007, pp 401–404

Power format requirements version 1.0, 25 Jan 2008. [Online] http://www.si2.org/?page=928

Rabaey JM, Pedram M (eds) (1995) Low power design methodologies, vol 336. Springer, New York

Rao RR, Devgan A, Blaauw D, Sylvester D (2004) Parametric yield estimation considering leakage variability. In: Proceedings of the design automation conference, July 2004, pp 442–447

Rosinger PM, Al-Hashimi BM, Nicolici N (2002) Scan architecture for shift and capture cycle power reductions. In: Proceedings of international symposium defect fault tolerance in VLSI systems, Nov. 2002, pp 129–137

Roy K, Prasad S (2000) Low-power CMOS VLSI circuit design. Wiley, New York. ISBN 0–471–11488-X

Roy K, Mukhopadhyay S, Mahmoodi-Meimand H (2003) Leakage current mechanisms and leakage reduction techniques in deep-submicrometer CMOS circuits. Proc IEEE 91(2):305–327

Sankaralingam R, Pouya B, Touba NA (2001) Reducing power dissipation during test using scan chain disable. In: Proceedings of VLSI test symposium, April–May 2001, pp 319–324

Savir J (1997) Scan latch design for delay test. In: Proceedings of the international test conference, Nov. 1997, pp 446–452

Si2 common power format specification version 1.1, 19 Sep 2008. [Online] http://www.si2.org/?page=811

Srivastava A, Sylvester D (2004) A general framework for probabilistic low-power design space exploration considering process variation. Proceedings of the international conference of computer aided design, Nov. 2004, pp 808–813

Tekumalla RC, Menon PR (1997) Delay testing with clock control: an alternative to enhanced scan. In: Proceedings of the international test conference, Nov. 1997, pp 454–462

Tiwari V, Malik S, Ashar P (1998) Guarded evaluation: pushing power management to logic synthesis/design. IEEE Trans Comput Aided Des Integr Circuits Syst 17(10):1051–1060

Tschanz JW, Kao JT, Narendra SG, Nair R, Antoniadis DA, Chandrakasan AP, De V (2002) Adaptive body bias for reducing impacts of die-to-die and within-die parameter variations on microprocessor frequency and leakage. IEEE J Solid-state Circuits 37(11):1396–1402

Unified power format (UPF) standard version 1.0, 22 Feb 2007, [Online] http://www.accellera.org/apps/group_public/download.php/887/upf.v1.0.pdf

Wang S, Gupta S (1998) ATPG for heat dissipation minimization during test application. *IEEE Trans Comput* 47(2):256–262

Wang S, Liu X, Chakradhar ST (2004) Hybrid delay scan: a low hardware overhead scan-based delay test technique for high fault coverage and compact test sets. In: Proceedings of the design, automation and test in Europe conference, Feb. 2004, pp 1296–1301

Wei L, Chen Z, Roy K, Johnson MC, Ye Y, De VK (1999) Design and optimization of dual threshold circuits for low-voltage low-power applications. IEEE Trans VLSI Syst 7(1):16–24

Whetsel L (2000) Adapting scan architectures for low power operation. In: Proceedings of the international test conference, Oct. 2000, pp 863–872

Xu G (2006) Thermal modeling of multi-core processors. In: Tenth intersociety conference on thermal and thermomechanical phenomena in electronics systems, pp 96–100

Yeo K-S, Roy K (2005) Low voltage, low power VLSI subsystems. McGraw Hill, New York

Yuan L, Qu G (2006) A combined gate replacement and input vector control approach for leakage current reduction IEEE Trans VLSI Syst 14(2):173–182

Zhang X, Roy K (2000) Power reduction in test-per-scan BIST. In: Proceedings of the international online testing workshop, July 2000, pp 133–138

Zorian Y (1993) A distributed BIST control scheme for complex VLSI devices. In: Proceedings of the IEEE VLSI test symposium, Apr. 1993, pp 4–9

Chapter 8
Test Strategies for Multivoltage Designs

Saqib Khursheed and Bashir M. Al-Hashimi

Abstract Reducing the power consumption of digital designs through the use of more than one Vdd value (multivoltage) is known and well practiced. Some manufacturing defects have Vdd dependency, which implies that defects can become active only at certain power supply setting, leading to reduced defect coverage. This chapter presents a coherent overview of recently reported research in testing strategies for multivoltage designs including defect modeling, test generation, and DFT solutions. The chapter also outlines number of worthy research problems that need to be addressed to develop high-quality and cost-effective test solutions for multi-Vdd designs.

8.1 Introduction

Minimizing power consumption through the use of low-power design techniques has been an active research area for nearly two decades, motivated by the portable and hand-held devices application market. The operating voltages needed for such designs are generated either through dedicated multiple power supplies on chip (Hamada et al. 1998) or through adaptive voltage scaling circuitry consisting of DC–DC converters and voltage-controlled oscillators (Lee and Sakurai 2000). These techniques operate gates or circuits not on the critical path of a design at lower operating voltage than those on the critical path thereby achieving low power without compromising performance. Commercial CAD tools support multi-Vdd design approach (Synopsys *galaxy*^{TM}) and for that reason it is normally employed in designs where power consumption is a key requirement. This chapter addresses the following general question, "Can existing test techniques be used to test multi-Vdd designs?" The simple answer is yes and to ensure high-defect coverage it is necessary to repeat the test at all operating voltages of the design since some defects show

S. Khursheed (✉) and B.M. Al-Hashimi
University of Southampton, Southampton, UK
e-mail: ssk06r@ecs.soton.ac.uk

P. Girard et al. (eds.), *Power-Aware Testing and Test Strategies for Low Power Devices*,
DOI 10.1007/978-1-4419-0928-2_8,

Vdd dependency. This may not be viable in designs where cost is of great importance as the case with hand-held devices market. Recently researchers have started to develop specific test solutions to multi-Vdd designs where the aim is to improve defect coverage without the need to repeat the test at all operating voltages of the design. Testing multi-Vdd designs is an orthogonal problem to very low voltage (VLV) testing (Hao and McCluskey 1993), which was proposed over a decade ago to improve reliability. It was shown that testing between $2V_t$ and $2.5V_t$, where V_t is the transistor threshold voltage, achieves high-defect coverage for resistive bridges. The differentiation is that in multi-Vdd designs there are a number of operating Vdds, in practice up to four, and the aim of multi-Vdd test is to determine the minimum number of voltage settings to ensure the highest level of defect coverage.

In this chapter, we outline recent findings for two major types of defects: resistive bridge and resistive open in the context of multi-Vdd designs. A nonresistive defect (e.g., a short) between an interconnect line and power supply (Vdd) or ground rail (Gnd) can be modeled using a stuck at fault model, which represents permanent failure of the line in terms of stuck-at 1 (short with Vdd) or stuck-at 0 (short with Gnd), respectively. Such type of failures do not show Vdd-dependent detectability[1] and therefore are not discussed in this chapter. Sections 8.2 and 8.3 discuss test techniques for resistive bridge and resistive open defects in the context of multi-Vdd designs. The DFT technique for devices employing multi-Vdd is discussed in Sect. 8.4, with the aim to achieve cost-effective test as well as reducing power dissipation during test. Section 8.5 provides a summary of emerging and new test research problems and finally, Sect. 8.6 concludes the chapter.

8.2 Test for Multivoltage Design: Bridge Defect

Resistive bridge represent a major class of defects for deep submicron (DSM) CMOS. It is due to an unwanted metal connection between two lines of the circuit, which deviates the circuit from its ideal behavior. A typical resistive bridge is shown in Fig. 8.1. A study on *resistive bridge distribution* is reported in Montanes et al. (1992) based on 14 wafers from different batches and production lines. The study shows that around 96% of bridges have a resistance value which is less than 1 kΩ. On the other hand, a physical defect between an interconnect line and power supply (Vdd) or ground rail (Gnd) is referred to as hard-short (bridge with 0-Ω resistance). It was shown in Khursheed et al. (2009b) that detectability of hard-short is irrespective of Vdd settings and therefore is not further discussed in this chapter.

This section discusses *modeling and test generation of resistive bridge* for multi-Vdd designs. Section 8.2.1 describes the analog and digital behavior of resistive bridge at single-voltage setting. This is further extended by showing

[1] Stuck-at fault model does not capture physical complexities at the fault site and therefore more complex fault models have evolved to improve testability of the design. For a comprehensive discussion on evolution of fault models see Delgado (2008).

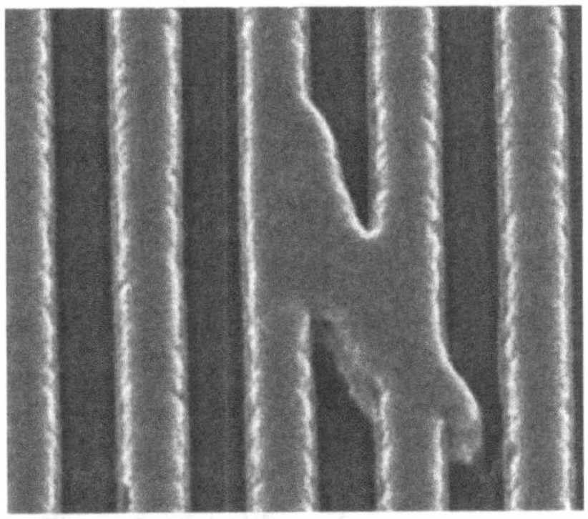

Fig. 8.1 Resistive bridge (Kundu et al. 2001)

Vdd-dependency of resistive bridge in Sect. 8.2.2. Finally, Sect. 8.2.3 provides a summary of recently reported research related to cost-effective testing of resistive bridge for multi-Vdd designs.

8.2.1 Resistive Bridge Behavior at Single-Vdd Setting

The resistance of a bridge is a continuous parameter which is not known in advance. A recent approach based on interval algebra (Engelke et al. 2004, 2006b) allowed treating the whole continuum of bridge resistance values R_{sh} from $0\,\Omega$ to ∞ by handling a finite number of discrete intervals. The key observation which enables this method is that a resistive bridge changes the voltages on the bridged lines from 0 V (logic-0) or Vdd (logic-1) to some intermediate values, which will be different for different R_{sh} values. The logic behavior of the physical defect can be expressed in terms of the logic values perceived by the gate inputs driven by the bridged nets based on their specific input threshold voltage.

A typical bridge fault scenario is illustrated in Fig. 8.2. D1 and D2 are the gates driving the bridged nets, while S1, S2, S3, and S4 are successor gates, i.e., gates having inputs driven by one of the bridged nets. The resistive bridge affects the logic behavior only when the two bridged nets are driven at opposite logic values. For example, consider the case when the output of D1 is driven high and the output of D2 is driven low. For illustration, we assume that the shown bridge R_{sh} affects only the output of D1, i.e., S1, S2, and S3 are affected by the resistive bridge. The dependence of the voltage level on the output of D1 (V_O) on the equivalent resistance of the physical bridge is shown in Fig. 8.3. The deviation of V_O from the ideal voltage level (Vdd) is highest for small values of R_{sh} and decreases for larger values of R_{sh}. To translate this analog behavior into the digital domain, the input threshold voltage levels V_{th1}, V_{th2} and V_{th3} of the successor gates S1, S2, and S3 have been added to the V_O plot. For each value of the bridge resistance R_{sh}, the logic values at inputs I_1, I_2, and I_3 can be determined by comparing V_O with the input threshold voltage of the corresponding input. These values are shown in the second part of Fig. 8.3.

Fig. 8.2 Example of a resistive bridge fault

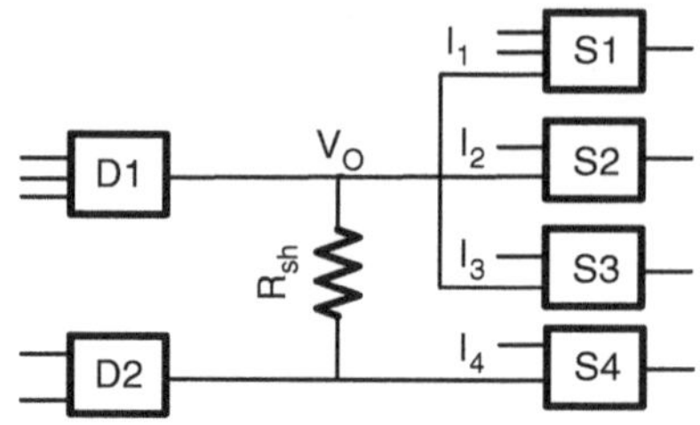

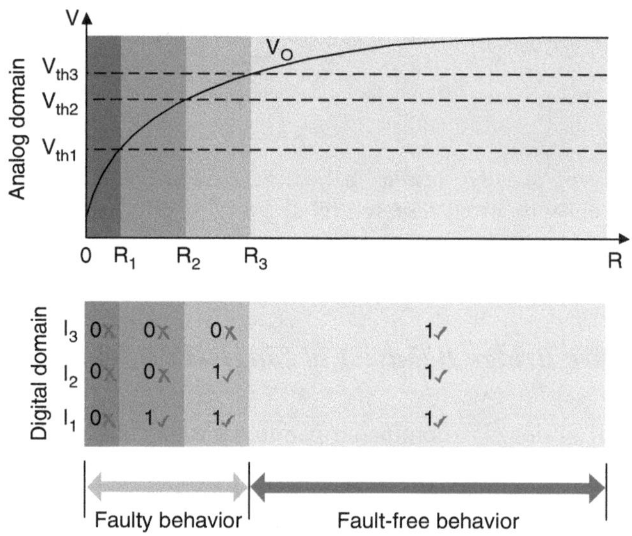

Fig. 8.3 Behavior of a bridge fault at a single-Vdd setting in analog and digital domains

Crosses are used to mark the faulty logic values and ticks to mark the correct ones. It can be seen that, for bridge with $R_{sh} > R_3$, the logic behavior at the fault site is fault-free (all inputs interpret the correct value), while for bridge with R_{sh} between 0 and R_3, one or more of the successor inputs are interpreting a faulty logic value. The R_{sh} value corresponding to R_3 is normally referred to as "critical resistance" as it represents the crossing point between faulty and correct logic behavior. Methods for determining the critical resistance have been presented in several publications (Sar-Dessai and Walker 1999; Engelke et al. 2006b).

A number of bridge resistance intervals can be identified based on the corresponding logic behavior. For example, all bridges with $R_{sh} \in [0, R_1]$ exhibit the same faulty behavior in the digital domain (all successor inputs interpret faulty logic value). Similarly, for bridges with $R_{sh} \in [R_1, R_2]$, successor gates S2 and S3 interpret the faulty value, while S1 interprets the correct value. Finally, for bridges with $R_{sh} \in [R_2, R_3]$ only S3 interprets a faulty value while the other two successor gates interpret the correct logic value. Consequently, each interval $[R_i, R_{i+1}]$ corresponds to a distinct logic behavior occurring at the bridge fault site. The logic behavior at the fault site can be captured using a data structure further referred to as logic state configuration (LSC), which can be looked at as logic fault model (Khursheed et al. 2008). The union of the resistance intervals corresponding to detectable faults forms the Global Analog Detectability Interval (G-ADI) (Engelke et al. 2006b). Basically, G-ADI represents the entire range of detectable physical defects. Given a test set TS, the Covered Analog Detectability Interval (C-ADI) represents the range of physical defects detected by TS. The C-ADI for a bridge defect is the union of one or more disjoint resistance intervals, the union of intervals corresponding to detectable faults (Renovell et al. 1996; Engelke et al. 2004, 2006a,b). The quality of a test set is estimated by measuring how much of the G-ADI has been covered by the C-ADI. When the C-ADI of test set TS is identical to the G-ADI of fault f, TS is said to achieve full fault coverage for f.

Several test generation methods for resistive bridge faults (RBF) have been proposed for a fixed supply voltage setting (Sar-Dessai and Walker 1999; Maeda and Kinoshita 2000; Shinogi et al. 2001; Chen et al. 2005; Engelke et al. 2006a). The method presented in Maeda and Kinoshita (2000) is to guarantee the application of all possible values at the bridge site without detailed electrical analysis. In Chen et al. (2005), the effect of a bridge on a node with fanout is modeled as a multiple line stuck-at fault. The study in Sar-Dessai and Walker (1999) identifies only the largest resistance interval and determines the corresponding test pattern. In contrast to Sar-Dessai and Walker (1999), the sectioning approach from Shinogi et al. (2001) considers all the sections (resistance intervals) $[R_i, R_{i+1}]$. For each section, the corresponding LSC (and associated faulty logical behavior) is identified. This avoids the need for dealing with the resistance intervals and improves the test quality compared with Sar-Dessai and Walker (1999), but the number of considered faults grows. In Engelke et al. (2006a), the authors combined the advantages of the interval based (Sar-Dessai and Walker 1999) and the sectioning approach (Shinogi et al. 2001) into a more efficient test generation procedure by targeting the section with the highest boundaries first. Interval-based fault simulation is then used to identify all other sections covered by the test pattern.

Prior research has analyzed the effect of varying the supply voltage on the defect coverage using pseudorandom tests (Engelke et al. 2004). The reported experimental results show that the fault coverage of a given test can vary both ways when the supply voltage is lowered, because not all faults can be covered using a single-Vdd setting during test. However, Engelke et al. (2004) suggests that applying the tests at a lower supply voltage in addition to the nominal can improve the fault coverage. This finding is further elaborated by Fig. 8.4. It shows the number of defects and respective resistance values, which cannot be detected (test escapes) at Vdd = 0.8 V [which would be a preferred Vdd for a 1.2-V process according to Renovell et al. (1996) and Engelke et al. (2004)]. The test escapes at 0.8 V, as shown in Fig. 8.4, are based on seven of the medium- and large-size ISCAS-85' and 89' benchmarks. The random spread of these defects across the resistance range suggests that to ensure high-defect coverage it will be necessary to test at more than one Vdd setting for 100% defect coverage, as motivated by Khursheed et al. (2008). In Sect. 8.2.2 we explain why it may be necessary to use more than one Vdd setting during test to ensure full bridge defect coverage for multi-Vdd designs.

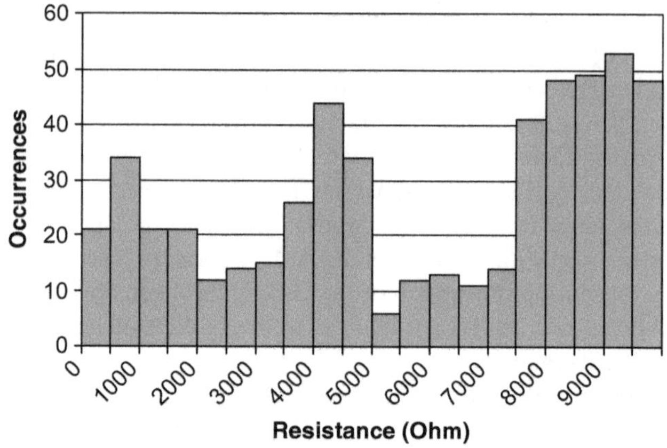

Fig. 8.4 Resistance values that cannot be detected at lowest Vdd setting (Khursheed et al. 2008)

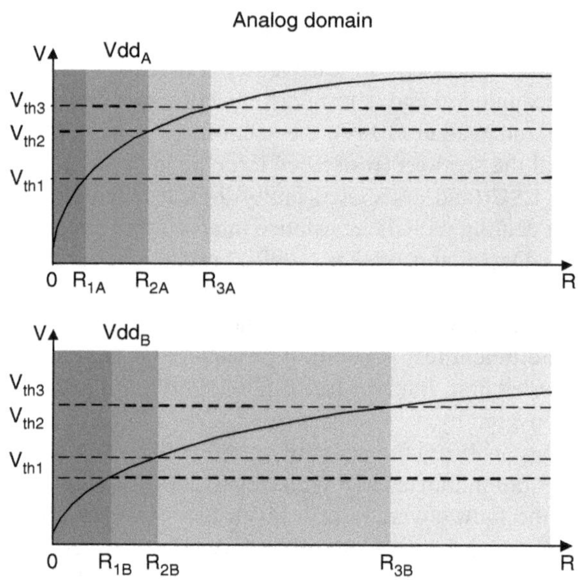

Fig. 8.5 Effect of supply voltage on bridge fault behavior: Analog domain (Khursheed et al. 2008)

8.2.2 Resistive Bridge Behavior at Multi-Vdd Settings

This section provides an analysis of the effect of varying supply voltage on bridge fault behavior. Figure 8.5 shows the relation between the voltage on the output of gate D1 (Fig. 8.2) and the bridge resistance for two different supply voltages Vdd_A and Vdd_B. The diagrams in Fig. 8.6 show how the analog behavior at the fault site translates into the digital domain. In this example, three distinct logic faults LF1, LF2, and LF3 could be identified for each Vdd setting. However, because the voltage

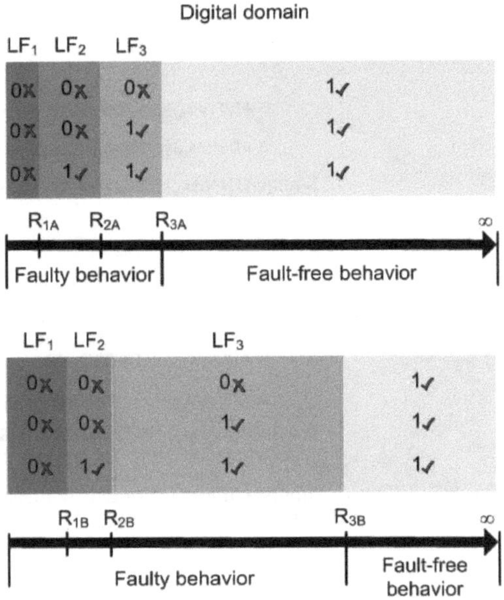

Fig. 8.6 Effect of supply voltage on bridge fault behavior: Digital domain (Khursheed et al. 2008)

level on the output of D1 does not scale linearly with the input threshold voltages of S1, S2, and S3 when changing the supply voltage (this has been validated through SPICE simulations), the resistance intervals corresponding to LF1, LF2, and LF3 differ from one supply voltage setting to another. This means that a test pattern targeting a particular logic fault will detect different ranges of physical defects when applied at different supply voltage settings. For example, at Vdd_A, a test pattern targeting LF3 will detect bridge with $R_{\mathrm{sh}} \in [R_{2A}, R_{3A}]$, while at Vdd_B it will detect a much wider range of physical bridge ($R_{\mathrm{sh}} \in [R_{2B}, R_{3B}]$). Analyzing this from a different perspective, a bridge with $R_{\mathrm{sh}} = R_{3B}$ will cause a logic fault at Vdd_B but not at Vdd_A. To demonstrate the need for using multiple Vdd settings during test we use the following two scenarios. In Case 1 (Fig. 8.7) all three logic faults LF1, LF2, and LF3 are nonredundant. Figure 8.7 shows the ranges of bridge resistance corresponding to faulty logic behavior for the two Vdd settings (basically the G-ADI sets corresponding to the two Vdd settings). Previous work on test generation for bridge faults (Engelke et al. 2006a) has used the concept of G-ADI assuming a fixed Vdd scenario. Ingelsson et al. (2007) has extended the concept of G-ADI to capture the dependence of the bridge fault behavior on the supply voltage by defining the multi-Vdd G-ADI as the union of Vdd specific G-ADIs for a given design

$$\text{G-ADI} = \bigcup \text{G-ADI}(\mathrm{Vdd}_i).$$

The overall G-ADI consists of the union of the two Vdd-specific G-ADI sets. It can be seen that G-ADI(Vdd_A) represents about 45% of the overall G-ADI while G-ADI(Vdd_B) fully covers the overall G-ADI. This means that a test set

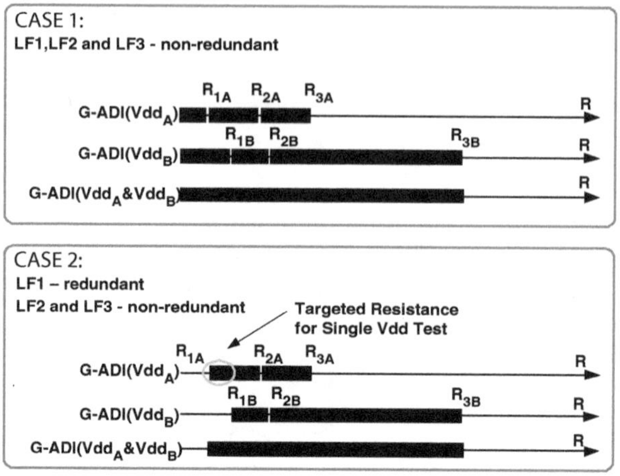

Fig. 8.7 Effect of supply voltage on bridge fault behavior: Observable bridge resistance ranges (Khursheed et al. 2008)

detecting LF1, LF2, and LF3 will achieve full bridge defect coverage when applied at Vdd_B. In Case 2 from Fig. 8.7, only LF2 and LF3 are nonredundant, which means that there is no test pattern which can detect LF1. In this case, G-ADI(Vdd_A) represents about 30% of the overall G-ADI while G-ADI(Vdd_B) represents about 90% of the overall G-ADI. This means that full bridge fault coverage cannot be achieved using a single-Vdd setting.

From this analysis it can be concluded that to achieve full G-ADI coverage in a variable Vdd system, it may be necessary to apply tests at several Vdd settings. Instead of repeating the same test at all Vdd settings, which would lead to long testing times and consequently would increase the manufacturing cost, it would be desirable to be able to determine for each Vdd settings only the test patterns which effectively contribute to the overall defect coverage.

It has been shown in Engelke et al. (2004) that the fault coverage of a test set targeting resistive bridge faults RBF can vary with the supply voltage used during test. This means that, depending on the operating Vdd setting, a given RBF may or may not affect the correct operation of the design. Consequently, to ensure high fault coverage for a design that needs to operate at a number of different Vdds, it may be necessary to perform testing at more than one Vdd to detect faults which manifest themselves only at particular Vdds. A multi-Vdd test generation (MVTG) methodology is presented in Khursheed et al. (2008), which computes a number of Vdd-specific test sets to achieve 100% defect coverage. In Khursheed et al. (2008), experiments are conducted using ISCAS-85' and 89' benchmark designs and fault list is compiled using coupling capacitance between neighboring nodes, these are most likely to form a bridge. Three Vdd settings are used for the experiment, i.e., 0.8 V, 1.0 V, and 1.2 V and the outcome is tabulated in Table 8.1. The first two columns show the benchmark designs along with the number of faults extracted for each design. In this experiment, Synopsys *TetraMAX*^{TM} is used to generate a

Table 8.1 Results of using Synopsys TetraMAX and multi-Vdd test generation (MVTG) as a combined test generation flow for RBF (Khursheed et al. 2008)

		TMAX		MVTG top-up			
		0.8 V		0.8 V	1.0 V	1.2 V	Tot.
Design	No. of RBF	DC	#tp	#tp	#tp	#tp	#tp
c1355	80	83	33	32			65
c1908	98	98	42	27			69
c2670	104	90	27	50			77
c3540	363	96	72	126	6	1	205
c7552	577	95	44	198		1	243
s838	34	88	17	17	2		36
s1488	435	96	82	82	2		166
s5378	305	95	60	123			183
s9234	223	89	48	92	2		142
s13207	358	95	60	89	5	1	155
s15850	943	98	56	144	4	5	209
s35932	1,170	96	33	89	36	66	224

test set for each design, which is then fault simulated at 0.8 V (since higher resistive bridge fault coverage is achieved at a lower Vdd). The defect coverage achieved and the number of test patterns in the TetraMAX test set are shown in the third main column of Table 8.1. Subsequently, MVTG (Khursheed et al. 2008) is used to generate top-up tests, targeting bridges that are not fully covered by the TetraMAX test set. It is therefore used to provide the remaining defect coverage up to 100%. The sizes of the test sets generated by the MVTG top-up run are given in the fourth column for each Vdd setting. Finally, the total test pattern count is shown in the last column of Table 8.1, marked as "Tot.". From test flow point of view, it is therefore suggested to use MVTG (Khursheed et al. 2008) as a postprocessing step to cover resistance intervals that remains uncovered by commercial ATPG tools.

8.2.3 *Cost-Effective Test for Resistive Bridge*

In Sect. 8.2.2, it has been shown that more than one Vdd setting is required to achieve 100% defect coverage of resistive bridging defects. Switching between different Vdd settings during test is not a trivial task, and therefore a large number of Vdd settings required during test can have a detrimental effect on the overall cost of test. Consequently, it would be desirable to keep the number of Vdd settings required during test to a minimum. By analyzing the scenario described in Case 2 (Fig. 8.7), it can be seen that full bridge defect coverage could be achieved using a single-Vdd setting (Vdd$_B$), if the logic fault (LF) corresponding to the resistance interval $[R_{1A}, R_{1B}]$ (shown separately in Fig. 8.7), LF1 in this case, would become detectable at Vdd$_B$. Based on this observation, two techniques are available in literature and are summarized in this section.

8.2.3.1 Test Point Insertion

The first method to reduce Vdd settings during test is by using *Test Point Insertion* (TPI) as proposed in Khursheed et al. (2008). Test points are used to provide additional controllability and observability at the fault-site to detect resistance intervals at the desired Vdd setting, which are otherwise redundant and therefore helps reducing the number of test Vdd(s). This can be understood using Fig. 8.7, which shows that marked resistance range is detectable only at Vdd_A. The TPI scheme proposed in Khursheed et al. (2008) is used to cover the resistance interval at desired Vdd (Vdd_B) by providing additional controllability and observability using test points. In this case, Vdd_B is desirable as it covers most amount of detectable resistance range as shown in Fig. 8.7. Experimental results presented in Khursheed et al. (2008) show that TPI can be used to reduce the number of Vdd settings during test, without affecting the defect coverage of the original test, thereby reducing test cost. One drawback with TPI scheme (Khursheed et al. 2008) is that it does not guarantee single-Vdd test and usually results in more than one test Vdd settings. Experimental results presented in Khursheed et al. (2008) and more recently in Khursheed et al. (2009a) show that TPI is unable to reduce test to single-Vdd setting for majority of circuits. This can be understood from the following explanation. In Fig. 8.2, the gates used for driving the bridge (D1, D2) and the driven gates (S1, S2, S3, S4), influence the number of test Vdd(s) in a circuit. For the same circuit, assume that D1 is driving high and D2 is driving low, the output of D2 (V_O) on the equivalent resistance of the physical bridge is shown in Fig. 8.8, which shows that higher resistance range is covered at 1.2 V (nonpreferred test Vdd) than at 0.8 V (preferred test Vdd). This means that 1.2 V becomes essential test Vdd and TPI includes it for 100% defect coverage, as resistance range covered at 1.2 V cannot be covered at 0.8 V.

8.2.3.2 Gate Sizing

Recently a new technique for reducing test cost of multi-Vdd designs with resistive bridging defect has been reported in Khursheed et al. (2009a). It targets resistive

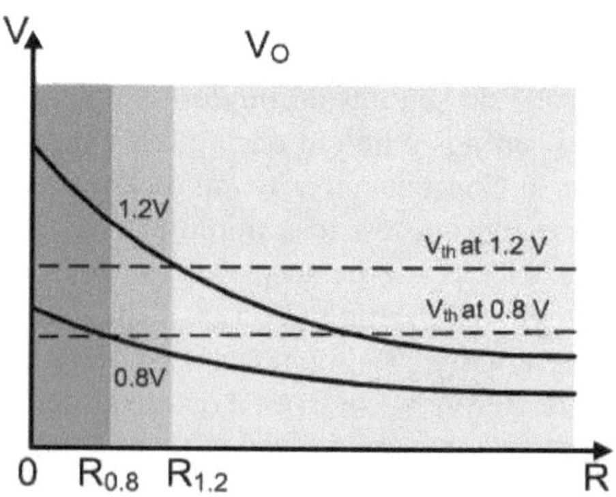

Fig. 8.8 Resistance range detection at different voltage settings

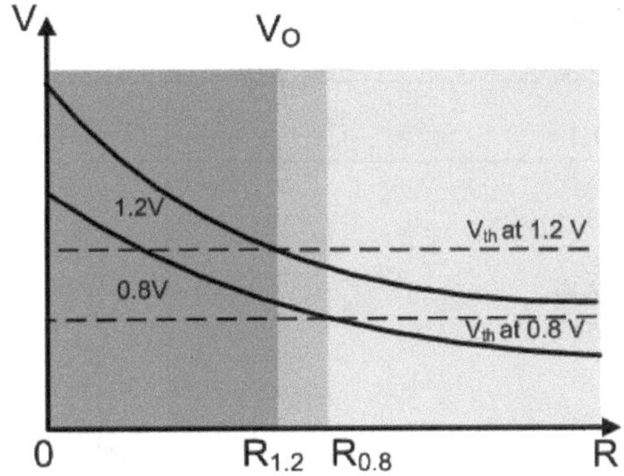

Fig. 8.9 Resistance range detection after adjusting the drive strength of the gates driving the bridge

bridge that cause faulty logic behavior to appear at a nondesired test Vdd setting and uses *Gate Sizing* (GS) to expose the same physical resistance at preferred test Vdd. This is achieved by adjusting the drive strengths of gates driving the bridge, such that higher resistance is exposed at the desired Vdd setting. The drive strength of the gates driving the bridged nets can be adjusted to increase the voltages on the bridged nets (V_O in Fig. 8.2). This increase in voltage level can help expose maximum resistance at the desired Vdd setting thereby reducing the number of test Vdd settings; additionally it can also be used to cover resistance intervals (such as the one marked in Fig. 8.7) at the desired Vdd setting. This concept is illustrated by Fig. 8.9, which shows same pair of bridged nets as shown in Fig. 8.8 (derived from Fig. 8.2, where D1 is driving high and D2 is driving low), i.e., the logic thresholds of the driven gates remain the same. In Fig. 8.9 it can be noticed that the voltage level V_O has increased such that $R_{0.8} > R_{1.2}$, by increasing the drive strength of the gates driving the bridge. This means that test generation will favor 0.8 V over 1.2 V, thereby reducing the number of test Vdd(s) and removing 1.2 V as a test Vdd and thus reducing total number of test Vdd settings. The drive current of a transistor I_{ds} is directly proportional to the gain factor β, which in turn is directly proportional to the W/L of the transistor. Thus replacing a gate with another having higher value of β (especially for transistors feeding the output) results in higher drive strength. This is feasible since, different versions of functionally equivalent gates are usually available in the gate library.

Experiments are conducted using ISCAS-85' and 89' full scan circuits, and results for TPI (Khursheed et al. 2008) and GS (Khursheed et al. 2009a) are tabulated in Table 8.2. The first two columns show the benchmark designs and respective gate count in each design. The third main column (labeled as Test Vdd(s)) tabulates total number of test Vdd setting(s) for each of the original design (labeled as Orig.) by TPI (Khursheed et al. 2008) (labeled, TPI) and by the GS technique (labeled, GS). As can be seen the GS technique is able to achieve 100% defect coverage at a single Vdd. This is unlike TPI, which requires two or more Vdd setting for most of the circuits to achieve the same defect coverage. Moreover, TPI is unable to reduce any test Vdd in case of c432 and c1908. The last main column of Table 8.2

Table 8.2 Results of gate sizing technique (GS) (Khursheed et al. 2009a) and its comparison with TPI (Khursheed et al. 2008)

		Test Vdd(s)			Gates	
CKT	No. of gates	Orig.	TPI	GS	GS	TPI
c432	93	All[a]	All	0.8 V	2	0
c1355	226	All	0.8 V	0.8 V	4	10
c1908	205	1.2 V, 0.8 V	1.2 V, 0.8 V	0.8 V	3	0
c2670	269	All	1.2 V, 0.8 V	0.8 V	6	19
c3540	439	All	1.0 V, 0.8 V	0.8 V	7	7
c7552	731	All	0.8 V	0.8 V	1	1
s344	62	1.2 V, 0.8 V	0.8 V	0.8 V	1	1
s382	74	1.2 V, 0.8 V	0.8 V	0.8 V	2	5
s386	63	All	1.2 V, 0.8 V	0.8 V	7	4
s838	149	All	0.8 V	0.8 V	14	28
s5378	578	All	1.0 V, 0.8 V	0.8 V	9	9
s9234	434	All	1.0 V, 0.8 V	0.8 V	6	2
s15850	1578	All	0.8 V	0.8 V	8	3

[a]All = 0.8 V, 1.0 V, 1.2 V

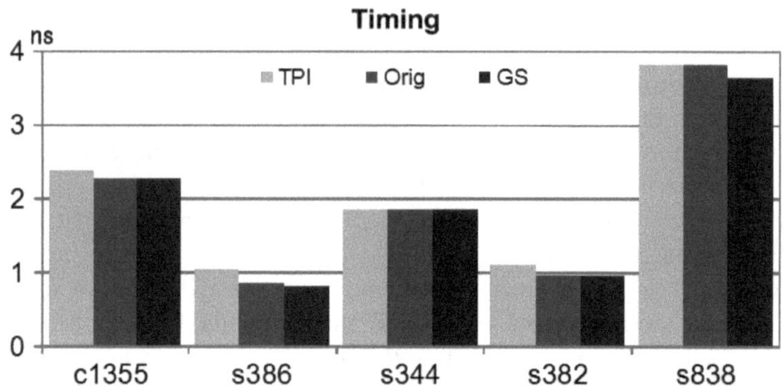

Fig. 8.10 Timing performance of TPI (Khursheed et al. 2008) and GS (Khursheed et al. 2009a) in comparison with the original design

(labeled as Gates) shows the number of gates replaced by GS technique and the number of test points (control/observation points) added by TPI.[2] The number of gates replaced by GS technique ranges from 1 to 14, while TPI has added up to 28 test points.

In another experiment reported in Khursheed et al. (2009a), the timing performance of the original design (Orig), is compared with the design altered by GS and by TPI techniques using Synopsys design compiler. Figure 8.10 shows the timing performance, as can be seen the GS technique has little affect on the timing

[2] The number of test points is the sum of control and observation points.

performance when compared to the original design. This is unlike the case with TPI, where the timing has increased because of test points in critical path. It should be noted that for some circuits the GS technique has reduced timing than the original design due to larger and faster gates. Thus GS technique represents an improvement over TPI, as it achieves 100% defect coverage at single test Vdd setting, while TPI mostly employs two or more test Vdd setting (Table 8.2). Furthermore, it has less cost of area, power, and timing overhead as compared with TPI. For further details refer to Khursheed et al. (2009a).

8.3 Test for Multivoltage Design: Open Defect

Section 8.2 considered test techniques for bridge defect, this section discusses test techniques for open defects, which is another dominant defect type commonly found in deep-submicron CMOS. It is due to unconnected nodes in a manufactured circuit that were connected in the original design and therefore deviates the circuit from ideal behavior. Open defects can be classified as full or strong opens with resistance greater than $10\,\text{M}\Omega$ and resistive or weak open with resistance less than $10\,\text{M}\Omega$ (Montanes et al. 2002). Full open causes logic failures that can be tested using static tests (test patterns applied without timing consideration). On the other hand, resistive open shows timing-dependent effects and therefore should be tested using delay tests. Figure 8.11 shows a cross section of resistive open defect. In this section, electrical characteristics of full open is discussed first, followed by resistive open.

8.3.1 Testing Full-Open Defect

Figure 8.12 shows *open defect distribution* in six different metal layers corresponding to 7,440 dies from 12 lots, manufactured in 180-nm CMOS process. As can

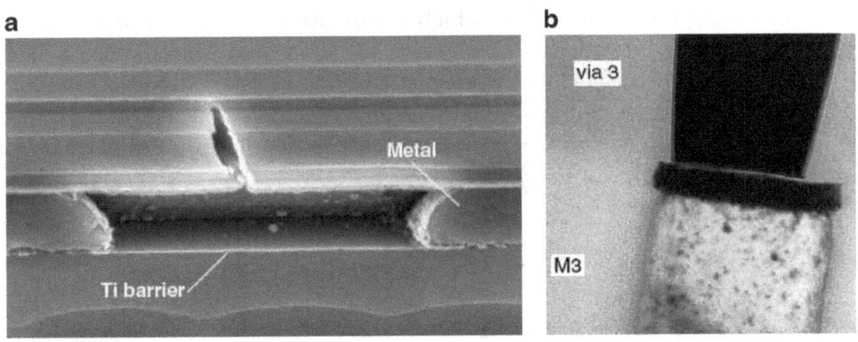

Fig. 8.11 Resistive or weak open defects: (**a**) Cross section of metal open line and (**b**) a resistive via (Montanes et al. 2002)

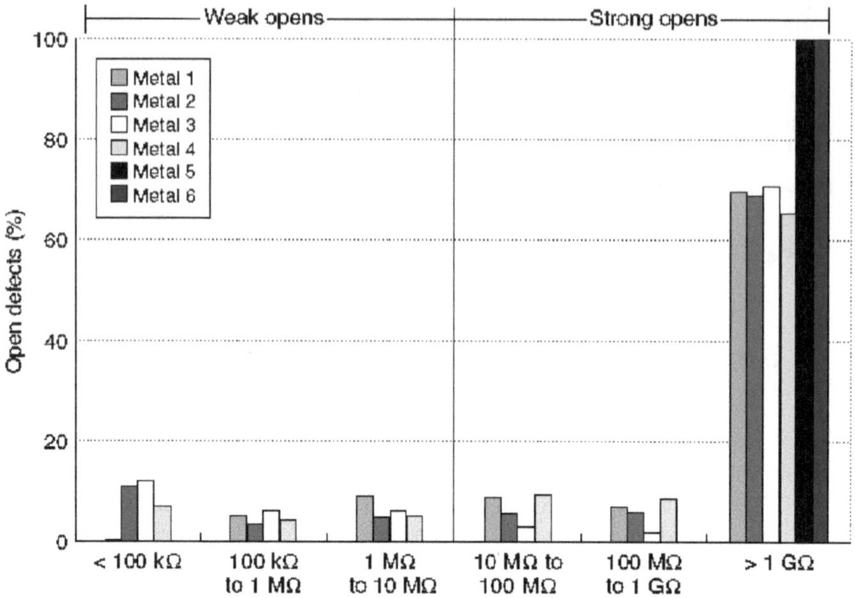

Fig. 8.12 Distribution of metal open resistances (Montanes et al. 2002)

be seen, the majority of open defects can be categorized as strong or full-open defects. Similar trend is reported for contact or via open (Montanes et al. 2002). The occurrence frequency of full-open defects is expected to increase in future technologies (Sreedhar et al. 2008; Arumi et al. 2008a). Two fault models are available in literature for modeling full-open defects, which can be categorized as *capacitance-based full-open fault model* (Henderson et al. 1991; Johnson 1994; Choudhury and Sangiovanni-Vincentelli 1995; Rafiq et al. 1998) and *leakage-aware full-open fault model* (Lo et al. 1997; Guindi and Najm 2003; Sreedhar et al. 2008; Arumi et al. 2008a). Several recent studies have used capacitance-based models (Gomez et al. 2005; Zou et al. 2006; Montanes et al. 2007; Spinner et al. 2008; Arumi et al. 2008b) for testing full-open defects, which use the following electrical characteristics (1) the capacitance between floating line (disconnected from the driver node) and its neighboring line(s), (2) the parasitic capacitance due to transistors (PMOS and NMOS connected to floating line) driven by the floating net, and (3) the trapped charge on the floating net. If F represents a floating net that is disconnected from its driver, then voltage V_F is given by Zou et al. (2006) and Ingelsson (2009):

$$V_F = \frac{C_{High}}{C_{High} + C_{Low}} V_{dd} + \frac{Q_{trap}}{C_{Gnd}}, \tag{8.1}$$

where V_F is voltage on the floating net, C_{High} and C_{Low} is capacitance due to neighboring lines driving high and low, respectively (including capacitance due to V_{dd} and Gnd), V_{dd} is the supply voltage, and Q_{trap}/C_{Gnd} represents the trapped charge on the floating net. From (8.1), it can be noticed that for detecting full-open defects, V_F can be induced such that voltage on the floating net is higher than the logic threshold L_{th} voltage of the gate input, i.e., $V_F > L_{th}$, thereby exciting a stuck-at 1 fault. Voltage on the floating net can be induced by using test patterns that result in setting the neighboring nets to desired logic value, thereby increasing the fraction $C_{High}/(C_{High} + C_{Low})$, as shown in (8.1). Similarly a stuck-at 0 fault can be induced on the floating net. The fault effect can then be propagated to any of the primary outputs for detection (Zou et al. 2006).

In nanometer CMOS ($\leq$90 nm), since the thickness of gate oxide is few tens of Å, it does not act as a strong insulator. This results in higher gate-tunneling leakage current in comparison to previous technologies (Sreedhar et al. 2008; Arumi et al. 2008a; Ingelsson 2009), and therefore affects the voltage on the floating net causing full-open defect. A floating net connected to a gate has a bistable input state (Sreedhar et al. 2008; Arumi et al. 2008a). In Sreedhar et al. (2008) an inverter synthesized using 45-nm technology was simulated with a floating input and the change in input voltage was observed. It was found that the voltage on the floating net increased from 0 to 0.17 V (due to gate leakage through the PMOS, as inverter output goes to logic high) and the input voltage reduced from 0.8 to 0.58 V (due to gate leakage through the NMOS, as inverter output goes to logic low). Furthermore, in Arumi et al. (2008a) an experiment is conducted using 0.18-μm technology with an open defect. It is shown that an interconnect open initially set to behave as stuck-at 1 [using (8.1) and procedure described above to set a particular logic value on an interconnect] changes to stuck-at 0 in approx. 2 s, due to *gate tunneling leakage* currents. Voltage behavior of the floating net is shown in Fig. 8.13. It is therefore concluded that for nanometer CMOS, gate tunneling leakage is a dominant player in setting the voltage on the floating net and the final steady-state value is independent of the initial state. Furthermore, it is predicted that the time period to reach the steady state will reduce in future technologies and will be in the order of hundreds of μs.

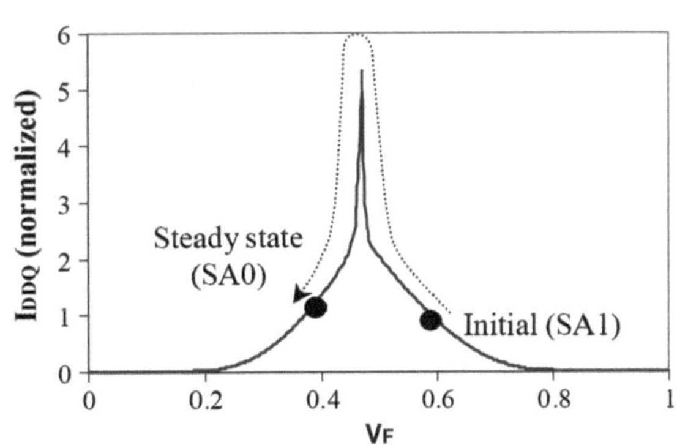

Fig. 8.13 Change in logic value due to gate tunneling leakage (Arumi et al. 2008a)

8.3.2 Testing Resistive Open Defect

This section summarizes recent research on test techniques for resistive interconnect open defect and the impact of voltage setting on their testability. Resistive open can be modeled as a resistor between two unconnected nodes, since it shows small inductive/capacitive component, which can be neglected for simplicity as used in Kruseman and Heiligers (2006) and Zain Ali et al. (2006). Figure 8.14 shows a typical *resistive open fault model*, where "D" and "S" represent the driver and successor gate, respectively.

Resistive open shows timing-dependent effects and therefore should be tested using delay tests. Delay fault testing is used to catch defects that create additional than expected delay and thereby cause a malfunction of the IC (Kruseman and Heiligers 2006). Using delay fault testing, a defect is detectable only when it causes longer delay than that of the longest path in a fault-free design. It was shown in Kruseman et al. (2004) that majority of tested paths show less than one-third delay in comparison to that of the longest path. Therefore, a defect in any of these shorter paths can only be detected if it causes higher delay than that of the longest path in the design.

In Kruseman and Heiligers (2006), the optimal test conditions for testing resistive open is analyzed for nonspeed-binned ICs, which are designed to meet timing under worst process and working conditions and typically have a logic depth of 30–70 gates. It is argued that for designs operating at few hundred MHz, one can expect to detect defects with resistance of $100\,k\Omega$ or more, while delay caused by smaller resistance defects are of the order of gate delays and does not cause additional delay even if they occur at the longest path. The paper analyzes two major sources of open defects, i.e., incompletely filled vias and partial breaks in the poly of the transistor (due to salicidation). Furthermore, it is argued that resistive open shows better detectability on silicon at elevated Vdd settings. This phenomenon is elaborated using two examples shown in Figs. 8.15 and 8.16 and discussed next. Figure 8.15 shows

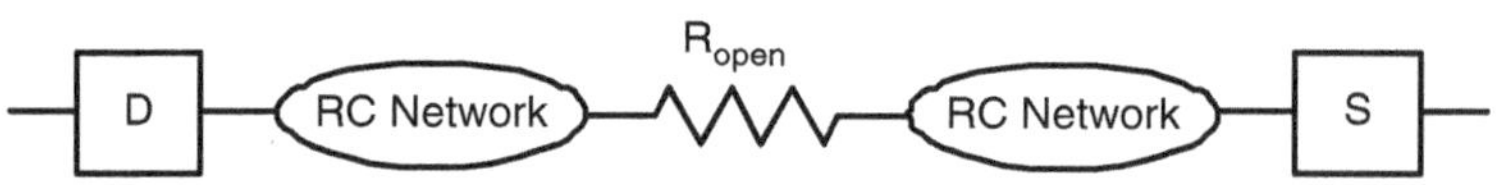

Fig. 8.14 Circuit model of resistive open defect

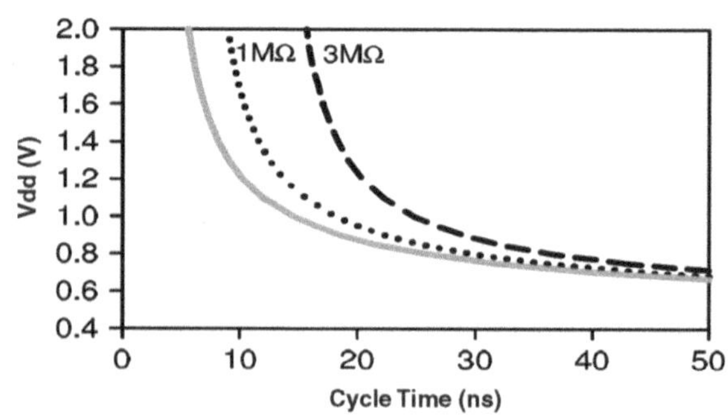

Fig. 8.15 Comparison of path delays due to resistive open defect in the longest path at different supply voltage settings. *Solid gray line* shows the fault-free design, while *dotted* and *dashed lines* show path delays using $1\,M\Omega$ and $3\,M\Omega$ in the longest path (Kruseman and Heiligers 2006)

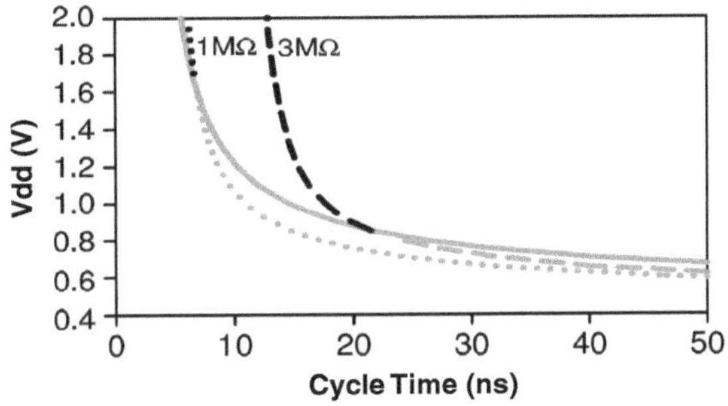

Fig. 8.16 Comparison of path delays due to resistive open defect in a short path at different supply voltage settings. The longest path is shown by a *solid gray line* (for the fault-free design), while *dotted* and *dashed lines* show path delays using 1-MΩ and 3-MΩ resistances in a shorter path (Kruseman and Heiligers 2006)

the delay caused by two different resistive opens (due to 1 MΩ and 3 MΩ) while considering these defects in the longest path and using different supply voltage settings (1.8 V being nominal supply voltage). The figure also shows the delay of the longest path in fault-free design (using solid gray line) and at various voltage settings. As can be seen, the defect-induced extra delay added to the expected delay is highest at elevated supply voltage (Vdd = 2.0 V) for both resistive open defects. Also, as expected, higher delay is observed at 3 MΩ than 1 MΩ. Figure 8.16 shows the effect of resistive open in a shorter path, with half the delay as the longest path in a fault-free design. Defects with same resistance values as Fig. 8.15 are inserted in the shorter path, and the delay is compared with that of the longest path (shown by solid gray line). As can be seen, delay due to 1-MΩ resistance show marginal detectability only at elevated Vdd setting (2.0 V), by causing higher delay than that of the longest path. It becomes undetectable at lower Vdd settings, as it shows lesser delay than that of the longest path. On the other hand, 3-MΩ defect resistance is best detectable at elevated Vdd (2.0 V) and becomes undetectable as Vdd setting is reduced further from 0.9 V. The behavior shown by these two examples (illustrated by Figs. 8.15 and 8.16) is commonly observed on silicon and is generalized using Fig. 8.17. As can be seen from Fig. 8.17, resistive open in general show better detectability at elevated Vdd setting and becomes undetectable at reduced Vdd. Finally, Kruseman and Heiligers (2006) show some cases where resistive open defects are better detectable at reduced Vdd setting.

Zain Ali et al. (2006) have also studied delay behavior for devices operating at multi-Vdd settings. Two types of defects are examined, i.e., transmission gate open and resistive open. Experiments are conducted using 0.35-μm using five (3.3, 3.0, 2.7, 2.5, and 2.0 V) discrete voltage settings on a four-level carry save adder (shown in Fig. 8.18). Each unit of carry save adder (for e.g., CSA-01) is made up of five transmission gates. The impact of transmission gate open is studied first, by inserting two NMOS open defects (one at a time) as shown in Fig. 8.18 (marked as "Fault A" and "Fault B"). The fault site and signal propagation path of inserted

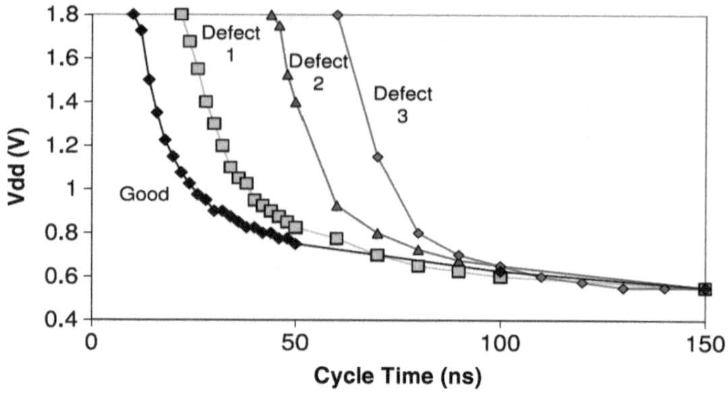

Fig. 8.17 Delay behavior of fault-free design (marked as "Good") in comparison to delay defect behavior due to three different defects (Kruseman and Heiligers 2006)

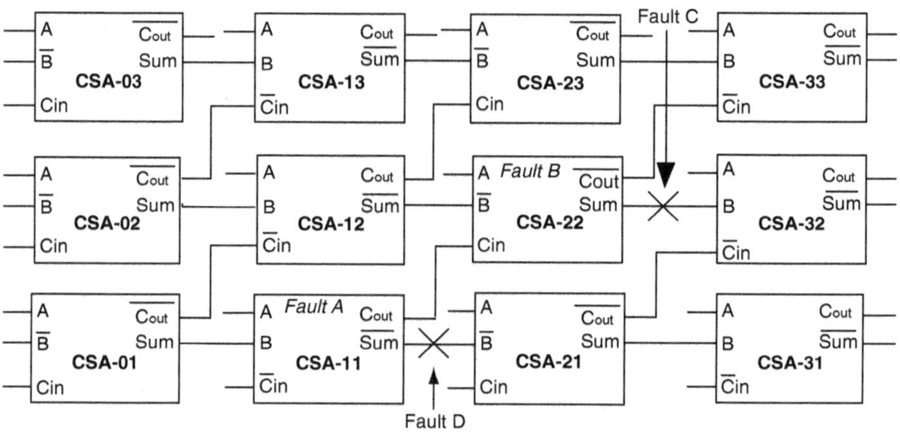

Fig. 8.18 Four-level carry-save adder, each adder cell is made of five transmission gates (Zain Ali et al. 2006)

defects is shown in Table 8.3. Gate delay ratio (GDR) and path delay ratio (PDR)[3] is calculated and results indicate that higher gate/path delay ratio is observed as Vdd setting is reduced and the two faults (transmission gate open) behave as stuck-at fault (SF) at lower Vdd settings. As expected, increased GDRs for both the faults result in higher PDRs at respective paths as well. Similar observations were reported in Chang and McCluskey (1996) using 0.6-μm and 0.8-μm technology and similar experimental setup. Study reported in Chang and McCluskey (1996) has suggested

[3] In Zain Ali et al. (2006), GDR (PDR) is calculated as a delay ratio between faulty and fault-free signal propagating gate (path) of a design.

Table 8.3 Signal propagating path for faults A and B Zain Ali et al. (2006)

Fault site		Signal propagating path
A	CSA-11 NMOS open	CSA-01(A) → CSA-11(B) → CSA-21(B) → CSA-32(Cin) → CSA-32(Cout)
B	CSA-22 NMOS open	CSA-01(A) → CSA-11(B) → CSA-22(Cin) → CSA-32(B) → CSA-32(Cout)

using $2V_t$ to $2.5V_t$ (VLV testing) for detecting defects due to transmission gate open, threshold voltage shift, and diminished-drive strength. This explains the SF behavior of transmission gate open at reduced Vdd settings.

The impact of interconnect resistive open is also studied in Zain Ali et al. (2006) by inserting two defects separately in the circuit, marked as "Fault C" and "Fault D" as shown in Fig. 8.18. For this experiment, three different resistance values (25 kΩ, 250 kΩ, and 1 MΩ) are used on both locations and results show that PDR increases with higher Vdd setting due to these two faults. As expected, PDR is more prominent for 1-MΩ resistance at elevated Vdd setting than the other two resistance values. These findings show that interconnect resistive opens are better detectable at elevated Vdd setting by delay test techniques. On the other hand, transmission gate opens are better detectable at lower Vdd settings. The application of delay test at single-Vdd setting reduces test cost by avoiding repetitive tests at other Vdd settings.

8.4 DFT for Low-Power Design

Sections 8.2 and 8.3 outlined test techniques for resistive bridge and resistive open for multiple-voltage designs. In this section, we summarize recent low-cost scan techniques for reducing power dissipation during test mode (Nicolici and Al-Hashimi 2003). These techniques are developed for devices employing multiple-voltage settings.

8.4.1 Multivoltage-Aware Scan

Designs that employ multiple voltage settings are divided into various voltage domains during physical placement of the design. Each voltage domain feeds various logic blocks and *level shifters* are used to communicate logic values across logic blocks operating under different voltage settings (Shi and Kapur 2004). The insertion of scan chains across logic block poses a challenge for scan chain ordering in multiple voltage designs due to two main reasons. First, it is desirable to reduce the number of level shifters required to transmit voltage levels from one scan chain to another, placed across different voltage domains. Second, power consumption during test can be reduced by fewer voltage domain crossing by the scan cells.

These challenges are met by multivoltage-aware scan cell ordering (Colle et al. 2005). The proposed methodology arranges scan cells based on respective voltage domains. This is achieved by scan cells ordering in such a way that scan cells operating under the same voltage levels are connected together. This in turn minimizes the number of level shifters that are otherwise required if scan cells are ordered without consideration of multivoltage designs. Furthermore, it reduces power dissipation by minimizing signal transmission in fewer voltage domain crossing. Experiments are conducted using industrial design with four voltage domains and it is shown that multivoltage-aware scan chain ordering shows 93% reduction in the number of level shifters, in comparison to scan chain ordering technique, which connects physically closer scan cells without considering its operating voltage. The proposed scheme has been implemented in Synopsys EDA tools and the DFT flow is shown in Fig. 8.19. As can be seen, DFT compiler recognizes the voltage/power domains and clusters the scan chains within the respective domains. The number of level shifters in the design are minimized by disabling voltage/power domain mixing, which is managed by "set scan configuration."

Recently a power-aware scan chain method is presented in Chickermane et al. (2008) for multi-Vdd designs. The method is implemented using Daisy-chaining scan approach to efficiently utilize expensive tester resources (bandwidth) and reduce test cost. The method avoids signal integrity issues during test by employing bypass multiplexers, which allows bypassing signals from power domains

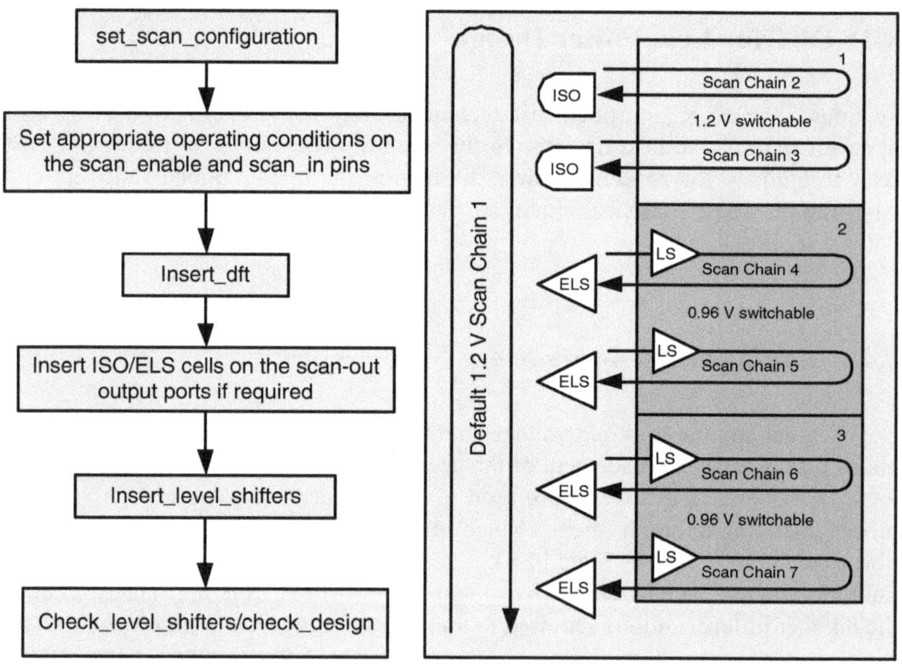

Fig. 8.19 DFT synthesis flow for multi-Vdd design using Synopsys design compiler (Baby and Sarathi 2008)

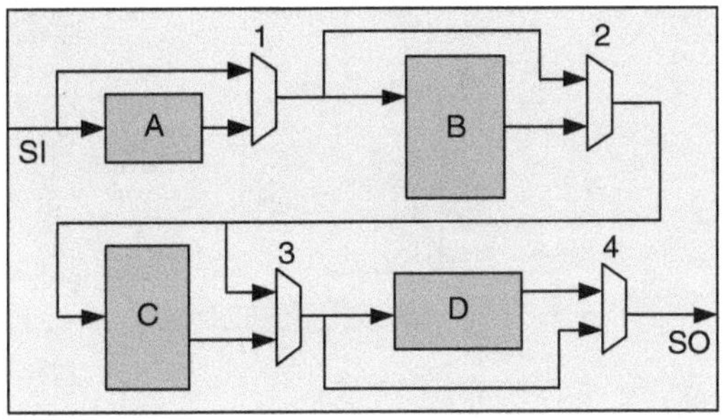

Fig. 8.20 Power-aware Daisy-chaining scan path (Chickermane et al. 2008)

that are switched off during test. Daisy-chain implementation along with bypass multiplexers (1, 2, 3, and 4) and four different power domains (A, B, C, and D) is shown in Fig. 8.20. As can be seen, bypass multiplexers allow testing of specific power domains in multi-Vdd environment. As an example, in a particular power mode, where power domains C and D are ON, while A and B are OFF, muxes 1 and 2 goes in bypass mode, while 3 and 4 are in pass-thru mode. This forms a scan chain between SI, 3, 4, and SO. The bypass multiplexers are placed on always-on power domain. This approach is implemented in Cadence *Encounter*^{TM} test tools.

8.4.2 Power-Managed Scan Using Adaptive Voltage Scaling

Reducing power dissipation during test has been an active area of research for nearly a decade and numerous techniques have been reported (Girard 2002; Bhunia et al. 2005). Recently an interesting technique that reduces both dynamic and leakage power during test through the use of adaptive voltage scaling Power Managed Scan (PMScan) has been reported (Devanathan et al. 2007). The presented methodology is motivated by three factors. First, it is known that dynamic power is proportional to V^2 (Weste and Eshraghian 1994) and gate leakage power is proportional to V^4 (Krishnarnurthy et al. 2002), where V is the operating voltage of the device. Therefore, reduction in supply voltage can significantly reduce total power (dynamic plus leakage) during test. Second, infrastructure for adaptive voltage scaling is widely deployed in modern microprocessors to reduce power consumption during functional mode. Therefore, it is suggested in Devanathan et al. (2007) to reuse voltage scaling infrastructure to reduce implementation (due to physical design and area) overheads. Third, scan-shift frequency is usually much slower than the operational frequency of the device, therefore scan-shift operation is ideal for voltage scaling during test.[4] Therefore, PMScan proposes voltage scaling during test to provide a

[4] Voltage scaling is widely used to reduce power consumption, while ensuring that timing requirements are met. It is therefore more effective for tasks that are less computationally intensive, i.e., tasks that can be completed at a slower speed.

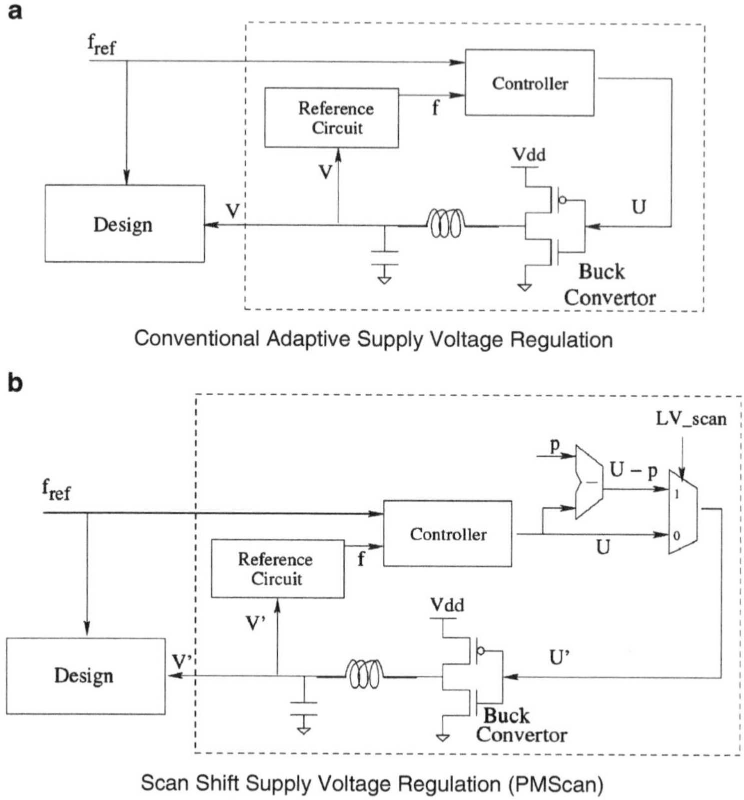

Fig. 8.21 Block diagram of adaptive supply voltage regulation in (**a**) conventional design, (**b**) PMScan (Devanathan et al. 2007)

trade-off between test application time and test power. This is achieved by modifying voltage regulation circuitry (used for adaptive voltage scaling) such that scan-shift operation meets acceptable timing, while supply voltage during scan shift is reduced. The voltage regulation circuitry changes the supply voltage to nominal during scan capture mode to ensure at-speed testing.

The conventional voltage scaling circuitry and the one proposed in Devanathan et al. (2007) are shown in Fig. 8.21. Figure 8.21a shows the conventional adaptive supply voltage circuitry showing the voltage regulation component in the dashed box. It uses feedback control and adjusts the supply voltage "V" using a DC–DC converter such that the delay of the circuit fits in one clock cycle of the desired clock frequency f_{ref}, which is usually generated using on-chip PLL. The reference circuit is made of a ring oscillator and determines the maximum delay of the design over process, voltage, and temperature variations. It determines the maximum frequency "f" corresponding to the voltage "V" provided to it. In Devanathan et al. (2007), the conventional voltage regulation design is modified for voltage scaling during scan-shift operation, as shown in Fig. 8.21b. It is designed such that when the signal

LV_scan = 1, the supply voltage "V" is lowered by "p." On the other hand when LV_scan = 0, the output "U" is applied to the multiplexer as in conventional design. Refer to Devanathan et al. (2007) for more details on design of such regulator.

Experiments are conducted using 90-nm library with nominal 1.1-V supply voltage using Synopsys *PrimePower*^{TM} for power analysis. The first experiment is conducted using seven different ISCAS 89 benchmarks using reduced Vdd (0.77 V) and at 25-MHz scan-shift frequency. Average dynamic, peak dynamic, and leakage power are compared between proposed PMScan technique with that of conventional scan (unaware of voltage scaling). It is shown that on average, PMScan reduces average dynamic power by about 44%, peak dynamic by 42%, leakage power by 91% contributing to overall total power by 64% in comparison with conventional scan. Moreover, it is shown that these results can be further improved by 5%, by using NOR-Gating scheme (Girard 2002)[5] along with PMScan. The second experiment analyzes test time and test power trade-off. It is conducted using an industrial design (with 9 million gates and 7 unwrapped cores), at three different voltage (1.1 V, 1.0 V, and 0.77 V) and scan-shift frequency (25 MHz, 75 MHz, and 125 MHz) settings. It is shown that for test application at 0.77 V and 125-MHz scan-shift frequency, test time reduces by 80%, while total power increases by 16%, in comparison with test application at 0.77 V with 25-MHz scan-shift frequency.

Another effective technique for reducing leakage power is by employing *state retention logic* (Keating et al. 2007). Recently a method to test state retention logic is proposed in Chakravadhanula et al. (2008). State retention logic is tested by scanning in test patterns, followed by powering down the logic block containing state retention logic, and then powering up again. This is followed by scanning out the test patterns, and is matched against the scanned in data for coherency.

8.5 Open Research Problems

Low-power design techniques present potential challenges to test and reliability of digital designs. At present there are continuing research efforts worldwide focusing on addressing these challenges. In the following three research problems are highlighted that need to be addressed to generate high-quality and cost-effective test solutions for reliable low-power designs.

8.5.1 Impact of Voltage and Process Variation on Test Quality

Previous sections have examined the impact of power supply variation on the behavior of manufacturing defects. It appears that test quality is also compromised

[5] NOR gate is used to halt unnecessary toggling of combinational logic (fed by scan flip-flop) during scan-shift operation.

due to another type of variation, i.e., due to fabrication process. While the impact of process variation on timing and power performance has been extensively investigated in the literature (Bhunia et al. 2007), its effect on test quality is an emerging area of research. In this section, we summarize two recent studies that take process variation into account using static and delay test techniques and motivate the need for joint *voltage and process variation* test.

In Ingelsson et al. (2008) and Ingelsson (2009), the impact of process variation on static test quality has been investigated for resistive bridge. It is shown that process variation has a negative impact on test quality of such defects leading to test escapes. A robustness matrix is developed to quantize the impact of process variation on test quality and a test generation method is developed to mitigate the impact of process variation and reduce test escapes. Experiments are conducted using ISCAS-85' and 89' benchmarks and synthesized using 45-nm CMOS technology. Results show that test generation method covers up to 18% more process-variation-induced logic faults than tests generated without consideration of process variation. In Lu et al. (2005), the influence of process variation on the longest path of the design has been investigated, while considering structural elements of the design (logic elements and interconnects). The method aims to reduce test cost without compromising on test quality, i.e., fault coverage. This is achieved by identifying minimum number of longest path candidates in polynomial time. Experiments conducted on ISCAS-85' and 89' circuits show that the number of testable paths are up to 6% of those found by Tani et al. (1998). In addition it is 300–3,000 times faster than the method proposed in Tani et al. (1998).

High-quality test for next generation multi-Vdd devices require improved static and delay test techniques capable of mitigating the impact of power supply and fabrication process variation. Such test techniques will need to be developed that will require realistic fault models, for both resistive bridge and resistive open, that mimic actual behavior at the physical level in the presence of voltage and process variation. Such fault models will be used for voltage and process variation aware test generation leading to higher test quality and therefore improve in-field product reliability of future multi-Vdd devices.

8.5.2 Diagnosis for Multivoltage Designs

Diagnosis is a systematic way to uniquely identify the defect causing malfunction in the circuit. It is critical to silicon debugging, yield analysis, and for improving subsequent manufacturing cycle. Recently diagnosis procedure for resistive bridge is investigated in Khursheed et al. (2009b) for ICs employing multiple-voltage setting. The diagnosis procedure (Khursheed et al. 2009b) is based on cause–effect diagnosis scheme (Abramovici et al. 1998) using a pass/fail dictionary (Pomeranz and Reddy 1992) to minimize memory storage. The proposed diagnosis algorithm combines information of resistance interval detection at all voltage settings and achieves overall higher diagnosis accuracy. Experiments are conducted using parametric fault model (Renovell et al. 1996), and ISCAS-85' and 89' benchmarks are synthesized on 120-nm technology. Experimental results show that the lowest Vdd setting achieves highest diagnosis accuracy for single-Vdd diagnosis, which is improved up to 38% by using multi-Vdd diagnosis. Furthermore, it establishes that multi-Vdd diagnosis is more effective for resistive bridge than for hard-shorts (bridge with 0-Ω resistance).

It is expected that future diagnosis strategies will need to employ process-variation-aware fault models to accurately diagnose resistive bridge and resistive open defects. Thereby accounting for test escapes due to process variation in nanometer CMOS and provide accurate diagnosis to DSM designs.

8.5.3 Voltage Scaling for Nanoscale SRAM

The above two open problems are related to test for low-power devices. Recent research indicates that low-power design also affects reliability of the device. One such work that determines optimal voltage setting to operate SRAMs in the presence of soft errors and gate-oxide degradation is presented in Chandra and Aitken (2009). Nanoscale SRAMs are vulnerable to soft errors and suffer from progressive gate-oxide degradation. Soft errors are faults induced by particle hit (alpha particle or neutrons), which can flip the stored data bit. These events are called single event upsets (SEU) and requires data content to be rewritten. SRAMs are especially vulnerable to SEU due to small node capacitance and small bit cell size.[6] On the other hand, gate oxide thickness is continuously decreasing with technology scaling in CMOS devices, which has resulted in increased gate tunneling currents. Increased gate tunneling currents result in progressive degradation of gate oxide, which is one of the most important reliability concern in current and future technologies. In Chandra and Aitken (2009), the optimal voltage setting to operate nanoscale SRAM in the presence of soft errors is investigated. This work has shown following three findings: For a given technology node (65 nm or 45 nm), higher voltage level results in higher immunity of SRAM cells against soft errors in the absence of gate-oxide degradation. On the other hand, gate tunneling currents increase with the increase in supply voltage, which in turn contributes to gate-oxide degradation. Therefore, an optimal voltage is formulated by an equation for operating nanoscale SRAMs in the presence of gate-oxide degradation and soft errors. The optimal voltage reduces with increasing level of gate-oxide degradation for nanoscale SRAMs.

It is expected that analytical models will be developed to achieve highest immunity against soft-errors for a given voltage setting value and gate-oxide degradation level, thereby improving reliability of nanoscale SRAMs in future technologies.

---

[6] Refer to Baumann (2005) for further reading on the effect of technology scaling and soft errors on memory and logic components of the circuit.

8.6 Summary and Conclusions

This chapter has presented an overview of recently reported research in testing strategies for multivoltage designs. Such strategies aim to reduce test cost and improve defect coverage of Vdd-dependent defects. The cost reduction has been obtained by using the least number (i.e., one) of voltage test setting for Vdd-dependent defects (resistive bridge and resistive open) by avoiding repetitive tests at several Vdd settings. For resistive bridge, the cost reduction is achieved by TPI and more recently by GS, which achieves 100% defect coverage at a single (lowest) test voltage. For resistive or full-open interconnect defect, elevated Vdd setting achieves better detectability using delay test and therefore repetitive tests at other voltage settings can be avoided.

Low-cost scan for multivoltage design is possible through various techniques. Some techniques focus on reducing implementation cost of scan chains in multivoltage environment through clustering scan chains according to their respective voltage domain thereby reducing the number of level shifters and also by employing power-aware scan that efficiently utilize expensive tester resources (bandwidth) and reduce test cost. Other technique achieves low-power test for multivoltage devices by reusing the existing functional infrastructure for voltage scaling to reduce power consumption leading to reduced cost. The chapter also outlines a number of worthy research problems that need to be addressed to develop high-quality and cost-effective test solutions for reliable low-power devices.

Acknowledgements The authors are thankful to Dr. Ilia Polian (Albert-Ludwigs-University of Freiburg) for useful comments and EPSRC (UK) for supporting this work under Grant EP/DO57663/1.

References

Abramovici M, Breuer MA, Friedman AD (1998) Digital systems testing and testable designs. IEEE, Piscataway, NJ

Arumi D, Rodriguez-Montaes R, Figueras J, Eichenberger S, Hora C, Kruseman B (2008a) Full open defects in nanometric CMOS. In: Proceedings of the VLSI test symposium, May 2008, pp 119–124

Arumi D, Rodriguez-Montanes R, Figueras J (2008b) Experimental characterization of CMOS interconnect open defects. IEEE Trans Comput Aided Des 27(1):123–136

Baby M, Sarathi V (2008) Advanced DFT implementation. (http://www.synopsys.com/news/pubs/insight/2008/art2_dftimplem_v3s4.html)

Baumann R (2005) Soft errors in advanced computer systems. IEEE Des Test Comput 22(3): 258–266

Bhunia S, Mahmoodi H, Ghosh D, Mukhopadhyay S, Roy K (2005) Low-power scan design using first-level supply gating. IEEE Trans VLSI Syst 13(3):384–395

Bhunia S, Mukhopadhyay S, Roy K (2007) Process variations and process-tolerant design. In: Proceedings of the international conference on VLSI design, Jan. 2007, pp 699–704

Chakravadhanula K, Chickermane V, Keller B, Gallagher P, Gregor S (2008) Test generation for state retention logic. In: Proceedings of the Asian test symposium, Nov. 2008, pp 237–242

Chandra V, Aitken R (2009) Impact of voltage scaling on nanoscale SRAM reliability. In: Proceedings of the design, automation and test in Europe (DATE) conference, April 2009
Chang JT-Y, McCluskey EJ (1996) Detecting delay flaws by very-low-voltage testing. In: Proceedings of the international test conference, Oct. 1996, pp 367–376
Chen G, Reddy S, Pomeranz I, Rajski J, Engelke P, Becker B (2005) An unified fault model and test generation procedure for interconnect open and bridges. In: Proceedings of the European test symposium, May 2005, pp 22–27
Chickermane V, Gallagher P, Sage J, Yuan P, Chakravadhanula K (2008) A power-aware test methodology for multi-supply multi-voltage designs, In: Proceedings of the international test conference, Oct. 2008, pp 1–10
Choudhury U, Sangiovanni-Vincentelli A (1995) Automatic generation of analytical models for interconnect capacitances. IEEE Trans Comput Aided Des 14(4):470–480
Colle AD, Ramnath S, Hirech M, Chebiyam S (2005) Power and design for test: a design automation perspective. J Low Power Electron 1(1):73–84
Delgado A (2008) Enhancement of defect diagnosis based on the analysis of CMOS DUT behaviour. PhD Thesis, July 2008
Devanathan VR, Ravikumar CP, Mehrotra R, Kamakoti V (2007) PMScan: A power-managed scan for simultaneous reduction of dynamic and leakage power during scan test. In: Proceedings of the international test conference, Oct. 2007, pp 1–9
Engelke P, Polian I, Renovell M, Seshadri B, Becker B (2004) The pros and cons of very-low-voltage testing: an analysis based on resistive bridging faults. In: Proceedings of the VLSI test symposium, April 2004, pp 171–178
Engelke P, Polian I, Renovell M, Becker B (2006a) Automatic test pattern generation for resistive bridging faults. J Electron Test Theory Appl 22(1):61–69
Engelke P, Polian I, Renovell M, Becker B (2006b) Simulating resistive bridging and stuck-at faults. IEEE Trans. Comput Aided Des 25(10):2181–2192
Girard P (2002) Survey of low-power testing of VLSI circuits. IEEE Des Test Comput 19(3):80–90
Gomez R, Giron A, Champac V (2005) Test of interconnection opens considering coupling signals. In: Proceedings of the international symposium on defect and fault tolerance in VLSI systems, Oct. 2005, pp 247–255
Guindi RS, Najm FN (2003) Design techniques for gate-leakage reduction in CMOS circuits, In: Proceedings of the international symposium on quality electronic design, March 2003, pp 61–65
Hamada M, Takahashi M, Arakida H, Chiba A, Terazawa T, Ishikawa T, Kanazawa M, Igarashi M, Usami K, Kuroda T (1998) A top-down low power design technique using clustered voltage scaling with variable supply-voltage scheme. In: Proceedings of the custom integrated circuits conference, May 1998, pp 495–498
Hao H, McCluskey EJ (1993) Very-low-voltage testing for weak CMOS logic ICs. In: Proceedings of the international test conference, Oct. 1993, pp 275–284
Henderson CL, Soden JM, Hawkins CF (1991) The behavior and testing implications of CMOS IC logic gate open circuits, In: Proceedings of the international test conference, Oct. 1991, pp 302–310
Ingelsson U (2009) Investigation into voltage and process variation-aware manufacturing test, PhD Thesis, University of Southampton
Ingelsson U, Rosinger P, Khursheed SS, Al-Hashimi BM, Harrod P (2007) Resistive bridging faults DFT with adaptive power management awareness. In: Proceedings of the Asian test symposium, Oct. 2007, pp 101–106
Ingelsson U, Al-Hashimi BM, Harrod P (2008) Variation aware analysis of bridging fault testing. In: Proceedings of the Asian test symposium, Nov. 2008, pp 206–211
Johnson S (1994) Residual charge on the faulty floating gate CMOS transistor. In: Proceedings of the international test conference, Oct. 1994, pp 555–561
Keating M, Flynn D, Aitken R, Gibbons A, Shi K (2007) Low power methodology manual: for system-on-chip design. Springer, New York

Khursheed S, Ingelsson U, Rosinger P, Al-Hashimi BM, Harrod P (2008) Bridging fault test method with adaptive power management awareness. In: IEEE Trans. Comput Aided Des 27(6):1117–1127

Khursheed S, Al-Hashimi BM, Harrod P (2009a) Test cost reduction for multiple-voltage designs with bridge defects through gate sizing. In: Proceedings of the design, automation and test in Europe (date) conference, April 2009

Khursheed S, Al-Hashimi BM, Reddy SM, Harrod P (2009b) Diagnosis of multiple-voltage design with bridge defect. IEEE Trans Comput Aided Des 28(3):406–416

Krishnarnurthy RK, Alvandpour A, De V, Borkar S (2002) High-performance and low-power challenges for sub-70 nm microprocessor circuits. In: Proceedings of the custom integrated circuits conference, May 2002, pp 125–128

Kruseman B, Heiligers M (2006) On test conditions for the detection of open defects. In: Proceedings of the design, automation and test in Europe (date) conference, March 2006, pp 896–901

Kruseman B, Majhi AK, Gronthoud G, Eichenberger S (2004) On hazard-free patterns for fine-delay fault testing. In: Proceedings of the international test conference, Oct. 2004, pp 213–222

Kundu S, Zachariah ST, Sengupta S, Galivanche R (2001) Test challenges in nanometer technologies. J Electron Test Theory Appl 17(3–4):209–218

Lee S, Sakurai T (2000) Run-time voltage hopping for low-power real-time systems. In: Proceedings of the design automation conference, June 2000, pp 806–809

Lo S-H, Buchanan DA, Taur Y, Wang W (1997) Quantum-mechanical modeling of electron tunneling current from the inversion layer of ultra-thin-oxide NMOSFET's. IEEE Electron Device Lett 18(5):209–211

Lu X, Li Z, Qiu W, Walker DMH, Shi W (2005) Longest path selection for delay test under process variation. IEEE Trans Comput Aided Des 24(12):1924–1929

Maeda T, Kinoshita K (2000) Precise test generation for resistive bridging faults of CMOS combinational circuits. In: Proceedings of the international test conference, Oct. 2000, pp 510–519

Montanes RR, Bruis EMJG, Figueras J (1992) Bridging defects resistance measurements in a CMOS process. Proceedings of the international test conference, Sept. 1992, pp 892–899

Montanes RR, de Gyvez JP, Volf P (2002) Resistance characterization for weak open defects. IEEE Des Test Comput 19(5):18–26

Montanes RR, Arumi D, Figueras J, Einchenberger S, Hora C, Kruseman B, Lousberg M, Majhi AK (2007) Diagnosis of full open defects in interconnecting lines. In: Proceedings of the VLSI test symposium, May 2007, pp 158–166

Nicolici N, Al-Hashimi BM (2003) Power-constrained testing of VLSI circuits. Kluwer, Dordrecht

Pomeranz I, Reddy SM (1992) On the generation of small dictionaries for fault location. In: Proceedings of the international conference on computer-aided design (ICCAD), Nov. 1992, pp 272–279

Rafiq S, Ivanov A, Tabatabaei S, Renovell M (1998) Testing for floating gates defects in CMOS circuits. In: Proceedings of the Asian test symposium, Dec. 1998, pp 228–236

Renovell M, Huc P, Bertrand Y (1996) Bridging fault coverage improvement by power supply control. In: Proceedings of the VLSI test symposium, April 1996, pp 338–343

Sar-Dessai VR, Walker DMH (1999) Resistive bridge fault modeling, simulation and test generation. In: Proceedings of the international test conference, Sept. 1999, pp 596–605

Shi C, Kapur R (2004) How power-aware test improves reliability and yield. http://www.eedesign.com/showArticlejhtml?articleID=47208594

Shinogi T, Kanbayashi T, Yoshikawa T, Tsuruoka S, Hayashi T (2001) Faulty resistance sectioning technique for resistive bridging fault ATPG systems. In: Proceedings of the Asian test symposium, Nov. 2001, pp 76–81

Spinner S, Polian I, Engelke P, Becker B, Keim M, Cheng WT (2008) Automatic test pattern generation for interconnect open defects. In: Proceedings of the VLSI test symposium, May 2008, pp 181–186

Sreedhar A, Sanyal A, Kundu S (2008) On modeling and testing of lithography related open faults in nano-CMOS circuits. In: Proceedings of the design, automation and test in Europe (date) conference, March 2008, pp 616–621

Tani S, Teramoto M, Fukazawa T, Matsuhiro K (1998) Efficient path selection for delay testing based on partial path evaluation. In: Proceedings of the VLSI test symposium, April 1998, pp 188–193

Weste NHE, Eshraghian K (1994) Principles of CMOS VLSI design: a systems perspective. Addison-Wesley, Reading, MA

Zain Ali NB, Zwolinski M, Al-Hashimi BM, Harrod P (2006) Dynamic voltage scaling aware delay fault testing. In: Proceedings of the European test symposium, May 2006, pp 15–20

Zou W, Cheng WT, Reddy SM (2006) Interconnect open defect diagnosis with physical information. In: Proceedings of the Asian test symposium, Nov. 2006, pp 203–209

Chapter 9
Test Strategies for Gated Clock Designs

Brion Keller and Krishna Chakravadhanula

Abstract One of the ways often used to design for low-power consumption during functional operation in CMOS devices is to gate off clocks to areas of logic not needed for the current state of operation. By gating off clocks to state elements that are known to not need updating, the dynamic switching current can be reduced compared with allowing state elements to update when you don't care what they contain. When clocks are gated, some amount of DFT is necessary to ensure ATPG can be used to create meaningful tests. This chapter describes some of the DFT approaches that can be applied so ATPG can deal with gated clocks. In addition, this chapter explores ways in which functional clock gating may be exploited to help reduce power during test.

9.1 Introduction

Functional use of clock gating has been utilized in sequential logic designs for decades. There exist numerous reasons for **gating of clock signals** within a design; however, for more than a decade, the ability to gate off clocks has been exploited as a means to reduce the **active logic switching** and thus reduce the **dynamic power consumption** of various CMOS devices (Benini et al. 1994; Nicolici and Wen 2007). By gating off the clock to state elements that are not actively participating in the current functional state operation, those state elements will not change even though functionally it may not matter whether they change value. If these state elements do not switch, then the logic they feed to will also not switch. In designs where only a small to modest fraction of the state elements may need to update for various functional operations, it may be possible to significantly reduce the active (or dynamic) power consumption of the device by gating off the clocks to areas and functional units that do not require being updated.

B. Keller (✉) and K. Chakravadhanula
Cadence Design Systems Inc., Endicott, NY, USA
e-mail: bkeller@stny.rr.com

P. Girard et al. (eds.), *Power-Aware Testing and Test Strategies for Low Power Devices*,
DOI 10.1007/978-1-4419-0928-2_9, © Springer Science+Business Media, LLC 2010

The use of clock gating to reduce dynamic power consumption has grown as we have seen the explosion in the use of battery-powered consumer electronics. Some battery-powered devices consume full power while turned on and little to no power while turned off. Other devices consume less power in certain modes of operation. For example, most cell phones today consume substantially more power while being actively used for a call than when they are simply monitoring for a call to be received. To achieve such a lower power operation in "standby" mode, it is clear that certain large units can have power shut off (e.g., the display screens and camera sensor). Shutting off power to large units is possible when there are **power shut-off switches** designed in for these units. Generally, power switches are designed for controlling power to large units that are either active or not based on relatively high level operating modes of the system. Power switches provide better power control than clock gating because power switches stop both the active and static/leakage power consumption of the affected logic (Chickermane et al. 2008). Because power switches are more complex to control and utilize in a design, they tend to be relegated to the high-level power mode controls while clock gating is used at lower levels of control; however, as more recent technologies show greatly higher quiescent (static) current drain, power shut-off switches may get utilized in ever lower levels with a finer granularity of power control. Most likely there will continue to be a mixture of power switches and clock gating used in future designs to help keep power consumption under control.

In the past, many logic designers tried to avoid using gating logic in the clock signal path because it could have an adverse impact on the clock skew that is so critically important to control within edge-sensitive designs. To get the same behavior without gating the clock signals, designers have used data path multiplexors (MUXes) such that when the state elements should not update, their current state is selected to be fed back to their data input – thus when the clock arrives, the state elements maintain their current state. This use of MUXing logic is often called **data gating** (Fig. 9.1) as opposed to **clock gating** (Figs. 9.2 and 9.3). For many digital designs, a significant portion of the power is consumed in the clock trees (some estimate it to be 30–50% of the dynamic power (Donno et al. 2004; Shen et al. 2007)), so there is additional benefit from a power consumption perspective from Clock

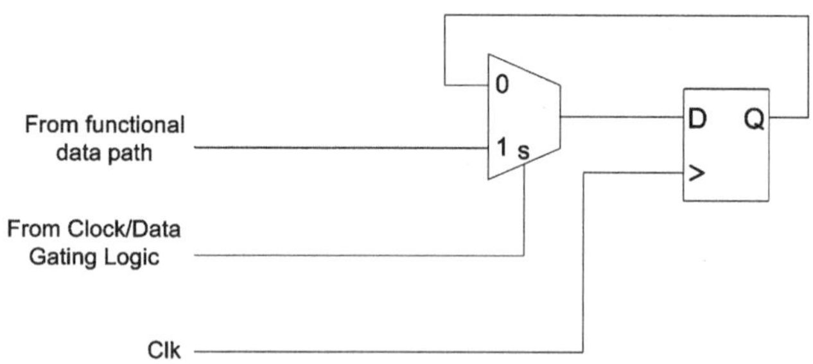

Fig. 9.1 Example of data gating implementing a "clock gating" equivalent behavior

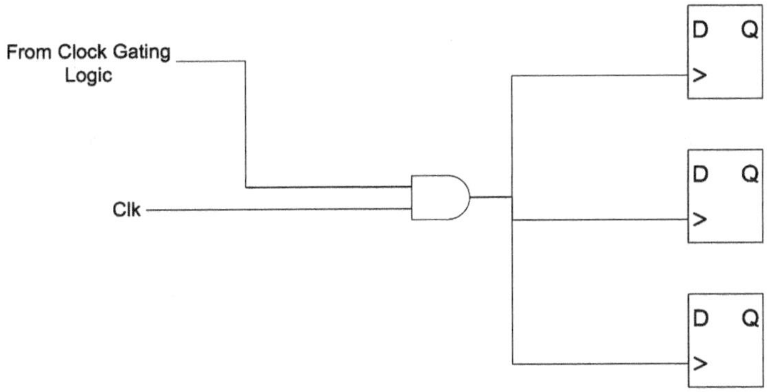

Fig. 9.2 Example of potentially glitchy clock gating

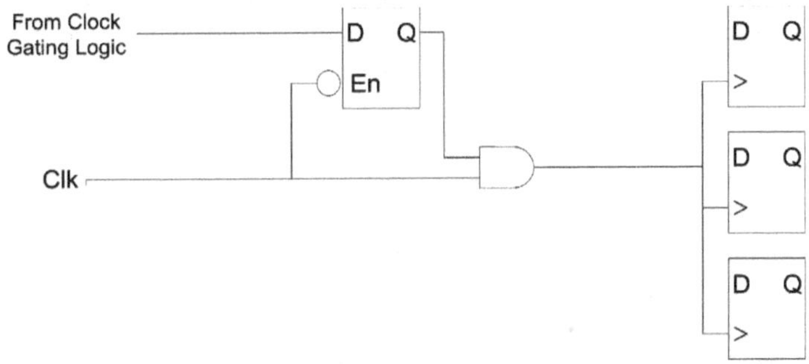

Fig. 9.3 Example of glitch-free clock gating

Gating as it additionally stops switching on the portion of the clock tree being gated off. One final advantage: clock gating can be applied at a normal fan-out point in the clock tree and all of the state elements down-stream from the clock gate will be affected; with data gating, a MUX must be inserted into the data path for each state element being affected by the gating – resulting in more total logic and thus more power consumption as well.

Even non-battery-operated devices are being made more power efficient as the world has become more energy conscious. Computer manufacturers have already begun marketing systems for their improved energy efficiency. As energy prices climb, it becomes a cost saving advantage to lower power consumption – including the costs to cool computer equipment.

All of this is leading to the high probability of having a lot of clock gating logic in future logic devices. It will become ever more important that testing be not only able to deal with clock gating, but even to take advantage of it when possible. The rest of this chapter is devoted to showing ways to deal with clock gating logic and also how to exploit it when trying to produce lower power consuming tests.

9.2 DFT for Clock Gating Logic

Functional clock gating can make it more difficult for ATPG software to create good tests. The following sections show some of the DFT techniques that can be used to help ATPG tools create high quality and efficient tests.

9.2.1 Safe Gating of Clocks in Edge Sensitive Designs

Before considering DFT for clock gating, it is useful to look into some basic ways of handling clock gating to ensure it will work well functionally. The clock gating shown in Fig. 9.2 depicts the Clk signal gated by some arbitrary logic function at an AND gate prior to driving the clock input to several rising edge flops. If the signal from the clock gating logic could possibly change on the rising edge of Clk, there will be a potential for a glitch, where initially the clock appears to get through and then is gated off. This is a poor way to implement clock gating.

To avoid the potential for **glitches** where the clock is gated, it is important to stabilize the signal feeding to the gate of the clock signal (the AND gate in Fig. 9.2). This can easily be done by inserting a D Latch (sometimes called a lock-up latch) on the gating signal and clocking that latch with the same clock that is being gated. The only requirement is to *ensure the latch updates on the phase of the clock when the clock will be at the controlling value at the gate input* to prevent glitches on the gating signal from affecting the clock signal after the gate. Figure 9.3 shows an example of safely implemented clock gating to avoid glitches. The inserted D Latch is enabled when Clk is 0 and in control at the AND gate – preventing any changes on the gating signal from causing any glitches.

In some cases it is also possible to insert a whole flop in the gating signal path instead of just a D Latch as long as the update occurs at the flop output when the clock is in control at the gate. Clock tree synthesis must account for clock gating and ensure the clock signal will get to the gate before any state element is updated by that clock. This naturally happens if the clock tree ensures the clock signal edges appear at state element clock inputs at the same time (within some tolerance for skew), including state elements that gate this same clock.

9.2.2 Edge Sensitive, MUXed Scan

Given that functional clock gating has been used for a long time, automated **synthesis of clock gating** logic is well supported by the synthesis tools currently available. Synthesis tools are sophisticated enough to handle timing exceptions during clock gating, insert clock gating with DFT, insert hierarchical clock gating, and also allow the user precise control over selection and insertion of clock gating logic. Synthesis tools can take into account timing exceptions on flops or on their pins like clock

Fig. 9.4 Example Verilog RTL that can result in a gated clock

```
input data_in;
reg   outdata;
input clk, enable;

always @ (posedge clk)
        if (enable)
            outdata  <= data_in;
```

and reset, and will synthesize different clock gate instances to flops having different exceptions (Mukherjee and Marek-Sadowska 2003; http://www.cadence.com; http://www.synopsys.com].

To enable the synthesis tool to infer clock gating structures, it is often a matter of writing the RTL in a particular way, one example of which is shown in Fig. 9.4.

The example Verilog RTL from Fig. 9.4 should result in the synthesis of a clock gate instance that gates the ***clk*** going to the flip-flop, and the ***enable*** signal is connected to the functional enable pin on the clock gate instance. The result could look similar to what is shown in Fig. 9.3, with synthesis using a clock gating cell that contains a lock-up latch on the enable signal. The inserted clock gate can be an instance of an integrated clock gate library cell, comprised of discrete library elements, or even a user-defined module. While this example shows gating a clock to a rising edge flop, in general either same or different clock gate modules can be used to drive rising and falling edge flops in the design.

In **MUXed Scan** designs, functional clocks are reused during test to scan load and unload the test data. Without some help, gated functional clocks cause the flops they are driving to not be controllable or observable during test, making the task of ATPG much harder and may lead to a loss of fault coverage. It is important that appropriate DFT be used to ensure gated clocks are controllable during scan operations.

A common DFT technique to make a gated functional clock controllable during test is to add some test control to the clock gating logic. The key is to ensure, when in the scan load/unload state, that the gated clock is forced to be enabled. This can be achieved by using the **scan enable** signal in the design to override the functional enable signal during the scan shift state. An example of this is shown in Fig. 9.5. Note that the **scan enable** signal is combined with the functional gating signal prior to the lock-up latch instead of after it as this ensures glitches on the scan enable are handled as well. If a lock-up flop is used, it may be wise to combine the scan enable with the gating signal after the lock-up; otherwise the effect of a change to the scan enable doesn't occur until after one or two edges of the clock have been applied. Without such test control logic, DFT rule checks would flag the gated clock as uncontrollable during scan, and prevent the flip-flops from being converted to scan flops.

The advantage of bypassing the clock gating using the scan enable signal is that it allows ATPG control over the clock gate during the capture operation. The scan

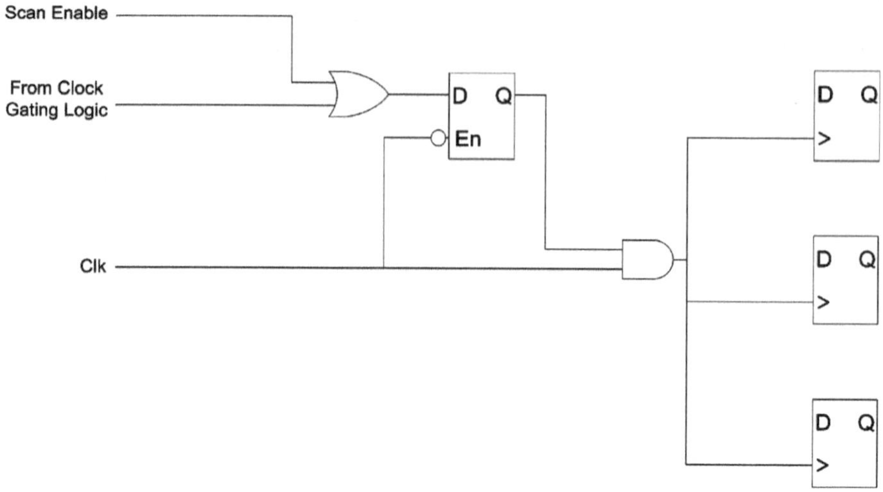

Fig. 9.5 Example showing how scan enable can be used to override clock gating

enable is active during the scan shift operation, but for most of the tests it will be at its inactive value during the capture clocking operation. It would also be possible to override the clock gating signal using a **test mode** control or constraint signal, but that would cause the clock gating logic to always be bypassed in that test mode. By not constraining ATPG to always bypass the clock gating, ATPG can then include the clock gating logic into the tests as needed. If a **test mode** (or **test enable**) signal is used to control the clock gate – as is sometimes seen – then the clock would be forced enabled at all times, possibly leading to unnecessary or excessive switching activity. As we will see later, it may be useful to allow ATPG the freedom to use the functional clock gates to help reduce switching activity during capture clocking.

If the functional clock enable (fed from the functional clock gating logic as shown in Fig. 9.5) is driven by a significant amount of logic, sometimes ATPG may not be able to generate the required value at the functional enable pin that would enable or disable the gated clock as needed. This scenario may also happen if there are constraints such that two separate clock gates cannot be turned off simultaneously. Figure 9.6 shows an example where a scannable flop, *DFT gate,* allows ATPG a simple means to turn off the gated clock without having to justify the off value on the functional clock gating logic path shown in the figure. If no flops controlled by the gate are required for detecting any faults in that test, *DFT gate* can be loaded with a logic-0 to turn off the clock. *DFT gate* is set to logic-1 for functional operation. Figure 9.7 shows a further enhancement where a *DFT Enable* flop can be used by ATPG to force the gated clock to be enabled. This can be useful if substantial effort would be required to enable the clock via the functional clock gating logic (the scan enable could be used, but that has other consequences that are usually undesirable). In this example, *DFT gate* is set to logic-1 and *DFT Enable* is set to logic-0 for functional operation.

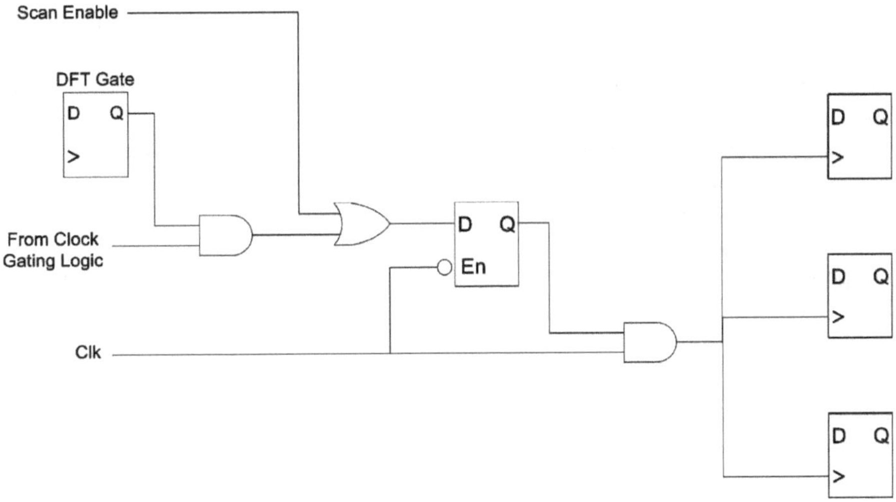

Fig. 9.6 Example of DFT to make it easy for ATPG to gate off a clock

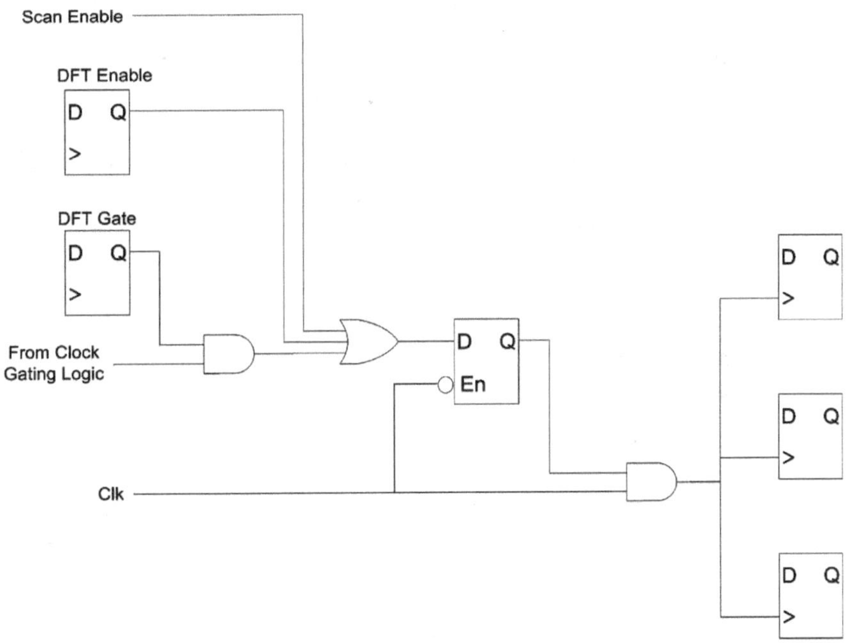

Fig. 9.7 Example DFT allowing ATPG to easily gate off or enable a gated clock

While the above-mentioned techniques have focused on improving controllability of the clock gating signal, an additional concern is the observability of the logic driving the functional clock enable. If there is a significant amount of logic driving the enable pin, ATPG may not be able to fully exercise and detect all faults in this logic. This can be resolved by using clock gate library cells that have observability logic built into them, or by adding test points externally [http://www.cadence.com]. To minimize area overhead, the observability logic can be shared across multiple clock gates and across multiple hierarchies.

Note that sometimes clock gating may not be necessary from a functional perspective. If the switching activity of the functional mode is very low, then there may not be any significant power savings in functional operation by using clock gating; however, clock gating may still be important from a test perspective to reduce capture power. This could conceivably lead to the insertion of clock gating for test use only – a form of DFT for reduced capture clock switching.

9.2.3 LSSD

Level sensitive scan design (LSSD) provides many useful features for creating safe tests. Since LSSD uses separate scan clocks, a gated functional clock cannot interfere with the ability to scan load and unload test data. Also, traditional LSSD clocking sometimes involves a separate test functional clock to test the functional logic paths. This LSSD test clock is typically gated by the functional clock (sometimes within a so-called clock splitter (Engel et al. 1996)), as shown in Fig. 9.8; the functional clocks (and any gating of them) are treated as data signals that gate the LSSD test C clock. In this example, the LSSD A, B, and C clocks all have an off state of 0 and it is expected that only one clock is pulsed or turned on at any time.

More modern LSSD clocking styles may utilize the level sensitive clocking just for the scanning operation – leaving the functional clock alone except to ensure it is held off/stable while scanning (Iyengar et al. 2006). In this form of LSSD, scan clocking is level sensitive while the functional clocks are typically edge sensitive and allow for functional, at-speed testing.

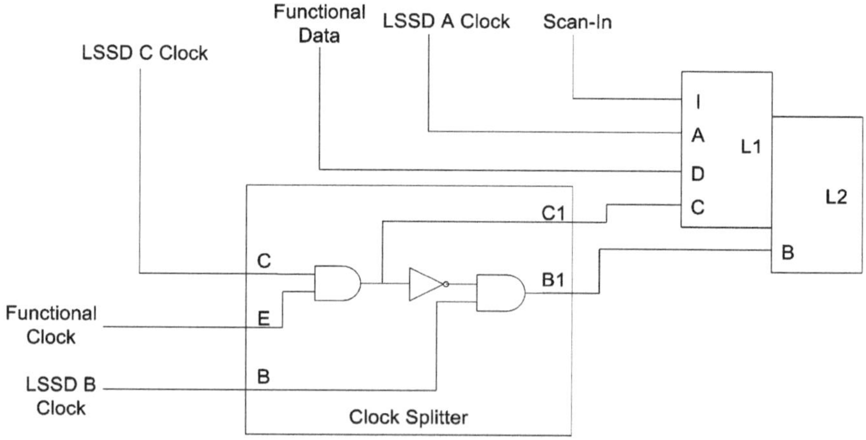

Fig. 9.8 Example LSSD clocking

In all forms of LSSD clocking, the scan is not affected by any gating of the functional clocks, so there is no DFT approach (beyond LSSD) required to deal with such gating, unlike what is required when using MUXed Scan. Other DFT considerations, such as DFT that can override the functional clock gating logic to force the clock to be gated off or to force the clock to get through, can be applicable to both LSSD and MUXed Scan.

9.2.4 Advanced DFT with On-Product Clock Generation (OPCG)

When devices run at functionally high frequencies (1 GHz or higher), or when the device is to be tested on a low-cost tester, the high-speed clocks needed to obtain high-quality delay tests have to be generated on-product. These clocks are usually created using **phase-locked loops** (PLLs) or similar structures that can accept a lower-frequency free running oscillator as input and output a higher-frequency oscillating signal for use on-chip. The PLLs are typically used functionally and can then also be utilized during test application by using the high-frequency oscillator to run a pulse generating state machine for each clock domain (Uzzaman et al. 2007).

The state machines for each clock domain are often programmable to produce 0, 1, 2, or even more pulses and then quiesce to allow the scan to occur. The example clock domain logic shown in Fig. 9.9 produces from 0 to 3 pulses, depending on how many 1s are loaded into the 3-bit program register (program load path not shown). The OSC input that runs the domain state machine is typically the output from a PLL or is divided down from the PLL to produce a lower frequency appropriate for the target domain. We mention the use of OPCG here because the pulse creating state machines act like clock gates at the very root of the clock tree for a clock domain. If the state machine is programmed to not produce any pulses, it is just as of the clock was simply gated off at the root of the clock tree. In fact, some OPCG program registers (as shown in Fig. 9.9) are simply a shift register that gates

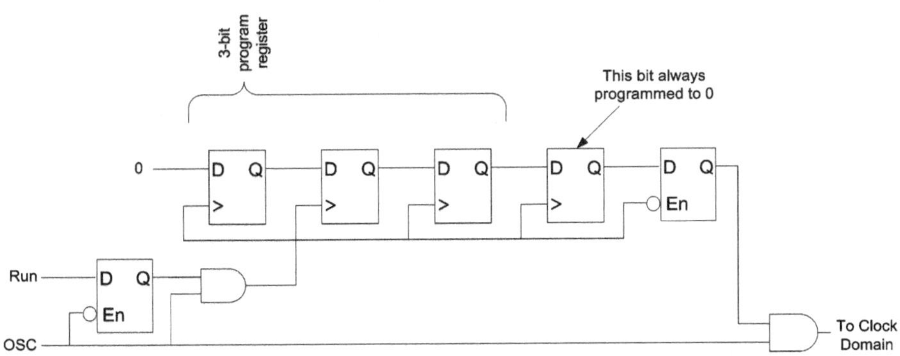

Fig. 9.9 Example OPCG clock generation logic for one domain

the clock once it is started, so to ATPG it looks like all domains may be getting clocked, but some are gated off depending on the values loaded into certain control registers.

A full investigation of OPCG is beyond the scope of this book, but it is useful to note that some aspects of OPCG can be looked at as holding off clocks, which could be exploited for lowering switching activity. Some approaches also include control of the scan enable by the OPCG state machine to allow switching into or out of scan to occur at speed and to enable launch-off-shift (LOS) style delay testing (Nadeau-Dostie et al. 2008); however, without some way to control the clocks outside of the scan operation, when capture clocks do occur, they may cause too much switching activity.

9.2.5 Overriding of Functional Clock Gating

The cone of logic feeding to the functional clock gate may be quite complex. If ATPG wants the clock to get through or be gated off, it may have to set several "care bits" to ensure the clock gate will be at the correct value. In these days when nearly all large chips have test compression logic utilized to improve test costs, it is not good if many care bits are required for a test (Touba 2006). If it is important that ATPG be able to gate off clocks to logic not participating in the current test, if doing so will add a lot of care bits to the existing test, then the test compression hardware may find it difficult to satisfy all the original test care bits plus the ones added to hold off some portions of the clock trees.

One solution to this problem is to add a test override to the functional gating logic that allows a much simpler means to force the clock off, force the clock enabled, or both conditions as necessary (see Figs. 9.6 and 9.7). When activated, this override mechanism will block the observability of the functional clock gating logic feeding that point (labeled "from clock gating logic" in Figs. 9.6 and 9.7); that only means we cannot utilize this override when trying to test for faults in that specific clock gating logic. The override signal itself should come from a scannable test signal. The clock-gate-override signals (e.g., the DFT Gate flop in Figs. 9.6 and 9.7) could be shared with several clock gates to help minimize the overhead for this DFT logic; however, you will lose flexibility if you share it with too many others as tests targeting many faults are more likely to include a fault in at least one of the areas shared with.

9.3 Taking Advantage of Clock Gating

Reduced power consumption during manufacturing test is becoming more important as chips are being designed for low power during functional operation. Of all power concerns during test, **instantaneous switching**, causing current spikes and inducing noise into the power rails, is the most insidious (Girard 2000; Li et al. 2007; Wang et al. 2006). When a clock is pulsed during a test, all flops controlled by that clock could potentially change values. Since most designs today are synchronous and edge clocked, there is a good possibility that many state elements will update on the same clock edge, which is as close to simultaneous as we can get.

One way that has been pursued to reduce switching during capture clock pulses is to try to make state element current values match their update values (Remersaro et al. 2006; Wen et al. 2007; Wu et al. 2007). This is done by modifying a test (that targets one or more faults) by specifying additional care bits in the test cube. This takes bits considered to be *don't-care* for the detection of the target faults and makes them *care* bits to reduce switching. In Wen et al. (2007), the don't-care bits are inferred from fully specified test cubes, but the use of them is the same. The effectiveness of modifying tests to make current and next states match for a significant number of state elements (or even for a set of important state elements that may feed to a large amount of logic) is very design dependent. There is no known relationship to the number of care bits needing to be added to a test to ensure that capture switching will be minimized. Even if you can do this by adding one care bit for every state element being considered, the number of care bits added to a test could be large. With **test compression** being used in so many of today's large chip designs, adding large numbers of care bits to tests will cause a lot of difficulty for the test compression to deal with such tests. Test compression exploits the typically high percentage of don't-care bits for each test on average, so greatly increasing the number of care bits is not going to make test compression work well.

Given that we would like to have low capture clock switching and tests that are compatible with test compression hardware (i.e., tests that contain a very low percentage of care bits), we need to find some way to greatly increase the number of state elements we can keep from switching for each care bit added to a test. A design that uses clock gating may provide just such a mechanism. If the clock gating is coarse-grained, the clock is gated off to large numbers of state elements at a time (see Fig. 9.10 gate A). If the gated off condition can be excited with just a few added care bits, it may be possible to turn off the clock to several thousand state elements with just a handful of added care bits. If there is moderately **fine-grained clock gating**, each clock gate may affect a few hundred state elements on average, which is still a good ratio of care bit control. If the design uses very-fine-grained clock gating (see Fig. 9.10 gate B), it is possible that individual (and independent) clock gates may control just a few tens of state elements, which might still provide a potential 20-to-1 factor or better of care bits to state elements held steady.

Clearly, if we can take advantage of existing functional clock gating, it may be able to keep capture clock switching minimized without adding large numbers of care bits to the tests. In (Illman et al. 2007, 2008; Czysz et al. 2008; Furukawa et al. 2008) it is suggested to exploit clock gating to help reduce switching during capture clocks. While the amount of capture power reduction achieved for a design depends on the number and granularity of clock gates present, all these techniques showed significant reduction in capture power across several designs. In Illman et al. (2007), several approaches for the use of clock gating to reduce switching are men-

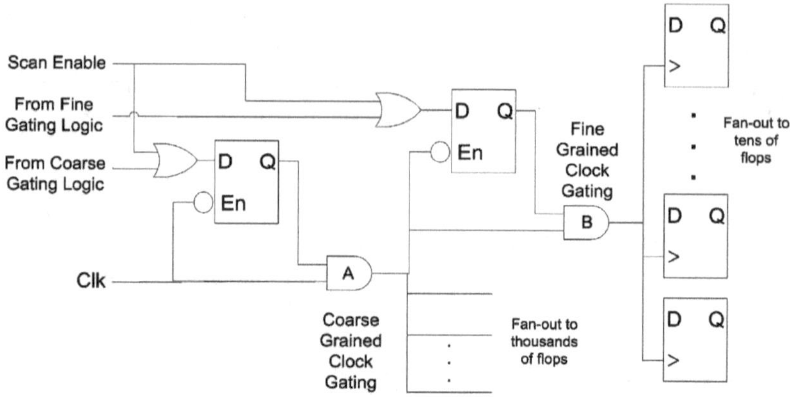

Fig. 9.10 Example of coarse and fine-grained clock gating. Gating that occurs higher up in the clock tree (gate A) affects a larger number of state elements compared with gating that occurs at lower levels of the clock tree (gate B)

tioned. In Illman et al. (2008), **default values** are used to help reduce switching during capture clocks. These "default values" are calculated for each clock gate up front during ATPG, and are the care bits that will force the clock off at a clock gate (Sect. 9.3.2). In Czysz et al. (2008). the ATPG notes and computes the care bits (clock control cubes) for gating off clocks to reduce capture switching in a compression environment. In CTX (Furukawa et al. 2008), the don't-care bits are inferred from post-ATPG fully specified test cubes and they are filled with 0 or 1 to allow exploitation of clock gating. To achieve further reduction in capture power switching, the clock gating approach is combined with the idea of making state element current values match their update values. A two stage process is used, in which first these don'-care bits are utilized to disable as many clock gates as possible, followed by analysis on flops that would cause a transition by capturing a value different than what they were loaded with. While making sure not to adversely affect the test data volume and fault coverage, transitions during capture clocks are further reduced by loading some of these flops with the same value they would capture.

There is also a side benefit to minimizing the number of state elements that switch during capture clocking: **scan cycle switching** can also be reduced. If the scan-load data are providing a substantial percentage of repeating values at the inputs to the scan chains, then there will be reduced scan cycle switching due to the scan-load data (Agarwal et al. 2008); however, after the capture cycles are applied, the **scan-unload switching** is really at the mercy of how the functional logic works. There is also a consideration for any inversion that may exist between scan bits in the chains. For example, if many flops tend to capture the same value (e.g., zero) during the capture cycle, if there is inversion between each bit along the scan chain, each shift cycle will cause transitions during scan out. When we avoid updating a large percentage (e.g., 80%) of state elements, these state elements will continue to contain values from the scan-load that induce low switching levels during scan.

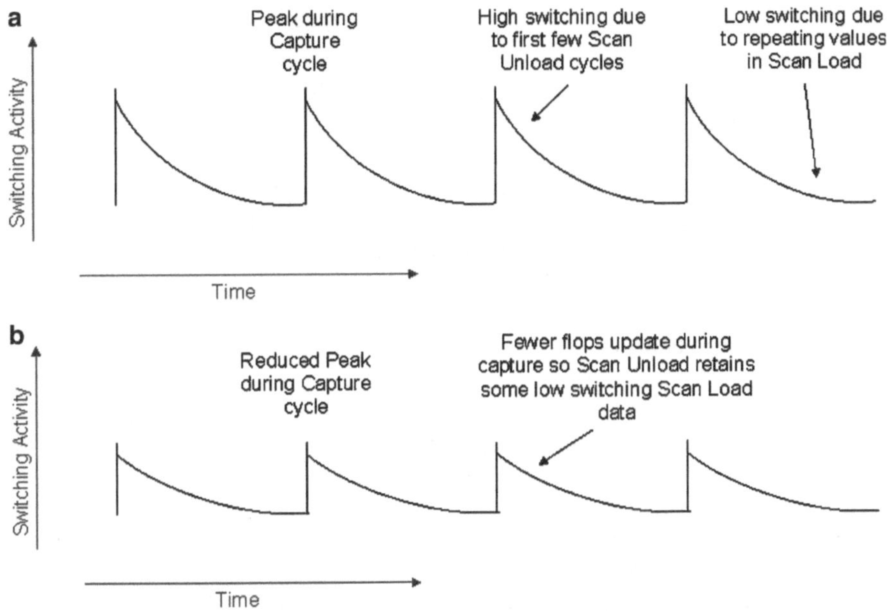

Fig. 9.11 Switching activity during test application. (**a**) Low switching scan load data, (**b**) low switching scan load combined with minimized switching during capture clocking

For example, suppose the scan load switching is held below 10% and capture clocking updates at most 20% of the flops. If we assume a random chance of switching during scan for the 20% that updated during capture (i.e., those 20% will have 50% – probability 0.5 – switching between just the values in those flops during scan out), then the scan out switching activity on the first few shift cycles should be:

$$(80\%) \times (10\%) + (20\%) \times (50\%) = 18\% \text{ or less}$$

The switching activity during test application resembles a *sawtooth* as shown in Fig. 9.11. If a large number of flops update during the capture cycle, not only do we see a peak during the capture cycle, but also high switching during the first few scan cycles as the captured data are shifted out. As the higher switching captured data are shifted out and lower switching scan-in data come into the chains, the switching activity should gradually fall to the scan-in switching level as scan cycles progress. The switching activity peaks again during the next capture cycle. Figure 9.11b illustrates how allowing only a few flops to update during capture clocking can benefit scan unload switching. After the capture clock most of the flops will retain the low switching data they were loaded with, hence the scan unload switching does not significantly exceed the scan load switching. The combined effect of reduced capture and scan unload switching causes the height of the sawtooth curve to be lower.

One thing needs to be emphasized: because DFT provides a means to bypass clock gating so that scan shifting can work in edge-triggered designs, it will be impossible to utilize the clock gating if ATPG is constrained to hold the clock gating in its bypass state. It is highly recommended that any clock gating bypass be enabled using a scan_enable signal that ATPG can change the value of rather than using a mode signal that is constrained to be constant. In the past, forcing the clock to be enabled all the time was often done to ensure each test would tend to clock as many state elements as possible – increasing the chance of observing and detecting more faults/defects per test; however, when trying to lower capture clock switching activity, this is no longer a recommended approach.

A note about reset clocks

Signals used to force many state elements to reset their values to some specific starting state (0 or 1), often referred to as reset clocks, are typically not gated. When these reset clocks are applied, depending on how many state elements are affected by the reset, you may get excessive switching activity. Functionally, when applying a reset sequence, the reset is held for some amount of time, which ensures the state elements being reset stabilize to the desired reset state; however, those state elements not being reset are typically not important and can be at any value – including that they may lose their prior value due to power supply noise from the excessive reset switching. During test application this can be a problem because ATPG can assume all nonreset state elements are still at their scanned-in values – but if there is excessive switching noise, some of these state elements can lose their values and would cause the test to fail. One way around this problem is for ATPG to create separate tests just for the reset clocks and these tests may expect unknown (X) values for all state elements not being reset. Another approach is to attempt to scan-load many of the state elements to their reset state prior to applying the reset clock so that only a portion of the state elements will switch. Any such reset clock test patterns will have to load all reset elements to the opposite of their reset values at some point in order to test that the reset is working correctly, but this can be done across several tests, allowing just a subset to switch in any one test.

9.3.1 Locating Where Clocks are Gated

To be able to take advantage of clock gates, we first have to locate them. It is often useful to trace forward from the root of all defined clock sources to identify all paths of clocks through the logic. This can be optimized to denote only paths that feed to clock inputs of state elements (including RAMs). Then, for all state elements of interest, trace back from the clock input along the clock path to locate points where the clock path from that point forward could be gated, i.e., forced to a steady state value.

While tracing to locate the clock gates, we can keep track of the number of flops controlled by each clock gate: The clock path should be traced back beyond any recognized clock gate so as to identify cases where a higher-level clock gate may exist (it would be a **coarse-grained clock gate** that controls many more flops than the fine-grained gates located further down the clock tree, closer to the flops). When there is such a hierarchy of clock gates, it may be useful to recognize it in order to utilize the higher-level gates when possible, ignoring the subordinate gates unless the higher level gate could not be utilized (Keller 2005). We have seen designs where fine-grained gating might affect 16 or 32 flops individually, but a higher level, coarse-grain gate could affect 10,000 or more flops. When test data compression is being used, it may not be worth bothering with clock gates that affect fewer than ten flops since adding care bits to test cubes needs to have a high payback per bit added.

Associated with each clock gate, it is useful to denote something that is a measure of how effective the gate is at reducing switching activity. One such measure is the simple number of state elements controlled by the gate. A slightly more accurate number might be a weighted count (similar to the Weighted Switching activity of (Czysz et al. 2008; Gerstendorfer and Wunderlich 1999)). Weighting each state element can be useful since a flop feeding to a single gate will likely produce less overall switching than a flop that feeds to 20 gates. There are many possible ways to weight the state elements, including use of signal probabilities (those closer to 0.5 will more likely switch than those skewed toward 0 or 1), node capacitance, and the number of gates affected by the state element (going through single input gates). All of these can be beneficial for establishing a reasonable metric for relative switching that allows meaningful comparisons of the expected effectiveness of each clock gate. We have found even assuming a weight of one for each state element is quite useful and often produces reasonable results.

It is also useful to track some information for each independent clock source, such as the number of flops it controls and what percentage of them can be gated off. When ATPG is looking to use a clock gate, it needs to know which clock is being gated, to avoid using a clock gate in a test that does not pulse the clock it gates. To reduce the number of tests generated, techniques like *multiclockcompaction* cause multiple clocks to be pulsed within the same test. This technique allows faults under different clock domains to be targeted within the same test. For example, in a design having three clocks (A, B, and C), turning on *multiclockcompaction* could cause both clocks A and B to pulse within the same generated test. Now to reduce the capture switching for this test, ATPG should utilize only the clock gates on clocks A and B, and ignore those on clock C. So many designs these days have tens if not hundreds of (internally generated) clock domains, it is important to know how many flops are driven by each domain and what portion of them can be gated off to avoid switching. This information might also be useful when deciding which clocks can be pulsed together to reduce the number of test patterns. It is recommended to avoid pulsing multiple clocks in the same test (tester cycle) if those clocks have no clock gating or limited clock gating to help reduce switching activity.

Note: Designs that avoid gating of clocks in favor of gating the data path (see Fig. 9.1) might still be able to take advantage of this concept of clock gating. It is more difficult to locate and identify the data gate logic equivalent to a clock gate, but this is conceptually still possible. Because gating to reduce functional power is most likely to use clock gating rather than data gating (in order to gain the benefit from reduced switching along the clock tree as well), it is not clear how many data gating designs will be seen in the future.

Note: Some designs use multiple, independent **power domains** (Chickermane et al. 2008). These may be referred to as multi-domain or multi-supply/multi-voltage designs. It is possible that different power domains have different capacities for handling switching activity. Different clock domains may in fact run in different power domains, which may allow them to operate independently from each other (switching in one power domain may have no impact on power supply noise in other domains). If it is acceptable to treat each power domain separately, then switching activity should also be tracked separately for each power domain. For example, a test might cause 20% switching of flops in the device being tested, but those 20% might be 80% of the flops within a single power domain. If that domain cannot handle such high switching, this test likely has a problem. It is important to be aware that tracking switching activity per power domain may be required for some designs.

9.3.2 Identifying "Default" Values

Once we have a list of clock gates, we can try to utilize them to help reduce switching. One simple way to do this is to identify care bit settings (scan flops and primary inputs) that can be set to force the clock off at the gate. If we identify a set of such care bits for each clock gate, we could define "default values" for these control bits. A default value is a value assignment to be made only if ATPG does not require a different value on that control input. This is also known as Preferred Fill (Remersaro et al. 2006).

The benefit of using default values for clock gating control is that the ATPG to identify what control bits are needed is done only once up front, so the overhead for this approach is fairly minimal. Some empirical results for certain designs have been reported (Illman et al. 2008). The potential down side of using default values is that the care bits for the current test may conflict with the default values to such an extent that enough clocks get through and the test has too much switching. If this happens, it may be better to compact test cubes less, with resultant larger test sets, in order to avoid the higher switching that occurs when clocks are not being gated off.

Figure 9.12 shows an example where a default set of care bits ($S1 = 0, S2 = 0$) are identified that will force the clock off at the clock gate. These default values are compared with the care bits of each of the three ATPG test cubes T1, T2, and T3 to see if they can be merged without any conflicts. In this example, the default values can be merged into test cubes T1 and T3 indicating that the clock can be gated off. Some don't-care bits in the test cubes are now replaced with new care bits (underlined) from the default values. For test cube T2, its care bits conflict with the default values, in which case the value chosen for X by the X-fill algorithms for reduced scan (or capture) switching will determine if the clock is gated off at that clock gate.

a

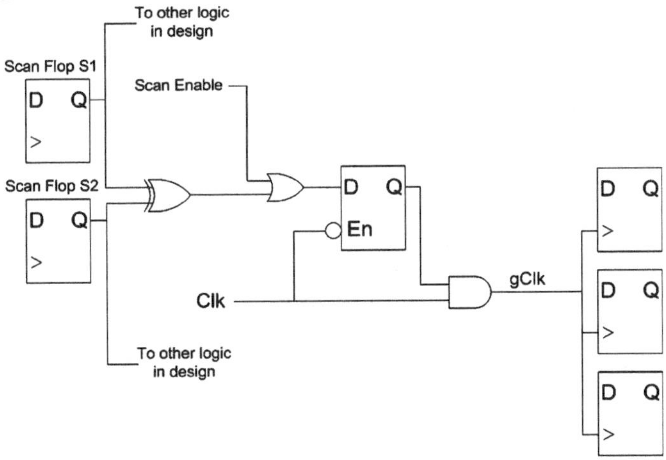

Gating logic controlled by two scan flops. Default values are S1=0, S2=0

b

Test Cubes	Scan flops in design					
	S0	**S1**	**S2**	S3	S4	S5
T1	X	**0**	**X**	1	0	X
T2	0	**1**	**X**	X	X	0
T3	X	**X**	**0**	1	0	X

Test cubes containing care bits from ATPG

c

Test Cubes	Scan flops in design						Merge "default values"?
	S0	**S1**	**S2**	S3	S4	S5	
T1	X	**0**	**0**	1	0	X	Yes
T2	0	**1**	**X**	X	X	0	Conflict
T3	X	**0**	**0**	1	0	X	Yes

Test cubes after adding "default values" to turn off clock gate

Fig. 9.12 Example of clock gate "default values" merged into ATPG test cubes. (**a**) Gating logic controlled by two scan flops. Default values are S1 = 0, S2 = 0. (**b**) Test cubes containing care bits from ATPG. (**c**) Test cubes after adding "default values" to turn off clock gate

When creating tests that have multiple capture clocks, such as for launch-off-capture (LOC) delay tests, it may be necessary to derive the default value care bits back through multiple time frames to ensure clocks are gated on each capture clock

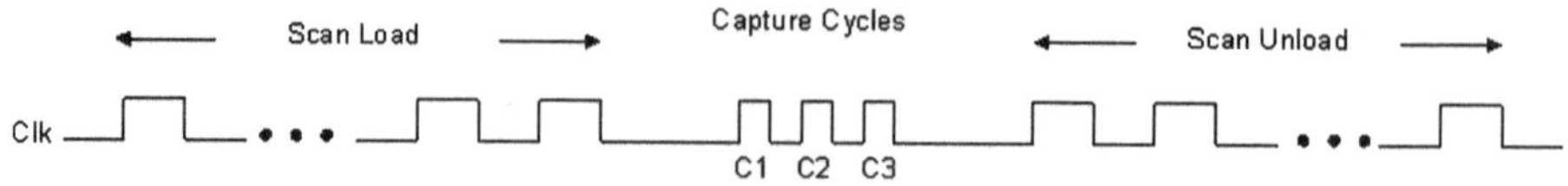

Fig. 9.13 Example of multiple capture clocks in launch-off-capture (LOC) delay tests

pulse. Because these default values need be derived only once up front, the multiple-time-frame ATPG should not be a big burden. Figure 9.13 shows an example delay test containing three capture clock pulses C1, C2, and C3. A three pulse test might be used to (1) create a transition, then (2) write the transition into a RAM, and then (3) read out of the RAM and capture into a scan flop. The three pulses are the time frames through which the default care bits have to be justified to ensure that the clocks are gated off to nonparticipating state elements for each pulse.

Although deriving clock gate default value care bits through multiple time frames makes sense to do, it may not be necessary. If the design is functionally low power, the functional logic operation may well tend toward low switching on consecutive applications of functional (capture) clock cycles. ATPG is almost guaranteed to create a circuit state that is out of normal functional state space on the scan load; however, as functional clock cycles are applied, the subsequent circuit states are likely to tend toward those that are closer to real functional operation states. If this is so, then if ATPG scan loads a state that causes low switching on the first capture clock cycle (pulse C1 in Figure 9.13), it is very likely that additional such functional clock cycles will also be low switching simply because the states are tending toward functional and those are designed to be low switching. This behavior will be circuit dependent and is not guaranteed, but it is at least plausible.

9.3.3 *Dynamically Augmenting a Test*

An alternative to the simplistic yet efficient default value approach is to look at the care bits in place for the current test from ATPG and obtain clock gates to control their clocks off with these values as a given. This requires modifying the test with values that have to be derived (using ATPG) dynamically for each unique test before it is fault simulated. This would be most appropriate to be applied after generating a test that targets multiple faults rather than for a test targeting a single fault that would be further enhanced to target additional faults. Once the clock gate care bits are added, it can significantly impede the testing of large chunks of logic, so it is desirable to do this only after all ATPG for faults on the test is complete.

Dynamically justifying the values to gate off clocks will work out better than the use of default values whenever there are multiple ways of setting the clock gate and only some of them conflict with the current test's care bits. This can be illustrated using the example in Fig. 9.12, where dynamic justification during test cube T2 would have determined that care bits $S1 = 1, S2 = 1$ are also a valid solution to gate the clock off. These care bits can merge with the care bits of test cube T2, thus ensuring that the clock can be gated off in a deterministic manner rather than relying on the fill algorithms. Also, when using **dynamically justified clock gating**, one can stop justifying additional clock gates once the test is known to have held the clock off to a sufficient percentage of the flops (Keller 2005). The default value approach may tend to produce tests that hold off clocks too much, resulting in perhaps too little switching on some tests.

As with the default values approach, the dynamic augmenting of a test could come up short if the test already has a lot of care bits set in areas that conflict with what is needed to gate off the clocks to significant parts of the circuit. To avoid this problem one can either avoid compacting the tests too much or actively monitor the clock gates affected by the test. Monitoring adds a fair amount of overhead, but allows detecting when a sufficient percentage of clock gates have been enabled by the current care bits in the test, so now is the time to apply care bits to gate off the areas still not controlled by the tests' care bits.

For tests with multiple capture clocks, e.g., LOC transition tests, adding care bits for clock gates may be needed for each time frame. As was mentioned before, if the capture clock cycles tend to bring the circuit state closer to functional states from the initial scan load, circuits designed for low switching in functional operation will tend to have low switching on subsequent clock cycles. This can help reduce the effort to augment the tests since, unlike with the default values approach, there will be ATPG applied to each test before it is sent to fault simulation to add clock gate care bits; the sequential ATPG required to justify these clock gate care bits back through multiple time frames could be expensive. If justifying the clock gate care bits only in the first time frame works, significant effort can be saved.

9.4 Summary and Conclusions

Functionally low power CMOS designs will tend to have a large percentage of their state elements controlled by gated clocks – even if they also utilize power switches and power modes. It is reasonable to take advantage of the clock gating to help reduce switching during test application – primarily to reduce switching activity during capture clocking, but also to help reduce switching when shifting out the captured results. It is important to ensure sufficient DFT is applied so that the clock gating will not cause problems for scan operations and yet ATPG should remain in control of the clock gates outside of the scan operations.

The use of "default" or "preferred" values for filling of don't-care bits in test cubes is a highly efficient and effective approach to reducing capture clock switching. The real advantage for utilizing the existing functional clock gating to do this is that it already exists for many low-powered designs and the care-bit to held state element ratio can be quite high, making this approach ideal for use with test compression.

It should be noted that for areas of the design where clocks may not be gated, any other approach that can keep switching low (e.g., the preferred values (Remersaro et al. 2006)) could be used in addition to clock gating default values. Just be aware that the care-bit to held state element ratio is not likely to be nearly as good.

By holding most state elements to the values they were scanned in to, we can further take advantage of any low switching values enabled by repeat-fill during the scan load. Although the state elements that change on the capture clocks may cause substantial switching during scan-out, if the percentage of updated state elements is kept low (perhaps to less than 20% of the scan flops), the scan unload switching will be kept under control.

Finally, tracking of switching activity may need to be done for each power domain on a multi-power domain design. This affects the way switching activity should be reported as well as how ATPG should attempt to reduce switching activity.

References

Agarwal K, Vooka S, Ravi S, Parekhji R, Gill AS (2008) Power analysis and reduction techniques for transition fault testing. Proc Asian Test Symp 403–408

Benini L, Siegel P, De Micheli G (1994) Saving power by synthesizing gated clocks for sequential circuits. IEEE Design Test Comp 11(4):32–41

Chickermane V, Gallagher P, Sage J, Yuan P, Chakravadhanula K (2008) A power-aware test methodology for multi-supply multi-voltage designs. Proc Int Test Conf, Paper 9.1

Czysz D, Kassab M, Lin X, Mrugalski G, Rajski J, Tyszer J (2008) Low power scan shift and capture in the EDT environment. Proc Int Test Conf, Paper 13.2

Donno M, Macii E, Mazzoni L (2004) Power-aware clock tree planning. Proc Int Symp Phys Design 138–147

Engel JJ, Guzowski TS, Hunt A, Lackey DE, Pickup LD, Proctor RA, Reynolds K, Rincon AM, Stauffer DR (1996) Design methodology for IBM ASIC products. IBM J Res Dev 40(4):387–406

Furukawa H, Wen X, Yamato Y, Kajihara S, Girard P, Wang LT, Tehranipoor M (2008) CTX: A clock-gating-based test relaxation and X-filling scheme for reducing yield loss risk in at-speed scan testing. Proc Asian Test Symp 397–402

Gerstendorfer S, Wunderlich HJ (1999) Minimized power consumption for scan-based BIST. Proc Int Test Conf 77–84

Girard P (2000) Low power testing of VLSI circuits: problems and solutions. Proc Int Symp Quality Electron Design 173–179

Illman R, Keller B, Bhatia S (2007) A review of power strategies for DFT and ATPG. Proc Eur Test Symp

Illman R, Keller B, Gallagher P (2008) ATPG power reduction using clock gate "default" constraints. Proc First Int Workshop Implications Low Power Design Test Reliability 45–46

Iyengar V, Grise G, Taylor M (2006) A flexible and scalable methodology for GHz-speed structural test. Proc Design Automat Conf 314–319

Keller B (2005) Clock gating support for low power test. Intern Cadence Specification Doc

Li B, Fang L, Hsiao MS (2007) Efficient power droop aware delay fault testing. Proc Int Test Conf, Paper 13.2

Mukherjee A, Marek-Sadowska M (2003) Clock and power gating with timing closure. IEEE Design Test Comp 20(3):32–39

Nadeau-Dostie B, Takeshita K, Cote JF (2008) Power-aware at-speed scan test methodology for circuits with synchronous clocks. Proc Int Test Conf, Paper 9.3

Nicolici N, Wen X (2007) Embedded tutorial on low power test. Eur Test Symp 202–210

Remersaro S, Lin X, Zhang Z, Reddy SM, Pomeranz I, Rajski J (2006) Preferred fill: A scalable method to reduce capture power for scan based designs. Proc Int Test Conf, Paper 32.2

Shen W, Cai Y, Hong X, Hu J (2007) Activity-aware registers placement for low power gated clock tree construction. Proc Int Symp VLSI 383–388

Touba NA (2006) Survey of test vector compression techniques. IEEE Design Test Comp 23(4):294–303

Uzzaman A, Li B, Snethen T, Keller B, Grise G (2007) Automated handling of programmable on-product clock generation (OPCG) circuitry for delay test vector generation. Proc Int Test Conf, Paper 17.3

Wang J, Walker DMH, Majhi A, Kruseman B, Gronthoud G, Villagra LE, van de Wiel P, Eichenberger S (2006) Power supply noise in delay testing. Proc Int Test Conf, Paper 17.3

Wen X, Miyase K, Kajihara S, Suzuki T, Yamato Y, Girard P, Ohsumi Y, Wang LT (2007) A novel scheme to reduce power supply noise for high-quality at-speed scan testing. Proc Int Test Conf, Paper 25.1

Wu MF, Hu KS, Huang JL (2007) An efficient peak power reduction technique for scan testing. Proc Asian Test Symp 111–114

http://www.cadence.com/products/ld/rtl_compiler, Encounter RTL Compiler, Cadence Design Systems, Inc

http://www.synopsys.com/Tools/Implementation/RTLSynthesis/Pages/DCUltra.aspx, Design Compiler, Synopsys, Inc

Chapter 10
Test of Power Management Structures

Mark Kassab and Mohammad Tehranipoor

Abstract Shrinking technology nodes offer higher levels of integration and better performance. However, they are accompanied by increased dynamic (switching) and static (leakage) power densities. As seen in previous chapters, a wide array of power management technologies is used to control dynamic and static power in integrated circuits. Those include clock gating and various types of power gating techniques. Power gating and multiple voltage supplies usually result in the use of special low-power cells such as state-retention registers, isolation cells, and level shifters. In addition to the challenges inherent in testing logic that can operate in multiple power modes, it is necessary to thoroughly test all the power management features including the clock gaters, power gaters (or switches), the logic that controls them, and the aforementioned low-power cells. Testing this logic will be presented in this chapter, as well as a method for validating the integrity of the power distribution networks.

10.1 Clock Gating Logic

Clock gating, as explained in previous chapters, is a widely used and relatively simple-to-implement method for effectively reducing dynamic power. By selectively shutting off a part of the clock tree, a clock gater can reduce dynamic power in both the logic driven by that clock as well as the clock tree itself. Clock gating is also used by synthesis tools to reduce design area. It is more power- and area-efficient, for example, to use clock gating than recirculating multiplexers when a large number of registers must conditionally hold their state (Keating et al. 2007; De Colle et al. 2005).

The manner in which clock gating logic is controlled during test has various implications on test, including: the testability of the functional clock gater control logic, the clock gater itself, dynamic power, and **automatic test pattern generation** (ATPG) pattern count. Those topics are covered in this section.

M. Kassab (✉)
Mentor Graphics Corp., Wilsonville, OR, USA
e-mail: mark_kassab@mentor.com

M. Tehranipoor
University of Connecticut, Storrs, CT, USA

P. Girard et al. (eds.), *Power-Aware Testing and Test Strategies for Low Power Devices*,
DOI 10.1007/978-1-4419-0928-2_10,

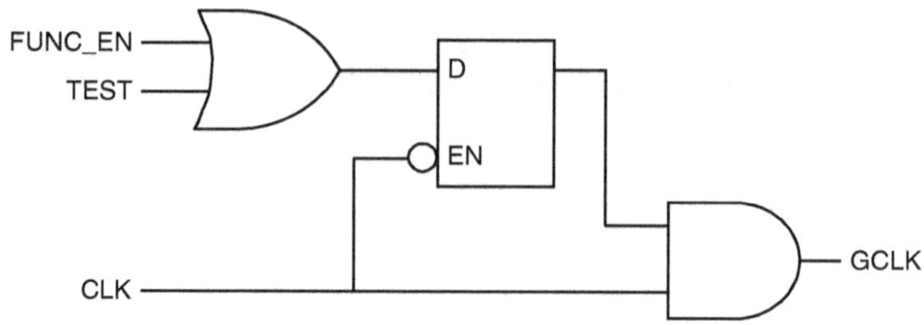

Fig. 10.1 Clock gater

10.1.1 Controlling Clock Gaters during Test

A typical clock gater cell is shown in Fig. 10.1. The latch prevents glitches on the enable signal (the data input of the latch) from propagating through the gater into the clock tree. It is necessary for any clock gater driving scan cells that are being used in a given test mode to be forced on during scan shifting. Therefore, a second enable signal (shown as TEST in Fig. 10.1) is OR-ed with the functional enable signal. It is used to override the functional signal and force the clock gater on when needed during test.

Synthesis tools typically provide the user with an option to control the test-mode pin (TEST signal in Fig. 10.1) using the *test enable* signal or *scan enable* signal. The *test enable* signal is asserted during the entire test session. Using it to control clock gaters results in the clock gaters being forced on during both the shift and capture cycles. The *scan enable* signal is asserted during shift, and almost always de-asserted during the capture cycle(s). Using it to control clock gaters during shift results in the gaters being controlled by the functional control logic during capture. Either option has its advantages and disadvantages.

10.1.2 Impact on Testability of the Clock Gater and its Control Logic

Like any other structure, the functional control logic has to be tested effectively to ensure correct functional behavior. It is usually controlled by scan cells. Controlling this logic or launching transitions for at-speed test can therefore be done by ATPG with no design changes. However, since the *test enable* signal is permanently asserted during test, using it to control clock gaters results in the functional control logic becoming unobservable and therefore untestable (De Colle et al. 2005). An observe point (OP in Fig. 10.2) must be added in this case to observe the functional control logic. Although the observe point allows the control logic to be tested for static faults, as with any use of observe points, the at-speed test may be inaccurate since the timing of the path ending at the observe point may differ from the functional path ending at the clock gater.

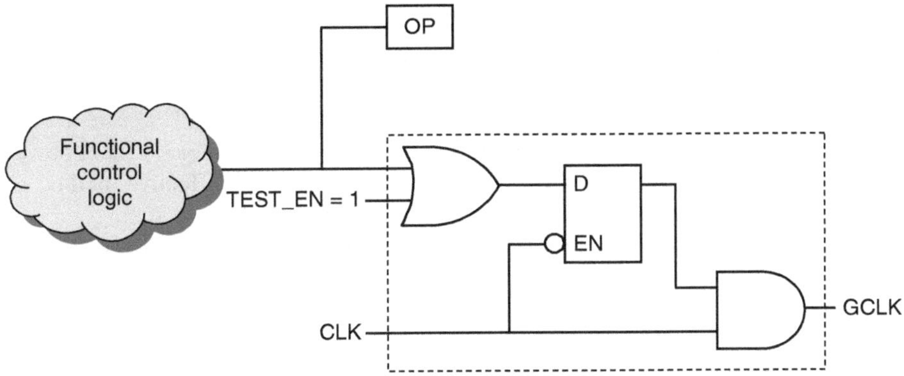

Fig. 10.2 Observing blocked control logic when using *test enable*

Even if the control logic is made observable by adding an observe point, the clock gater itself is still largely untestable. A stuck-at-1 fault on the latch output, for example, will not be tested although the presence of such a defect can be catastrophic to the functional operation.

Driving the clock gater test mode signal using *scan enable* instead of *test enable* greatly improves the testability of the clock gater and its control logic, and eliminates the need for observe points in most cases. With the test signal de-asserted during the capture cycles, the clock gater and its control logic operate in a similar manner to the functional mode of operation. A fault that propagates to the input of the clock gater can often be easily observed. If the clock gater is expected to pulse but does not due to the fault, or vice versa, one of the many scan cells driven by the gater will likely unload a different value than the one expected. Therefore, any functionally irredundant faults in the clock gater and clock gater control logic can be tested, and the paths used for at-speed test will match those used during functional operation.

10.1.3 Impact on Power and Pattern Count

When *test enable* is used to control clock gaters, all clock gaters in the design or design partition under test are forced on. All state elements therefore get clocked during the capture cycle if the root clocks are pulsed. This further exacerbates the high switching activity during test compared to the functional mode, which has been discussed in previous chapters. One positive side effect, if the design can support the increased switching activity, is that pattern count is often lower when all state elements are capable of capturing since ATPG can test more faults with each scan pattern.

The converse is true when *scan enable* is used. Switching activity during capture is reduced since not all clock gaters are on. It may not even be possible to turn them on simultaneously due to mutual exclusivities in the control logic. This leads to the switching activity being closer to functional levels. In fact, ATPG can be constrained to turn off clock gaters through the functional control logic to meet power constraints (Czysz et al. 2008). This has been shown to be an effective method of capture power reduction, and one that is compatible with on-chip scan compression methods since relatively few scan cells have to be controlled to disable a large number of flip-flops from getting clocked.

10.2 Power Control Logic

In designs that employ power gating strategies for reducing leakage power, the **power management unit** (PMU) controls power modes and orchestrates safe transition sequences between power modes. It also presents new test challenges. A defect in this unit may not affect the functional operation of the design, yet can affect its power consumption. The power status of a gate also introduces a new dimension to structural test pattern generation. Since power is not explicitly represented in the netlists that ATPG tools use as input, logic involved in power gating is not adequately tested by conventional test strategies. In addition, DFT changes are necessary to enable scan test in the presence of power gating, as well as to facilitate testing of the power control logic in the PMU.

10.2.1 Role of Power Control Logic

The PMU (Fig. 10.3) controls the various power domains that use power gating (Chickermane et al. 2008). In each domain, it can control the following signals:

1. Power enable: This signal enables or disables the header/footer transistors used to connect the logic domain to V_{DD}/V_{SS} and therefore to power on or off a given domain. Its inverse is commonly referred to as the sleep signal. There may be multiple signals used to independently control groups of switches. This can be used to control inrush current when a domain is powered on and avoid severe voltage droops.
2. Isolation: If in a given power mode, a domain is powered off and it feeds another domain that is powered on with no special handling, a large short circuit current can result due to floating logic voltage levels feeding the powered-on domain. To avoid this, special isolation cells are inserted between the domains and activated in such power modes to block the two domains and provide valid voltage levels into the enabled domain. The isolation cells are usually functionally simple gates such as AND or OR gates, where one of the inputs is the output of the first domain and the second input is driven by the isolation signal from the PMU. Isolation

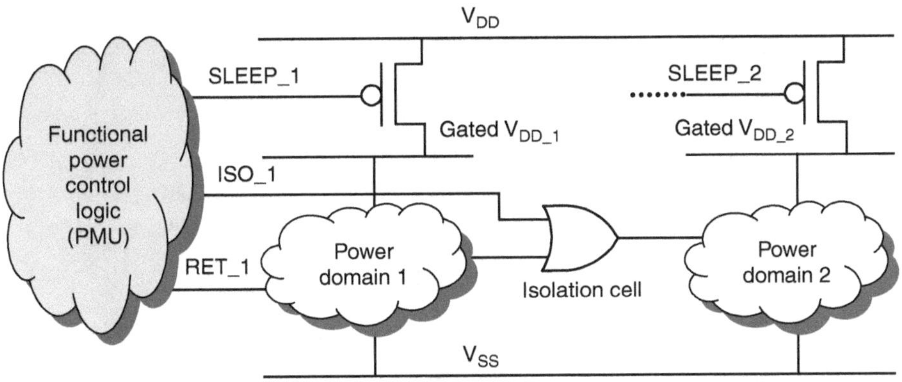

Fig. 10.3 Power management unit (PMU)

is obviously active when the isolation signal has the controlling value of the isolation cell and the output of the isolation cell is constant. Latches are also sometimes used for isolation. The latch is made transparent when the source domain is powered on, while the latch enable is de-asserted when isolation is required.

3. Retention: One of the main design challenges introduced by **power shut-off** (PSO) is that state elements lose their state when the power in that domain is gated off. Consequently, some or all of the state elements may be replaced by state retention registers (SRRs) (Zyuban and Kosonocky 2002) that are capable of retaining their values through a power-off cycle while consuming low leakage power. The main cost is larger area overhead. There are operational differences between different types of retention cells depending on their technology. Commonly, to retain the value in a state element, either a retention signal must be asserted throughout the power-off cycle, or the value is saved by pulsing a *save* signal prior to PSO and restored by pulsing a *restore* signal after power is restored. The PMU is responsible for controlling those control signal(s).

10.2.2 Power Control during Shift

As seen in Sect. 10.1, the shift clock(s) of the block(s) under test in a given test mode must be enabled during shift when using clock gating. Similarly, the aforementioned functional power control signals must be overridden and fixed during shift as shown in Fig. 10.4. Power domains that are being scanned in a given test mode must obviously be powered on throughout the shift process. The state retention mode and isolation cells must be disabled for the domains powered up and involved in shifting. While it would be simpler to bypass all the power management and power up the entire device during test, this can result in exceeding the power limits of the design and may not be a viable option. The power mode (or state) used in a given

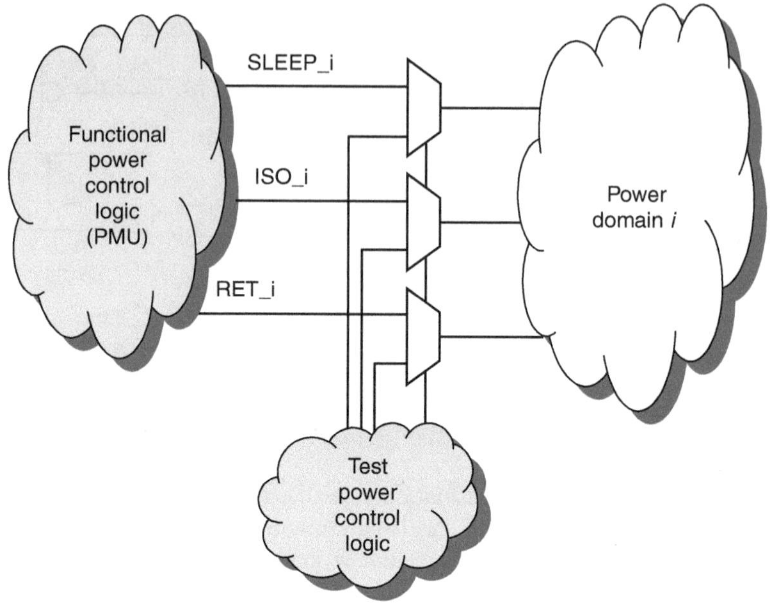

Fig. 10.4 Overriding power control during test

test session should ideally match a valid functional power mode as specified by one of the power intent formats: **unified power format** (UPF) or **common power format** (CPF). In addition, control over the power modes will be needed to test the low-power cells as will be seen.

In power domains that are not being tested in a given test mode, it is preferable to force the power off, or even on if necessary, than to allow it to be driven by the functional control logic during shift. If not fixed, the power status may keep switching during shift and enter invalid power modes. The current drawn as a result may exceed design specifications and even lead to invalid operation of the logic under test.

10.2.3 Power Control during Capture

Consider testing of the power domains – the logic that may be powered off. A requirement for testing faults in that logic is that the gates involved in the test be powered on during the capture cycle(s). If the power is controlled through its functional operation in the capture cycle after having been overridden during shift, the following considerations must be made:

- The power status and its control must be known to and handled by the ATPG tool so that the relevant logic is powered on when a value is required on a gate. For example, the fault site must be forced to a binary value. The gate attached to the fault site is clearly relevant for the test and must be powered on when this fault is being tested. Similarly, other gates that control and observe the fault site need specific binary values and must be powered on. ATPG must have knowledge of how to control the power status of the different domains such that those gates that are needed for the test are powered on. One way to pass this knowledge to ATPG is through UPF or CPF, if the ATPG tool can process this abstraction of power information and make use of it during pattern generation. The information in UPF/CPF enables the ATPG tool to determine how to turn on/off a domain, and whether a domain is on or off at a given time. Another method to account for the power status during ATPG, if using an ATPG tool that has no understanding of power, is by remodeling within the design and library cells. For example, the model of a scan cell with no retention capability can be done by adding an input pin to the model to represent the power status of the domain that includes this scan cell. This pin would have a value of 1 when the cell is powered on and a value of 0 when the cell is powered off. The cell model would additionally be changed such that when this power signal has a value of 0 (power off), the state of the scan cell becomes X (unknown). When the power signal has a value of 1 (power on), the cell operates normally.

- If a state element is in a domain that gets powered off in any capture cycle without retention being used, this cell loses its state, which can include values scanned in or captured fault effects. Again, this needs to be modeled for and used by ATPG.
- If any power mode transitions occur during the capture cycles, the transition may span multiple cycles, especially if fast cycles are being applied during at-speed test. For example, assume that in a two-cycle test, the power control signal for a domain is de-asserted (domain is off) in the first cycle, and asserted (domain is on) in the second cycle. If there is insufficient time between the two cycles for the power domain to completely power up, the power domain cannot be assumed to be operating reliably in the second cycle even though its power control signal is asserted. Timing will therefore need to be taken into account; power transitions cannot be assumed to occur within one cycle as with clock gating.

It is therefore more straightforward to keep the power mode used during shift fixed and in effect during capture. In other words, the same power mode is then used for the entire test session when testing faults in the power domain(s). A design would need to be tested in multiple test modes to avoid powering on the entire device simultaneously. We refer the reader to the test generation and DFT techniques for multi-voltage designs covered in Chap. 8. For testing low-power cells such as retention, it will be necessary to allow the power mode to change during the capture phase. This will be further discussed in Sect. 10.4.

10.2.4 Testing the Power Control Logic

If the functional power control logic is blocked and overridden during shift and capture when most of the logic is being tested, special handling is required to test it.

The power control logic is usually driven by scan cells and belongs to the always-on power domain (always powered on), so controllability of this logic is straightforward with no design changes required. Nodes in this logic can be controlled to a 0 or 1 value by loading the scan cells driving the logic, just as with conventional structural test. However, this logic cannot be observed directly since it drives the power control signals and not scan cells. Observing the logic can be done in one of two ways:

1. By observing the logic being controlled by the control logic. For example, consider the SLEEP_i signal in Fig. 10.4. To observe a stuck-at-0 on this signal, one would have to allow the functional control logic to control the power in the domain, force the signal to 1, and observe whether the power domain is powered off (excepted) or on (unexpected).
2. By inserting dedicated observe points (Fig. 10.5) that are usually observed through scan.

Using the power domain or low-power components being controlled (Method 1) involves observing faults in the controller through that logic and ultimately into scan cells (or primary outputs). The logic that controls the power switches can be

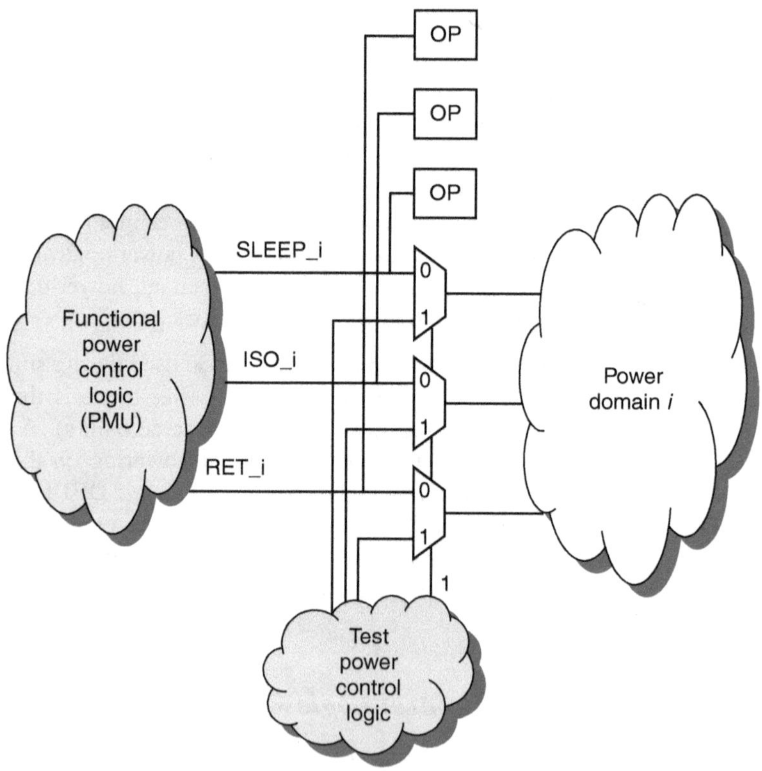

Fig. 10.5 Observing the control logic through observe points

observed by observing the virtual (or gated) V_{DD} or V_{SS}, as shown in Sect. 10.3. In other words, the SLEEP_i signal is observed by detecting whether the power domain is powered on or off. Logic that controls the retention and isolation cells can be observed in a similar manner to the tests for those power cells, which are discussed in Sect. 10.4.

This observation method minimizes additional test logic and therefore area overhead. It allows testing (including at-speed test) of the functional control logic through its functional path. However, it also has some negative consequences. Diagnostic resolution can be compromised since it becomes more difficult to distinguish defects in the low-power components from defects in the control logic of those components. For example, it may be difficult to differentiate a defect in the power switch from a defect in the logic controlling the power switch. In addition, the control logic would need to be tested in a separate test session since the control signals cannot be overridden during capture as shown in Fig. 10.4 when using this observation method. The ATPG complexity is also considerably higher, unless the scan cells in all but the always-on power domain are masked during this test session.

A simpler solution is to add observe points to the outputs of the control logic (Method 2). This allows the functional control logic to be tested while those signals are overridden downstream and the power domains are being tested. For example, it is then possible to control and observe the functional power control logic while the test power control logic overrides the power controls and forces all power domains to be powered on so they can be tested simultaneously. However, this method does not allow testing of the functional path between the functional control logic and the power switches.

10.3 Power Switches

To reduce power dissipation, especially leakage power dissipation introduced by shrinking process technologies, power switches are commonly used in modern low-power designs. To enable the power gating functionality, different parts of the design are equipped with one or more power switches. Figure 10.6 shows an example of

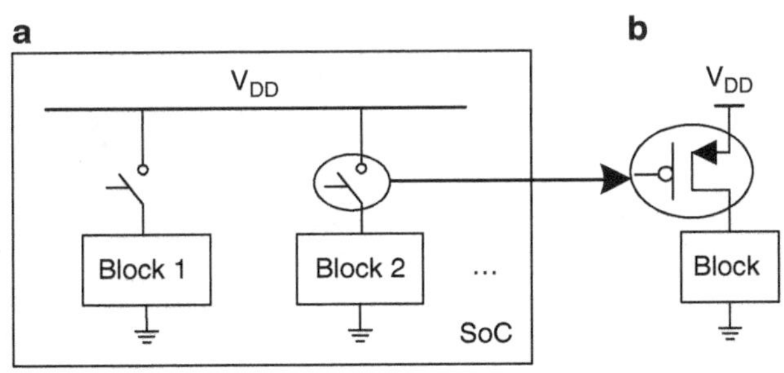

Fig. 10.6 A power gating scheme in an SoC design

the power gating implementation on an SoC. According to the functionality and activity of circuit blocks in the SoC, a block or several blocks can be individually powered off through the power switches. The static power dissipation on these powered-off blocks will therefore be minimized, reducing the overall power consumption of the SoC.

10.3.1 Types of Power Switches

Several techniques have been proposed for implementing power gating (Kao et al. 1998; Kosonockyet al. 2001; Kim et al. 2003). Figure 10.7 shows four examples of power switch configurations. In Fig. 10.7a, a PMOS transistor is used as a *Header Switch* and controlled by a dedicated control signal *P_sig*. When *P_sig* equals logic "0," the power-gating transistor conducts and the circuit block is connected to V_{DD} to be able to operate. When *P_sig* equals logic "1," the power-gating transistor will switch off the power supply to the circuit block, drastically reducing its leakage power dissipation. Note that in practice, the power gating transistors are also referred to as sleep transistors. In Fig. 10.7b, an NMOS transistor is used instead of the PMOS transistor as a *Footer Switch*. Setting the control signal N_sig = "1" creates the power-on condition, while N_sig = "0" creates the power-off condition. Figure 10.7c presents an example of a *Symmetric Power Switch* architecture. In this case, there are two control signals *P_sig* and *N_sig*. Generally, *N_sig* is the inverse of *P_sig* such that the two power gating transistors can switch on and off

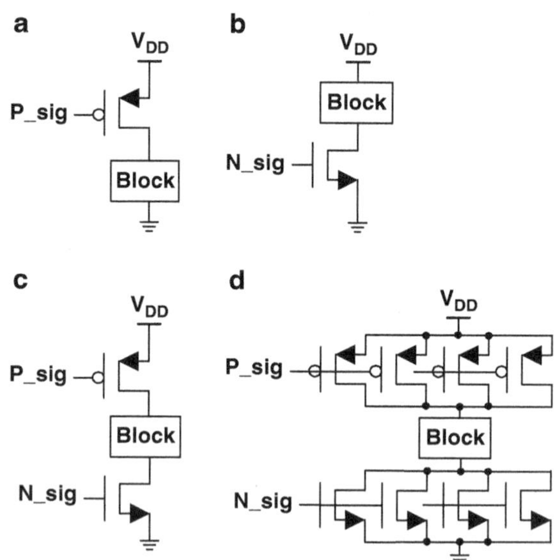

Fig. 10.7 Examples of power switch types

simultaneously. To supply sufficient current to the circuit block, the sleep transistors are designed to be very large if there exists only one transistor for power gating. Considering layout, design for manufacturability, and limiting inrush current when switching on a power domain, it is common in practice to design several transistor segments (named *Segmented Power Switch*) for power gating as showing in Fig. 10.7d. There may be several segments constituting a switch, and each segment can contain one or more transistors. All the transistors in one segment share the same drain, gate, and source. All the segments share the same drain and source. Together, they can provide the necessary supply current to the power domain, and allow switching of the segments in a staggered manner. This strategy will also benefit the physical placement procedure.

10.3.2 Testing of Power Switches

The insertion of power-gating transistors also introduces testing problems. Due to imperfection in the fabrication process causing manufacturing defects, the power-gating transistors may not work properly. For example, a short between the source and drain of the power-gating transistor will cause the switch to be permanently on, rendering the transistor useless for reducing leakage power in the connected domain. Conversely, an open between the source and drain of the power-gating transistor will cause the switch to be permanently off, in which case the connected logic would no longer function. If a number of transistors work in parallel as power switches similar to the ones shown in Fig. 10.7d, it is necessary to ensure that all the transistors are working correctly. If some of the switches are permanently on, the power-reduction ability of the device is impaired. If some are permanently switched off, the remaining transistors have to provide more current. This can affect the device's timing. The current overload may even impact the life-time of the switch. Therefore, it is extremely important to verify the correct functionality of the power switches and validate that they switch on when the control signal is enabled, and switch off when the control signal is disabled.

10.3.3 Methodologies for Testing Power Switches

Method 1: Test for header switch with comparator

In recent years, several methods have been proposed for testing power switches. While this section presents methods for testing header switches, similar methods can be used for testing footer switches. (Goel et al. 2006) proposed using a comparator to test the power switches. The basic idea of the method is to use an XOR gate as the comparator to compare the logic level value of the core's power supply V_{core} with the logic value of the *standby* signal, as shown in Fig. 10.8a.

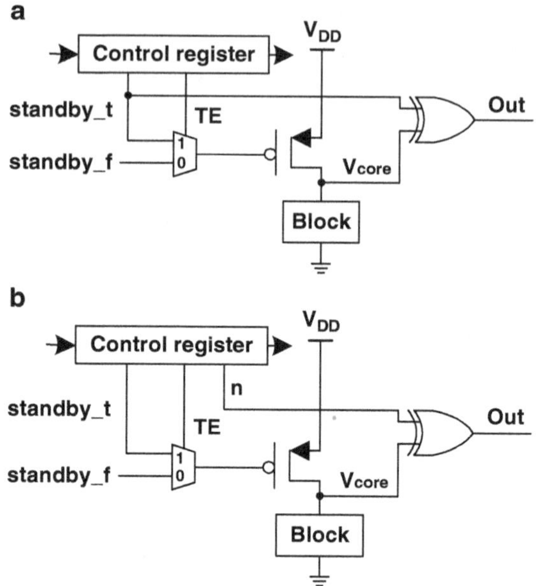

Fig. 10.8 Test circuits for power switches

The signal *standby_f* is the functional signal to control the operation of the power switch, while the signal *standby_t* is the test signal for the same operation. The signal *standby_t* can be provided by means of a shift register. The test-enable signal *TE* must be set to 1 so that the multiplexer placed in front of the power switch selects the signal *standby_t*. To test the power switch with the proposed circuitry, two patterns are needed:

Pattern 1: $TE = 1, standby\_t =$ "1." This pattern should turn off the power switch. V_{core} should be much lower than VDD, ideally "0." Therefore, the output of the XOR gate should be "1" for a correctly operating power switch. By observing the value of the *Out* signal, one can check whether there is a permanent short fault on the power gating transistor.

Pattern 2: $TE = 1, standby\_t =$ "0." This pattern will turn on the power switch. V_{core} should be equal to VDD. Therefore, the output of the XOR gate should be "1" for a correctly operating power switch. By observing the value of the *Out* signal, one can check whether there is a permanent open fault on the power gating transistor.

However, two problems must be considered for this test method. First, the order of the test patterns is essential. Generally, pattern 1 should be applied first. If pattern 2 is first applied, node V_{core} will be charged to VDD and will need a long time for a complete discharge before applying pattern 1. Secondly, the value of the *Out* signal is important. Note that for both patterns, the *Out* signal is always "1" for correct power switch functionality. In this case, if there is a stuck-at 1 fault at the output of the XOR gate, it will mask the power switch faults and make them undetectable.

To circumvent this issue, it is preferable to differentiate the control signal of the power switch and the input of the XOR comparator as shown in Fig. 10.8b.

Controlling the *n* input of the XOR gate enables testing of both the power switch and the XOR gate. This allows fully testing the functionality of the power gating transistors and the test circuitry.

Method 2: Test for header switch with logic gate tree

The disadvantage of Method 1 is the number of input control signals and output observation signals. Since two input control signals (except test enable signal *TE*) and one output observation signal are needed as shown in Fig. 10.8b, the number of additional signals may be huge considering that there is a large number of power gating transistors to be tested in an SoC design. An alternate method is to test these power switches with the same input control signals, and add a multiplexer to select the desired output signal. However, some extra control signals are needed for the signal-selection multiplexer.

Tsai et al. (2008) proposed a logic gate tree method to test power switches, which will minimize the number of input control and output observation signals, as shown in Fig. 10.9a.

Figure 10.9b shows a table containing possible input test patterns and the expected outputs. This pattern set first turns all transistors on in cycle 1 to see whether

a

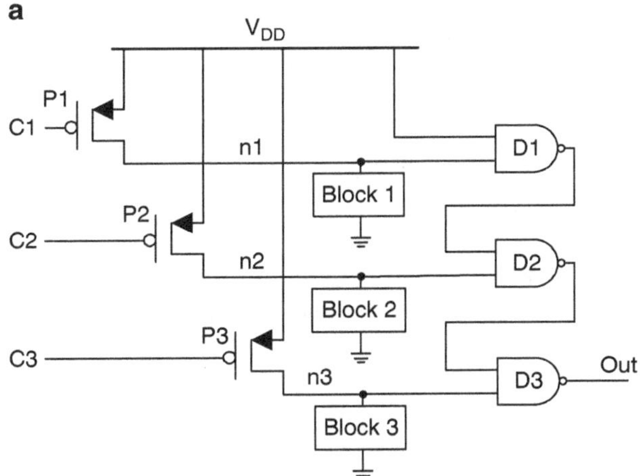

b

Cycle	C1	C2	C3	Out
1	0	0	0	0
2	1	0	0	1
3	1	1	0	0
4	1	1	1	1
5	1	1	0	0
6	1	0	0	1
7	0	0	0	0

Fig. 10.9 NAND gate tree and patterns for testing power switches

there is a permanent switch-off fault on power gating transistor P1. In cycle 2, it turns off P1 by setting C1 to logic "1" to test whether there is a permanent switch-on fault on P1 and whether there is a permanent switch-off fault on P2. Then in cycle 3, it turns off P2 by setting C2 to logic "1" to test whether there is a permanent switch-on fault on P2 and whether there is a permanent switch-off fault on P3. This procedure is iterated until all faults on power-gating transistors are tested.

Unfortunately, there is still a test pattern problem for this method as well. Note that there is no path to discharge nodes $n1, n2$, and $n3$ efficiently; they can only be discharged by leakage. Therefore, sufficient time is necessary between cycles to make sure the test procedure works correctly.

Method 3: Test for symmetric and segmented switch

The previous methods can also be extended for the *symmetric* and *segmented switch* test. For example, consider the comparator-based method proposed in Goel et al. (2006) to test the header and footer switches sequentially as shown in Fig. 10.10. Note that in a symmetric structure, there will be different control register and test enable (*TE*) signals for testing the header switch and the footer switch.

The *segmented power switch* can be tested segment by segment. Figure 10.11a shows the test circuitry for a two-segment power switch. This design contains two segments and each segment contains two transistors. The two segments are controlled by control signal $S1$ and $S2$, respectively. The test pattern set and expected outputs are shown in Fig. 10.11b. In cycle 1, all the segments are turned off to see whether there is a permanent switch-on fault in segment 1 or segment 2. Next, in cycle 2, segment 1 is turned on to see whether there is a permanent switch-off fault in segment 1. In cycle 3, all the segments are turned off to discharge the V_{core} node, preparing to test segment 2 in cycle 4. In this cycle, the switch-off fault in segment 2 is tested.

Method 4: Parametric testing of power switches

A new architecture for parametric testing of micro power switches was presented in Souef et al. (2008). The architecture utilizes the DFT included in the

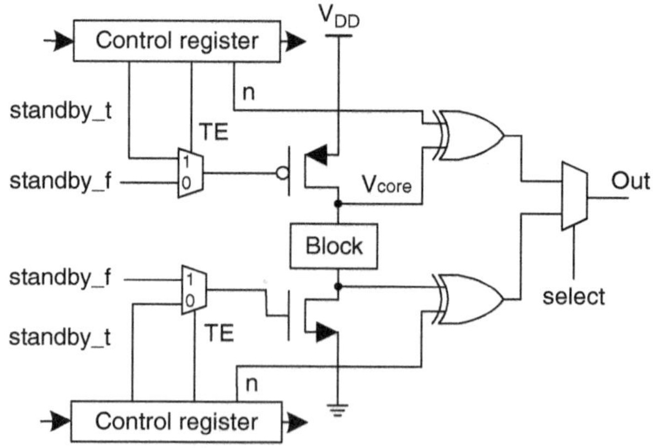

Fig. 10.10 Test circuitry for a symmetrical power switch

a

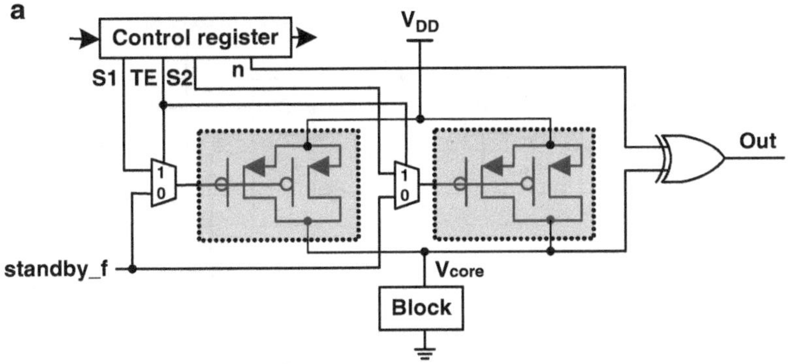

b

Cycle	S1	S2	TE	Out
1	1	1	1	1
2	0	1	1	0
3	1	1	1	1
4	1	0	1	0

Fig. 10.11 Test circuitry and test patterns for segmented power switch

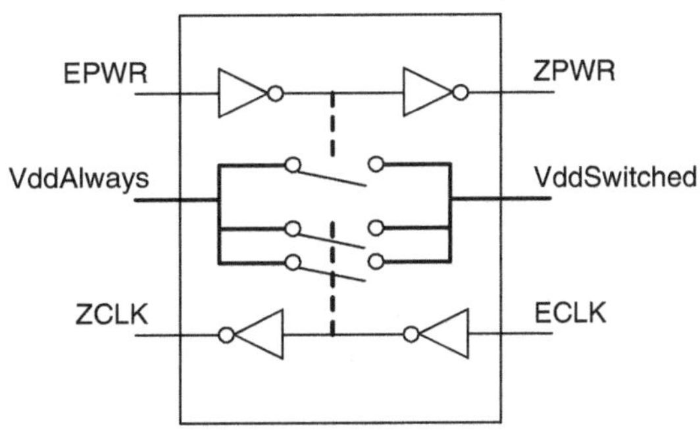

Fig. 10.12 A schematic of a micro switch

device-under-test to measure the resistivity of the micro power switches and verify its correctness. Figure 10.12 shows an example micro switch cell that is composed of two control input signals *EPWR* and *ECLK*. The control signals are buffered using two internal inverter cells on *ZPWR* and *ZCLK*, respectively. The input power supply *VddAlways* feeds the output power supply *VddSwitched* by means of internal transistors represented as tiny switches in the diagram. The first control signal *EPWR* turns on or off a single transistor, giving a highly resistive path between *VddAlways* and *VddSwitched*. The second control signal *ECLK* turns on or off multiple transistors, giving a low resistive path between *VddAlways* and *VddSwitched*. For the target design in Souef et al. (2008), the authors used a micro switch cell that contained ten identical transistors for which one transistor is controlled by *EPWR* signal and the other nine transistors are controlled by the *ECLK* signal.

When using micro switches, it is mandatory to use many of them to properly power the block. The micro switches are daisy chained to control the power-on slew

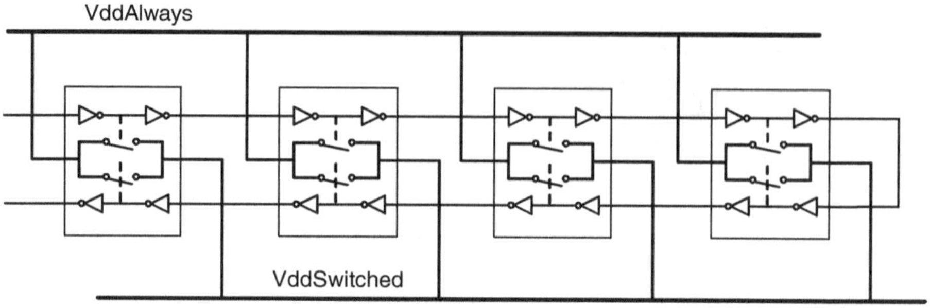

Fig. 10.13 A schematic for a micro switch chain

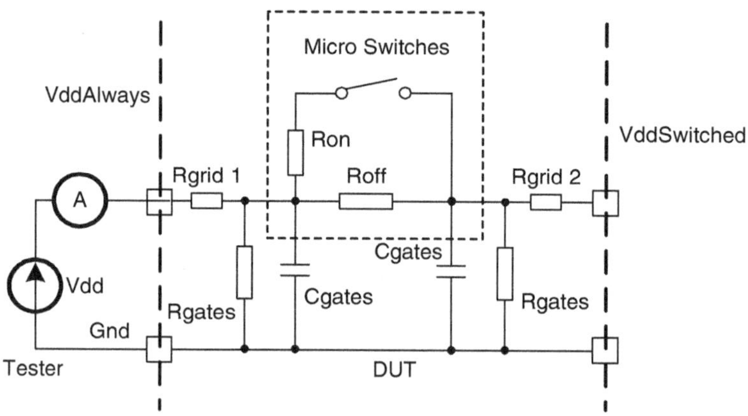

Fig. 10.14 Test environment modeling

rate. The control signals open the switches in a cascaded manner. The first switch activates the second one, which launches the third one, and so on. Figure 10.13 shows the schematic for a micro switch chain.

The objective for performing a parametric test is to detect fine resistive defects introduced during the fabrication process that would make the chip not perform as expected. To extract the resistivity of the micro switches that are embedded inside a chip, a model of the environment is built that comprises the tester and the chip. Figure 10.14 shows the simplified model that was used in Souef et al. (2008).

As shown in Fig. 10.12, the micro switches have two positions:

- Closed switch: It is a closed position for the switch where the resistivity is minimal (only a few ohms). Hereafter, it is referred to as *Ron*.
- Open switch: It is an open position for the switch where the resistivity is maximal (several mega ohms). Hereafter, it is referred to as *Roff*.

The micro switch is modeled with an ideal switch and two resistors, *Ron* and *Roff*. The chip has two different power domains which are the *VddAlways* power domain and the *VddSwitched* power domain. The ground is common for both and is called *Gnd*. The two power domains are assumed to be accessible externally from chip power pads. Inside the chip, each power grid resistance is modeled by resistors in series with the power pads named *Rgridl* and *Rgrid2*. These resistances are important because their value, a few ohms each, will not be negligible during *Ron* measurement. Each power domain is modeled by a simple pair of resistive and capacitive values. The resistance models the equivalent leakage of the domain (in the range of kilo ohms) and the capacitance models the equivalent gate charge capacitance. Each RC network is connected in parallel between the V_{dd} pad and the ground. They are called *RgateVddAlways* and *CgateVddAlways* for the *VddAlways* part, and *RgateVddSwitched* and *CgateVddSwitched* for the *VddSwitched* part.

The above model can be used to determine the *Roff* and *Ron* equivalent resistance of the micro switch clusters. This model relies on static behavior; it does not model the transitions that would hardly be detectable on production testers. This means that the capacitances will not be used, and the measurement on the tester will need to wait until the system settles.

Using the proposed test architecture in Souef et al. (2008), *Ron* and *Roff* values are measured per cluster. After measuring the resistances, a simple analysis can be made to determine whether the power switches are working properly. For example, an excessively low *Roff* resistance will create a higher leakage current, preventing the chip from achieving the autonomy specifications (idle or playtime). Parts showing high leakage need to be screened and rejected on the production line.

The same analysis can be made for *Ron*. An excessively high resistive value on a cluster indicates a defect in one of the micro switches of the cluster. An excessively high *Ron* resistance will affect the performance of the core in the *VddSwitched* power domain. This abnormally high resistance will introduce a voltage drop and the design might not be able to perform at the required speed. Therefore, parts with excessively high *Ron* values must be rejected during test.

10.3.4 Testing Problems and Possible Solution

Methods 1, 2, and 3 digitally test open and closed states of the power switches. Although effective, properly discharging the test node will be a major issue for all of them. Method 4 seems to be able to effectively measure switch open and closed equivalent resistances, *Ron* and *Roff*. This method, however, requires an ATE with sophisticated current measurement capability and must deal with variations in resistances in power distribution network especially in sub-65 *nm* technology designs.

New methods are required to effectively discharge the test node when using Methods 1–3. Since there is no efficient discharge path, sufficient time is necessary between the test pattern applications to discharge the test node via leakage. A discharge path is needed for faster power switch testing similar to the one shown in Fig. 10.15 (Peng and Tehranipoor 2008). However, a test vector pair is needed for

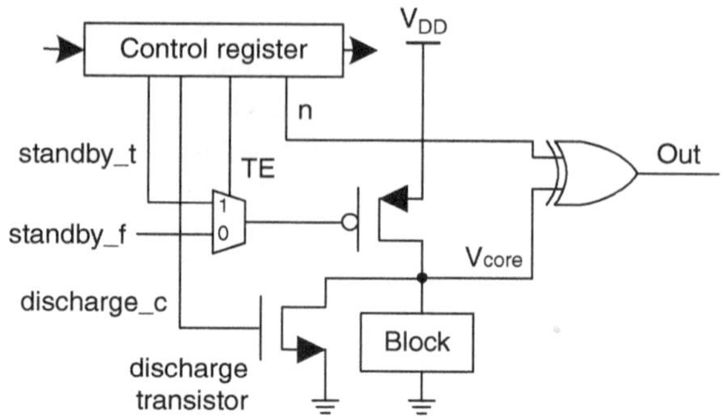

Fig. 10.15 Power switch test circuitry with discharge path

this testing method. With the first test vector, all the power switches must be turned off, and the discharge transistor should be turned on. The second test vector will turn off the discharge transistor and apply the proper pattern to the power switches. This will make the power switch testing structure slightly more complex but it will significantly speed up the test procedure.

10.4 Low-Power Cells

Designs with PSO or multi-voltage domains include a number of special cells such as SRRs, isolation cells, and level shifters. Those cells logically behave like regular gates during normal structural tests, and are therefore tested by conventional fault models. However, their low-power features create additional test requirements. This section will discuss the additional tests required to ensure correct operation of those cells.

10.4.1 State Retention Registers

The ability of **state retention registers** (SRRs) to retain their state when the power domain is powered off needs to be validated during test. Conventional structural tests, if applied while a given power domain is forced on, cannot validate state retention since the domain is not powered off during the capture cycle.

At a minimum, a test such as the following must be applied to test the retention capability. The power domain(s) containing the SRRs under test is powered on and data is shifted in. The ability of each register to retain both a 0 and 1 must be validated, so at least two scan tests are needed. After shift, retention is activated, and the power is cycled while controlling isolation as needed. Retention is then disabled such that the data is restored from the retention latch into the master flip-flop or latch, and scan is unloaded for comparison. Scan cells with retention capability are checked to ensure they unload the value loaded into them. Scan cells without retention capability that are in the domains powered off are masked since their unload value is unknown. The following summarizes the sequence used for each test pattern:

1. Shift in value *v*.
2. Enable retention, or pulse retention *save* clock.
3. Enable isolation cells if this power domain feeds another that will remain on in the next step.
4. Power the domain off.
5. Power the domain on.
6. Disable isolation cells.
7. Disable retention, or pulse retention *restore* clock.
8. Shift out, expect value *v*.

To cycle power during the capture cycles, the ATPG tool must be able to control the power mode. While this can be done by controlling the functional enable logic during capture for those special tests, it is more common and convenient that the test logic inserted to control power modes is used for that purpose. The power mode controlled by the test logic is typically changed either through a JTAG operation, or by changing primary inputs that are allowed to control the power in this test mode.

In addition to testing the retention capability, it is advisable to also test the robustness of the retention capability. The retained value should not usually be affected by application of the state element's clock, or its asynchronous set/reset. If that is the case, and it is possible for those signals to be asserted during retention, then the ability of the retained value to be unaffected by such events needs to be validated. This can be done by retaining a value, then attempting to load the register with the opposite value either through the data port or through the set/reset. The test must be repeated for each clock that can change the value of the register, and for both retained 0 and 1. Up to four tests are needed for a flip-flop with asynchronous controls:

1. Load and retain 0. Apply 1 to data input and pulse clock. Expect 0 after restore.
2. Load and retain 0. Pulse set port. Expect 0 after restore.
3. Load and retain 1. Apply 0 to data input and pulse clock. Expect 1 after restore.
4. Load and retain 1. Pulse reset port. Expect 1 after restore.

The power domain is not powered off during those tests. Note that tests (1) and (2) can be applied in one scan load by successively attempting to load a 1 through the data port, then pulsing the set signal (or vice versa). Similarly for tests (3) and (4).

10.4.2 Isolation Cells

Since the purpose of an isolation cell is to provide a reliable voltage level and corresponding binary value into a powered-on power domain when the source power domain is off, isolation cells must be tested under those conditions as well and not only when both power domains are on.

Isolation cells that are combinational gates such as AND or OR will only have one value on their output during isolation, that being the value when one of their inputs (the isolation enable signal) is forced to the gate's controlling value. So when isolation is enabled, the output of an AND gate would be 0 and that of an OR gate would be 1. In addition to the tests normally applied to those gates when the power domains are on, it is sufficient to add one fault on the output of the isolation cell with a constraint that the input power domain must be powered off. If the isolation cell is a latch and can hold a 0 or 1 during isolation, then the cell needs to be tested for both output stuck-at faults. In both cases, of course, the source power domain must also be powered off when output of the isolation cell is observed by capturing the stuck-at fault effect.

10.4.3 Level Shifters

Level shifter cells are inserted between power domains that operate at different voltage levels, and serve to convert the voltage levels up or down as needed by the driven domain. Functionally, the level shifter is often a buffer. In other cases, the isolation cell also performs the level shifting function.

Level shifters are the simplest low-power cells to test. If the two power domains operate at fixed voltages, the level shifter is adequately tested by the conventional static and at-speed tests. If voltage scaling is used such that the two domains can operate at different voltage levels, the faults on the level shifter must be tested at the different operating conditions. The power modes specified by UPF or CPF can be analyzed to determine the different voltage levels at which those two domains can operate and therefore the different operating conditions at which the tests need to be repeated.

10.5 Power Distribution Network

As described in Chap. 2, the **power distribution network** (PDN) delivers power and ground voltages from power pads in a wire-bond package or C4 bumps in a flip-chip package to all cells in a chip. A robust PDN is essential to ensure correct and reliable operation of modern high-performance VLSI circuits. As technology scales, designs are becoming increasingly sensitive to power supply noise impacting signal and power integrity. Power supply noise refers to the noise on the power and ground distribution network, which reduces the effective supply voltage levels reaching gates in a circuit. In general, high average currents cause large ohmic IR voltage drops (Tang and Friedman 2001) and the transient currents cause large inductive Ldi/dt ground bounce (Tang and Friedman 2000) in the PDN s (Zhu 2004). The main effects of IR voltage drop and ground bounce are on circuit timing and signal integrity. Since power supply reduction will slow down the gate transition, IR drop and ground bounce affect setup and hold times as well as clock skew (Kelley 2006), which could potentially result in silicon failure. The PDN must be designed effectively to minimize the voltage drops and maintain the local supply voltage within specified noise margins.

The sensitivity of circuit delay to the power supply noise increases due to (1) scaling of supply voltage and (2) limited scaling of voltage threshold (Tang and Friedman 2001). Experiments show that a 2.4× increase in gate delay can be observed during simulation with a 12.5% decrease in supply voltage at a 130-nm technology node (Ahmadi and Najm 2003). It has also been shown that a 1% voltage change in a 90-nm technology design causes approximately a 4% change in gate delay (Kelley 2006). From these examples, it is seen that the impact of power supply noise on gate/circuit delays is becoming increasingly significant with technology scaling.

As functional clock frequencies and gate densities increase, the simultaneous switching activity of the chip also increases. This results in higher peak and average currents concentrated in regions of the chip with higher gate density, stressing the PDN that is supplying that region. A large amount of power has to be distributed to all gates and devices across the entire chip through a hierarchy of up to 12 metal layers. This trend creates a major challenge for PDN design, test, and failure analysis (Lin and Chang 2001).

Unlike logic cells, power/ground vias and lines are not accessible by primary inputs or scan flip-flops. Due to the complexity of PDNs and lack of controllability and observability, it is extremely difficult to test and diagnose PDN manufacturing defects.

The PDN reliability challenges caused by technology scaling have been discussed in McPherson (2006). It is a common practice to use 20–40% of the metal resources to build a high-density PDN in modern high-performance microprocessors (Anderson et al. 2001; Tsai 2001). Since the PDN utilizes such a large portion of area in a design, based on inductive fault analysis (IFA) (Shen et al. 1985), the probability of defects occurring on its vias and interconnects can be quite high especially considering that the width of lower level metal layers in modern PDNs is only about 2× the width of circuit interconnects.

There is a fundamental difference between gate defect and PDN defect behavior in a circuit. A gate defect can cause a gate to malfunction and impact the circuit function or timing behavior. A PDN defect, however, may not necessarily result in a gate/circuit functional or timing failure. Contrary to the common assumption that defects in PDNs can typically be detected by implication during functional or structural tests, only a small percentage of defects that result in catastrophic failures may be detected during manufacturing test. (Ma et al. 2008).

In general, defects in PDNs could cause two types of problems in an integrated circuit:

1. Functional/timing problem: Such problems are likely to be detected during manufacturing test.
2. Reliability problem: In this case, the chip under test passes the functional/structural tests but an in-field failure occurs when the applied input pattern

maximizes the effect of the PDN defect on circuit timing, or causes a malfunction in certain gates powered by the PDN.

New test pattern generation methods must be developed to effectively deal with open and resistive-open defects in PDNs. Such patterns will be useful during pre-tapeout stages of the design process where the PDN robustness can be verified. They can also be used for manufacturing test to detect potential PDN-induced timing/functional failures. By using such patterns for dynamic power analysis, the designers would be able to redesign the PDN by inserting additional vias or power wire sizing to eliminate the impact of a potential defect, if one exists after fabrication, resulting in reduced escapes and higher yield. Since defects in power/ground vias and power wires adversely affect yield in addition to severely impacting reliability, time-to-market, and profitability, there is a need to detect these defects and identify their locations in the PDN to improve design quality and in-field reliability.

10.5.1 PDN Structures

Power and ground distribution networks are typically designed hierarchically, from block level to chip level. The block-level PDN, which is also called the local PDN, is designed either in a fully customized way (usually for hard macros) or using an automated router to uniformly arrange the power/ground over standard cells. In high-performance digital ICs, a grid structured network is widely used for global PDN design, while the structure for the local PDN can be different from block to block. In a typical integrated circuit, the lower the metal layer, the smaller the width and pitch of the lines (Mezhiba and Friedman 2004).

Compared to global PDNs which are routed on higher metal layers with wider lines and redundant vias, local PDNs are more prone to be affected by spot defects, process variations, and electro-migration. In state-of-the-art SoC circuits, such as microprocessors, hundreds of millions of power/ground vias and wires are used to deliver power and ground to all cells. Given a uniform distribution of spot defects, according to IFA proposed in Shen et al. (1985), area-intensive routing such as local PDNs will result in more defects. As technology scales, more open/resistive-open vias and wires are expected to occur in local PDNs. In addition to the difficulty of testing PDNs, their restricted accessibility makes it difficult to localize detected defects.

To further clarify the main objective for PDN testing, a grid-structured PDN can be used (see Fig. 10.16) to analyze the impact of possible open defects on the circuit performance and functionality (Ma et al. 2008). As shown in Fig. 10.16, vertical power straps on Metal 6 and horizontal power straps on Metal 5 build up a grid structured network which can distribute power across the entire chip. Ground straps are not shown for sake of simplicity. The lowest level power/ground (P/G) lines on Metal 1 run horizontally as power rails. Standard cells are arranged in rows and connected to Metal 1 P/G wires with two adjacent rows sharing the same power

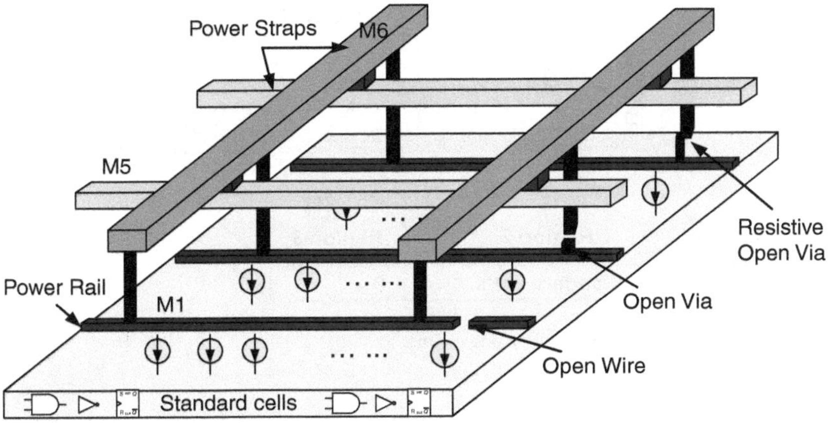

Fig. 10.16 A representation of power distribution network and potential open defects. The ground straps are not shown for simplicity

line. Each P/G rail in Metal 1 is connected by vias to the P/G lines in Metal 6 at the overlap sites, respectively. Metal 1 connects to the Metal 6 PDN through stacked vias. Open/resistive-open defects on power wires and vias have also been shown in Fig. 10.16. Since significantly wider wires are used in the global PDN, the probability of an open defect is extremely low. Thus, only defects on the local PDN (e.g., Metal 1) are considered. At this level, the power line width is about 2× that of circuit interconnects connecting logic cells, P/G vias connecting logic cells to P/G lines, as well as vias connecting the upper-metal layers to power lines in Metal 1. Note that in a stacked via, the via connecting to Metal 1 is the smallest, which is less than a quarter of the power line width.

10.5.2 *Open Defects in PDNs*

While power lines tend to exhibit similar defects as any other interconnect in a design, the impact of these defects on the circuit depends on the power grid design. Most defects in different PDN designs will result in similar faulty behavior. As an example, an open defect on the power line or power via will weaken the power network and result in an increased IR drop, changes in delays in the neighboring cells, and possibly multi-path delay faults; i.e., more than one path may be impacted by the extra delay induced by the IR drop increase. If one of multiple redundant power vias from an upper-layer metal to a lower-layer metal is broken, the network is still connected, but weakened. As a result, the PDN will not be able to supply as much power to the underlying cells, potentially causing timing and functional failures if the current demand for that region becomes too high. Similar behavior may result from shorted vias and shorted power line defects. For instance, a local drop in voltage will occur around the short, but farther away there may be no perceptible difference.

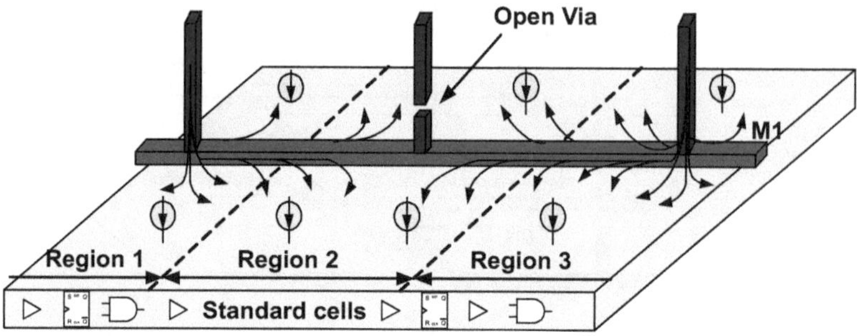

Fig. 10.17 Open via defect on Metal 1 power line

In Fig. 10.17, an open via defect with its two nearby regions is shown. Due to the open via defect, gates underneath the open via (region 2) cannot draw current through the power via in that region (region 2) from the upper-layer PDN. The cells in region 2 will likely draw current from neighboring vias; i.e., vias in regions 1 and 3 as shown with arrows in Fig. 10.17. This will increase the current flowing through the power vias and wires in the two neighboring regions, causing increased IR drop in the neighboring regions. For the cells in the region with the open via, since they need to draw current from farther power vias, their power pin resistance increases and thus they also suffer from increased IR drop. As for open wire defects, in the region with an open defect, only cells that are separated from the nearest power vias will experience an IR drop increase. Cells in the neighboring region of one side of the open defect will suffer from increased IR drop as well due to the increased current drawn through their Metal 1 power rail. Such increased IR drop in neighboring regions due to open defects may result in functional or timing failures, especially when there is already a large amount of switching activity in any of the neighboring regions (region 1 or region 3). In case of resistive opens on vias or P/G wires, the increased resistance will increase the IR drop in the region where the defect exists. Therefore, similar behavior is expected (Ma et al. 2008).

10.5.3 Pattern Generation Procedure

To address the issues mentioned, test patterns must be generated to target and detect open defects in the PDN since relying on incidental detection may negatively impact reliability and **defective parts per million** (DPPM) rates. As shown in Fig. 10.18, the pattern generation procedure consists of three major steps: (1) region sorting; (2) pattern generation; and (3) pattern validation. After the physical design, the post-layout netlist file and the design exchange format (DEF) file are generated. The DEF file, which contains physical placement information of the elements in the circuit, is used to define regions in Step 1 (Ma et al. 2008).

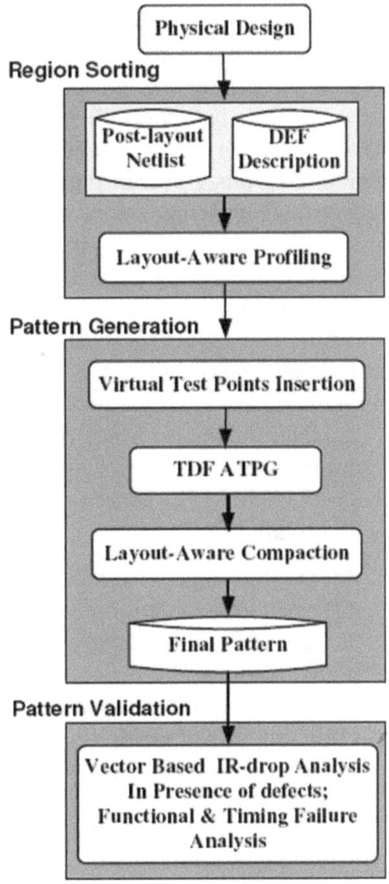

Fig. 10.18 Pattern generation flow for detecting open defects on PDNs

Step 1: Region sorting

Since there are numerous vias connected to power/ground rails in Metal 1, targeting all vias and interconnects would be very time consuming and impractical. To increase the processing speed and reduce the calculation effort, regions are targeted rather than individual vias. This allows a single pattern to detect many potential open defects at one time. Since some regions of the chip are more susceptible to IR-drop issues created by opens than others, a region sorting method is integrated into the pattern generation procedure to further reduce the CPU run time. Thus, only regions where PDN defects could potentially cause functional failures (IR-drop hotspots) or timing failures are considered. To perform region sorting, the design is divided into regions based on the upper-layer PDN structure, which is generated during physical synthesis.

For a chip comprising a rectangular power ring with k vertical and l horizontal power straps, the design is divided based on the intersection of the straps/rings as midpoints for each region in the design, which is similar to power points (power pads in wire-bond and C4 bumps in flip-chips) in very large designs (Ma et al. 2008).

Step 2: Pattern generation

In the second step of the procedure, the goal is to generate a pattern that can exacerbate any defects present in the target region of a local PDN. To highlight these defects, patterns will be generated that introduce large switching activity both in the faulty-via centered region (region 2 in Fig. 10.17) and the two adjacent regions (regions 1 & 3 in Fig. 10.17). By increasing switching activity in all three regions and applying test patterns generated by the procedure shown in Fig. 10.18, open/resistive-open defects in the local PDN that make the PDN nonrobust can be detected by observing the timing failure of the chip under test.

To maximize the switching activity in the selected three regions, transition delay fault (TDF) ATPG is used. Note that the switching activity created in the target regions should not exceed the threshold set by the designer based on functional operation during the power network distribution synthesis step. With only three regions targeted during the pattern generation process, the remainder of the circuit will have minimum switching since scan cells on those other regions are filled with values that reduce switching. To generate patterns with greater switching in the target region, two steps are taken:

1. All the flip-flops in the three regions are considered as observation points during TDF pattern generation.
2. Virtual test points are inserted at the output of gates in these three targeted regions. Outputs of all gates in the three regions are considered as fault sites. The virtual test points provide new observation points to (1) reduce the amount of effort the ATPG needs to propagate the transition to an observation point, (2) increase the number of transitions, and (3) reduce the number of care-bits in the pattern when generating one pattern per fault site. This temporary netlist with virtual test points is used by ATPG to generate test patterns.

This method can generate one pattern using TDF ATPG for each fault site in the selected regions. The procedure treats each net as a fault site. In this case, p vectors will be generated, where p is the total number of ATPG-testable TDF fault sites in the targeted regions. These patterns are then compacted using a layout-aware compaction algorithm to generate a single pattern for the target regions. The compaction algorithm counts the switching activity in the selected region introduced by each vector and only compacts those vectors that will increase the switching activity in the targeted regions. Once the switching activity of the compacted pattern has reached the user-defined upper threshold, the compaction program will stop compacting the vectors. The compaction is a simple, but layout-aware, greedy algorithm that checks the bit-compatibility of each two consecutive vectors in the vector set and compacts them. The launch-off-capture (LOC) method can be used to generate TDF patterns using any commercial ATPG tool.

Step 3: Pattern validation

To validate the effectiveness of the generated pattern in Step 2, open vias or open wires can be intentionally inserted in the PDN in the targeted region. In the presence of an open defect, a large amount of current will be drawn from neighboring vias. If the total current drawn from power vias in the neighboring regions (e.g. regions

1 and 3 in Fig. 10.17) is greater than the threshold, it will likely result in a timing failure (if the critical path going through these three regions is critical) or functional failure (if the voltage drop on a gate in these regions is very large) during test.

If the design fails, then the targeted region is identified to have an open defect since the generated pattern should not fail a fault-free design. If the design still works properly, then the PDN is considered as robust even in presence of open defects. This robustness can also imply that such a via is redundant in the design and can potentially be removed to save area. A simple way to perform the pattern validation is to compare the worst IR drop with and without simulated open defects in the selected regions. The pattern generated in Step 2 can be used as input vector and vector-based IR-drop analysis can be conducted.

10.6 Summary and Conclusions

Clock gaters, power switches, level shifters, and isolations cells are commonly used in practice during the design of low-power digital integrated circuits. Although effective at reducing static and dynamic power consumption, they pose serious challenges to the test engineers. This chapter provided insight into the challenges and methodologies for testing power management cells and control units. It also presented an effective methodology for testing the PDN faults and verifying the integrity of the PDN during pre-tapeout design validation. It is recommended to employ a combination of the proposed methods depending on the cells used for power management to reduce the in-field failures and increase circuit reliability.

Acknowledgments The authors wish to thank the following for their insightful feedback and discussions: Kun-Han Tsai and Greg Aldrich of Mentor Graphics, Teresa McLaurin of ARM, Andy Halliday of AMD, and Ke Peng and Junxia Ma of the University of Connecticut.

References

Ahmadi R, Najm FN (2003) Timing analysis in presence of power supply and ground voltage variations. Proc IEEE/ACM Int Conf Comp Aided Design 176–183.

Anderson CJ, Petrovick J, Keaty JM, Warnock J, Nussbaum G, Tendier JM, Carter C, Chu S, Clabes J, DiLullo J, Dudley P, Harvey P, Krauter B, LeBlanc J, Lu P-F, McCredie B, Plum G, Restle PJ, Runyon S, Scheuermann M, Schmidt S, Wagoner J, Weiss R, Weitzel S, Zoric B (2001) Physical design of a fourth-generation POWER GHz microprocessor, in Proc. IEEE Int. Solid-State Circuits Conf 232–233, February 2001.

Anderson CJ et al (2001) Physical design of a fourth-generation power GHz microprocessor. Proc IEEE Int Solid-State Circuits Conf 232–233.

Chickermane V, Gallagher P, Sage J, Yuan P, Chakravadhanula K (2008) A power-aware test methodology for multi-supply multi-voltage designs. Proc Int Test Conf, Paper 9.1.

Czysz D, Kassab M, Lin X, Mrugalski G, Rajski J, Tyszer J (2008) Low power scan shift and capture in the EDT environment. Proc Int Test Conf, Paper 13.2.

De Colle A, Ramnath S, Hirech M, Chebiyam S (2005) Power and design for test: A design automation perspective. J Low Power Electron 1(1):73–84.

Goel S, Meiger M, Gyvey J (2006) Testing and diagnosis of power switches in SOCs. Proc Eur Test Symp 145–150.

Kao J, Narendra S, Chandrakasan A (1998) MTCMOS hierarchical sizing based on mutual exclusive discharge patterns. Proc Design Automation Conf 495–500.

Keating M, Flynn D, Aitken R, Gibbons A, Shi K (2007) Low power methodology manual for system-on-chip design. Springer, New York.

Kelley K(2006) Using first encounter and voltage storm to optimize peak IR drop or power mesh area. CDNLivehttp://www.cadence.com/rl/Resources/conference_papers/lptp_cdnlive2006sv_kelly_IRDrop.pdf.

Kim S, Kosonocky S, Knebel D (2003) Understanding and minimizing ground bounce during mode transition of power gating structures. Proc Int Symp Low Power Electronic Design 22–25.

Kosonocky S, Immediato M, Cottrell P, Hook T, Mann R, Brown J (2001) Enhanced multi-threshold (MTCMOS) circuits using variable well bias. Proc Int Symp Low Power Electronic Design 165–169.

Lin S, Chang N (2001) Challenges in power-ground integrity. Proc IEEE/ACM Int Conf Comp Aided Design 651–654.

Ma J, Lee J, Tehranipoor M, Wen X, Crouch A (2008) Identification of IR-drop hot-spots in defective power distribution network using TDF ATPG. Proc Workshop Defect Data Driven Testing (D3T).

McPherson JW (2006) Reliability challenges for 45nm and beyond. Proc Design Automation Conf 176–181.

Mezhiba AV, Friedman EG (2004) Power distribution networks in high speed integrated circuits. Kluwer Academic, Dordrecht.

Peng K, Tehranipoor M (2008) An effective test method for power switches in digital ICs. Technical Report, CADT-12–01–2008.

Shen JP, Maly W, Ferguson FJ (1985) Inductive fault analysis of MOS integrated circuits. IEEE Design Test Comp 2(6):13–26.

Souef L, Eychenne C, Alie E (2008) Architecture for testing multi-voltage domain SOC. Proc Int Test Conf, Paper 16.1.

Tang KT, Friedman EG (2000) On-chip delta-i noise in the power distribution networks of high speed CMOS integrated circuits. Proc IEEE Int ASIC/SOC Conf 53–57.

Tang KT, Friedman EG (2001) Estimation of transient voltage fluctuations in the CMOS-based power distribution network. Proc IEEE Int Symp Circuits Sys 5:463–466.

Tsai LC (2001) A 1 GHz PA-RISC processor. Proc IEEE Int Solid-State Circuits Conf 322–323.

Tsai Y, Hsin-Chu (2008) Method and apparatus for testing power switches using a logic gate tree. US Patent 7,394,241, 1 July 2008.

Zhu QK (2004) Power distribution network design for VLSI, Wiley, New York.

Zyuban V, Kosonocky SV (2002) Low power integrated scan-retention mechanism. Proc Int Symp Low Power Electronics and Design 98–102.

Chapter 11
EDA Solution for Power-Aware Design-for-Test

Mokhtar Hirech

Abstract Previous chapters of this book covered various techniques for testing low-power devices. The objective of this chapter is to help design-for-test experts understand challenges related to implementing these techniques in EDA flows. EDA tools have been constantly challenged with problems related to integrating DFT insertion with logical/physical synthesis and timing closure. Power gating techniques add power as a significant dimension to the complexity. DFT insertion tools must be made power-aware so that DFT logic such as scan can be correctly architected across different power domains and voltage islands. At the same time, any DFT inserted logic must have the minimum possible impact on power consumption. The user has, now more than ever, to be well prepared to make trade-off decisions as each technique brings its new set of constraints and implementation costs. In this chapter, we describe the challenges facing the EDA industry, discuss some existing solutions, and finally, we propose some future directions.

11.1 Introduction

Satisfying power dissipation requirements has become a critical component for design closure in a large number of integrated circuits. It is now in the designer's mind since the start of a design project. Traditional design-for-low-power techniques such as clock gating are no longer sufficient. Designers are forced to radically change their design practices by explicitly implementing more intrusive power management schemes. Designs are now architected based on challenging power consumption profiles, making use of multi-voltage domains, power domain shutoff, and multi-threshold cells. The creation of technology libraries requires more effort as cells need to be characterized for different voltages, and power and ground pins have to be explicitly visible in the logical domain. Every point tool in a design flow must now

M. Hirech (✉)
Synopsys Inc., Mountain View, CA, USA
e-mail: hirech@synopsys.com

P. Girard et al. (eds.), *Power-Aware Testing and Test Strategies for Low Power Devices*,
DOI 10.1007/978-1-4419-0928-2_11, © Springer Science+Business Media, LLC 2010

consider low-power intent as an unavoidable constraint from the user. Analysis tools take guidance from the power intent, but do not change the power intent, whereas implementation tools update and refine the power intent, thus modifying the design.

When it comes to test automation, the over-the-wall testing strategy has become a dead path. Test strategy is no longer defined solely by test experts. It has to be defined at the very early stage by a team of designers and test experts. If this task is not planned adequately, it could be very challenging, not to say impossible, to achieve good quality of test within an acceptable power consumption budget. Power-aware ATPG techniques have been and continue to be helpful in generating power-optimized patterns. However, their efficiency is sometimes limited, and not always predictable. Predictability, a key differentiator in today's flows, can only be achieved through DFT. In other words, what you get is what you plan for.

Some new low-power challenges facing the DFT engineer include the following:

- Power dissipation is so important that designs are now architected in such a way that, at a give time, only few parts are alive in functional mode of operation. These devices cannot be tested with all parts switched on at the same time and testing one part at a time will result in significant test application time.
- DFT logic such as scan has to be carefully architected across power domains. Otherwise, the outcome may be inefficient due to an excessive number of level shifters and isolation cells, or to unusable scan chains.
- It is difficult, and sometimes impractical, for ATPG tools to determine when power domains are switched on and off during test, particularly if this is intended at the pattern level. To manage in-rush current, power switches are usually powered on in sequence. If the ATPG is to power on one power domain, it has to account for the related latency in the test protocols.
- The test engineer must be able to test the power management structures themselves. While level shifters, isolation cells, and retention states are easy to test, power switches are very challenging due to the analog nature of their faulty behavior.
- With multi-voltage designs, the location of DFT blocks such as scan compression must be carefully considered. If the block is not placed in the same power domain as the configured scan chains, additional level shifters and isolation cells may be required.
- Sometimes power switch cells are simple placeholders in the logical domain. Their implementation details are made available only later on in the physical domain.
- DFT insertion tools must be fully integrated in design flows to make sure that power management structures, added by DFT, are preserved by subsequent design optimizations.

In this chapter, we will review test automation aspects of design flows for low power, describe test automation objectives, review challenges facing EDA from the test automation perspective, describe some solutions, and close with future development alternatives.

11.2 Design Flows for Power Management

Before power dissipation became a critical constraint, there was only a single Vdd/Vss power supply pair in a design. All power supply connections were implicit, and every cell in the design was powered on all the time, at a constant voltage. EDA tools were successfully used on very complex designs. The real challenge, back then, was in timing closure, area optimization, and routing congestion.

Today, advanced low-power designs with power gating and multi-voltage domains break many assumptions that were built into EDA tools. The power view must be considered orthogonal to traditional logical, physical, and test views (Goering 2008). A same logic design may have many possible power implementations. To guarantee high productivity within flows and inter-operability between flows, a concise specification of structural power aspects is being standardized. Two competing format exist. One is CPF (for common power format) (CPF 2007) and the other is UPF (for unified power format) (UPF 2007). The intention here is not to compare one format against the other, but to use one format as an example to show how structural power aspects are defined. We explore Accellera's UPF because the author has been involved in that effort and because UPF has been standardized as IEEE 1801 (Brophy 2008; Std-1801 2009). UPF offers consistent semantics for implementation and verification tools. As shown in Fig. 11.1, synthesis, simulation, formal verification, and physical implementation all rely on this single format. Section 11.2.2 illustrates the usage of UPF on an example design.

11.2.1 Multi-voltage and Power Gating Context

A typical advanced low-power design may comprise the following design concepts and capabilities: (1) Multiple power supplies, multiple voltage islands, and possibly

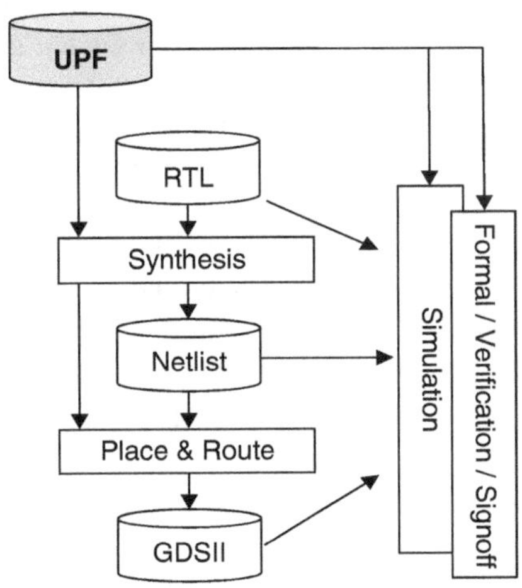

Fig. 11.1 UPF-based low-power design flow

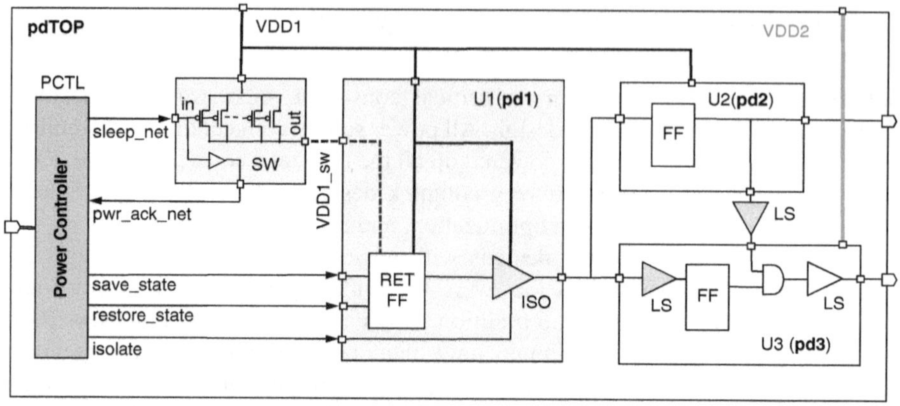

Fig. 11.2 Multi-voltage and power gating concepts

different power domains of the chip; (2) Power-down of selective power domains while ensuring proper isolation between shutdown and live parts, as well as ensuring proper retention of flip-flop states; (3) Supply voltage scaling/switching, together with frequency scaling/switching, across multiple scenarios (operation modes); (4) Clock-gating of flip-flops; (5) Mapping of technology cells from libraries with different threshold voltage, and so on.

Figure 11.2 shows a schematic illustration of a typical low-power design where dedicated cells (isolation cells, power switches, level shifters, and retention registers) are used along with a power controller. In this example, we have four power domains: pdTOP is defined for the top-level design and it is always powered on, pd1 which includes instance U1 is designed to be switched off when not in use, pd2 which includes instance U2 and pd3 which includes U3 are both always powered on. Instances U1, U2, and the power controller logic operate at the same supply voltage, VDD1, as the top-level design. However, instance U3 has a different supply voltage, VDD2. Implementing this configuration would require the following power management structures: a power switch cell SW that is needed to power on/off power domain pd1; a state retention strategy in pd1 so that a state in U1 could be saved before pd1 is shut off and then quickly retrieved when pd1 is back on again; an isolation strategy to ensure that signals out of pd1 do not float when pd1 is powered down. Isolation cell outputs are safely clamped to a known logic value using an isolation enable signal which prevents noise and other issues from propagating through active power domains; a level shifter strategy that translates the voltage swing of signals to compatible value is required between power domains that operates under different voltage supplies; and finally, a power controller circuitry that needs to regulate all of these power management structures.

11.2.2 *Unified Power Format*

UPF is a joint work between major EDA vendors and well-established semiconductor companies developed under Accellera. The initial work resulted in version 1.0 that is now fully implemented in EDA tools. UPF is the basis of the IEEE 1801 standard. The UPF standard is dedicated to the specification of the structural aspects of power intent (Brophy 2008). Operating environment details (process, temperature, operating voltage data, and leakage power calculation) are not part of UPF. They need to be provided separately.

The following is an example UPF description for the design of Fig. 11.2. This example is for illustration purposes only. It shows the steps and commands used to define power management structures and legal power states of the design. In this example, courier font is used for UPF keywords.

11.2.2.1 Creation of Power Domains

The creation of power domains is the first step in the UPF description. The user creates a power domain as a collection of design objects that operate with the same power supply nets. In this example, we create four power domains.

```
create_power_domain pdTOP
create_power_domain pd1 -elements U1
create_power_domain pd2 -elements U2
create_power_domain pd3 -elements U3
```

11.2.2.2 Top-Level Connections

The second step is to create power supply ports and power supply nets, and to define their association to the power domains.

```
create_supply_port VDD1 -domain pdTOP
create_supply_port VDD2 -domain pd3
create_supply_net VDD1 -domain pd1-reuse
create_supply_net VDD1 -domain pd2 -reuse
create_supply_net VDD2 -domain pd3
create_supply_net VDD1_sw -domain pdTOP
create_supply_net VDD1_sw -domain pd1 -reuse
connect_supply_net VDD1 -ports VDD1
connect_supply_net VDD2 -ports VDD2
```

11.2.2.3 Primary Power Nets

This step is to establish associations between power domains and primary power and ground nets.

```
set_domain_supply_net pdTOP -primary_power_net VDD1
-primary_ground_net VSS
set_domain_supply_net pd1 -primary_power_net VDD1_sw
-primary_ground_net VSS
set_domain_supply_net pd2 -primary_power_net VDD1
-primary_ground_net VSS
set_domain_supply_net pd3 -primary_power_net VDD2
-primary_ground_net VSS
```

11.2.2.4 Creation and Mapping of Power Switch Cell

This step is for directives that create and map the power switch cell SW.

```
create_power_switch S*W -domain pdTOP
-input_supply_port {in VDD1}
-output_supply_port {out VDD1_sw}
-control_port {sleep sleep_net}
-ack_port {pwr_ack pwr_ack_net}
-on_state {my_on_state in {sleep}}
map_power_switch SW -domain pdTOP -lib_cells SWLIBCELL
```

11.2.2.5 Definition of Isolation Strategy and Isolation Control

The following specification commands define the isolation strategy and the isolation control for power domain pd1 as it is meant to be switched on and off during functional operation of the design.

```
set_isolation iso_pd1 -domain pd1
-isolation_power_net VDD1
-isolation_ground_net VSS
-clamp_value 1
-applies_to outputs
set_isolation_control iso_pd1 -domain pd1
-isolation_signal PCTL/isolate
-isolation_sense high
-location self
```

11.2.2.6 Retention Strategy and Retention Control in pd1

Retention strategy and retention control are established for power domain pd1 to make sure its state is saved before it is shut-off and then restored when it goes back on again. The mapping of the retention cell to a technology library cell is explicitly specified.

```
set_retention ret_pd1 -domain pd1
-retention_power_net VDD1
-retention_ground_net VSS
set_retention_control ret_pd1 -domain pd1
-save_signal {PCTL/save_state high}
-restore_signal {PCTL/restore_state high}
map_retention_cell ret_pd1 -domain pd1 -lib_cell_type RETLIB-
CELL
```

11.2.2.7 Power State Table

This section specifies the legal states of the design. In state s0, all power domains are on. In state s1, power domain pd1 is off while all other domains are on.

```
add_port_state VDD1 -state {HV 1.08}
add_port_state VDD2 -state {LV 0.9}
add_port_state SW/out -state {HV 1.08} -state OFF off}
create_pst my_pst -supplies {VDD1 VDD1_sw VDD2}
add_pst_state s0 -pst my_pst -state {HV HV LV}
add_pst_state s1 -pst my_pst -state {HV OFF LV}
```

11.2.2.8 Level Shifter Strategy

This is to specify level shifter strategy for signals that flow between different power domains. Each power domain could have different strategies that dictate the type, location, and applicability to interface signals.

```
set_level_shifter ls_pd1 -domain pd1 -applies_to outputs
-location self
set_level_shifter ls_pd3 -domain pd3 -applies_to both
-location self
set_level_shifter ls_pd2 -domain pd2 -applies_to outputs
-location parent
```

11.3 Test Automation Objectives

As power consumption becomes a mainstream concern, power design intent and requirements must be considered at each stage in the design flow. Each verification or implementation tool in the flow must understand the designer's intent and honor the requirements. In this section, we describe test automation goals that are required in advanced low-power designs.

11.3.1 Quality of Results

Test engineers, pressed by increasingly stringent test requirements, continue to strive for key traditional objectives: low pattern count, the highest possible fault coverage, and the shortest test application time. Testing advanced low-power designs adds an important new dimension to an already complicated task. Today, the goal is to achieve the traditional objectives while also keeping power under control. Both DFT and ATPG tools must be made power friendly and must adjust to the new conflicting requirements.

Test tools must handle designs with various power management implementations. For DFT tools, this includes the following requirements: (1) DFT architecture must be made power-aware; (2) DFT techniques should be carefully designed to reduce power consumption during test to acceptable margins; and (3) DFT should facilitate access to power management structures. For ATPG tools, this includes the following requirements: (1) ATPG tools have to be made power-aware when dealing with power domains during pattern generation; (2) They must generate patterns for the power management circuitry; and (3) They must produce power optimized patterns for acceptable power consumption during test. In the context of advanced low-power designs, the user must be prepared to make the necessary trade-off decisions while the test tools are enhanced to help the user make informed decisions. The objective is to manage power consumption while achieving acceptable test application times and pattern counts. There is no flexibility as far as test coverage is concerned.

11.3.2 DFT Requirements in Mission Mode

In mission mode, DFT logic must be transparent. For formal verification, the transparency means that DFT logic must be disabled so equivalence checking properly verifies the design before and after DFT insertion. In the context of low-power devices, the transparency means that, ideally, DFT logic should not have any negative impact on power consumption during functional operation. But, in practice, there is almost always some impact caused, for example, by activity on test dedicated clocks and internal power in test logic. The objective is to minimize that impact.

11.3.3 Integration into Design Flows

Designers optimize designs based on a variety of constraints such as timing, placement, layout, area, power, testability etc. To build powerful EDA solutions through efficient one-pass methodologies (Beausang et al. 1996), these various optimization engines must be built over a common synthesis platform. This enables designers to implement their designs from RTL-to-GDSII without costly iterations or managing multiple netlist and constraint formats.

Building power and DFT solutions on a common synthesis platform enables optimal implementation of power management and DFT structures leading to high-quality manufacturing diagnostics and working silicon with low-test costs.

More specifically, since DFT has to also add level shifter cells and isolation cells, it is more efficient and practical to use the same synthesis engines for this work. DFT tools do not need to know all power intent described details, such as power supply nets, and other synthesis concepts such as "always ON" logic which refers to active gates inside powered down domains. An efficient integration with synthesis means that DFT insertion would transparently use these synthesis provided functionalities. The role of DFT tools will then become more focused on improving the test related value. The challenge of the integration to synthesis is to make sure that any gates inserted by DFT tools should not create timing violations or cause congestion issues in physical synthesis, and they have to be preserved by post-DFT synthesis optimizations of the netlist.

11.4 Integration of Power Management Techniques in Design-for-Test Synthesis Flows

As part of the synthesis-based flows (De Colle et al. 2005; DFTMAX 2008), test automation products need to understand power related constraints and management structures. For a DFT product, this translates into the following considerations: (1) Each step in the DFT insertion process must be made low-power aware; (2) Additional work has to be done in order to test the power management structures themselves; (3) The tool must allow the user make the best trade-offs between DFT architecture options and their impact on power management structure needs. On the other hand, an ATPG tool: (1) Must be guided by a power budget, usually in term of toggling activity; (2) It needs to support the power management structures themselves; (3) Finally, it also has to help the user make trade-off decisions in the area of pattern count, test application time and power consumption. This section describes the new capabilities required for test automation.

As shown in Fig. 11.3, DFT synthesis involves a multi-step process. The test protocol creation helps the user define test protocols. Usually the user provides an initialization sequence, and the tool completes the protocol based on user specification of test control signals, clocks, and reset signals. The test design rules checking phase

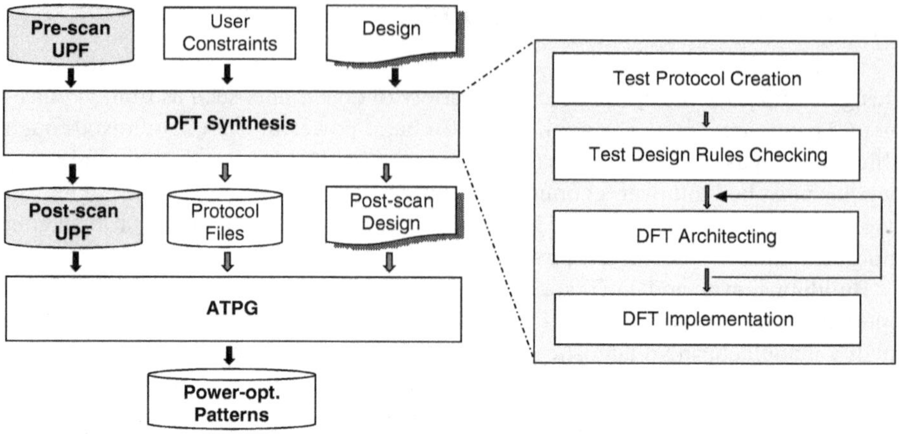

Fig. 11.3 The key components of DFT synthesis

analyzes the design, and based on the test protocol, it checks for critical issues that will negatively impact the testability of the design. Typical issues include clock and reset controllability. The user needs to fix any critical issues before moving to the DFT architecting phase. This phase does not change the design. Therefore, it allows the user to explore many DFT architectures based on variations of constraint specifications. The DFT implementation is the final step of the process. It realizes the DFT architecture in the form it was previewed in the exploration phase. This results in DFT insertion and design optimizations that take care of any design constraint violations introduced by DFT on global control signals.

11.4.1 DFT for Low-Power Rules

The implementation of DFT techniques rely on a set of test design rules. These rules are design requirements that could be generic or specific to the type of DFT technique being implemented. When complied with, this enables standard ATPG tools to generate high coverage test vectors. Best design practices require full controllability to clock, reset, and test control signals. These basic rules form a common DFT foundation. Additional architecture-specific rules then distinguish a variety of distinct DFT methodologies. Scan compression, for example, is very sensitive to X-generators, while logic BIST requires busses to be 1-hot during test. Further detailed descriptions are found in Keating et al., (2007).

A test design rule checking program is a key component of a DFT synthesis flow. Given a design and a DFT technique to be implemented on that design, the tool checks the design against the relevant rules, and catches violations. Critical violations have to be fixed, otherwise this can lead to lower or even unacceptable quality of test coverage. If those violations are not fixed early in the design flow, the user will incur costly design iterations before achieving a testable design.

Power gating creates additional challenges for DFT. These challenges include the following:

- Ensure DFT logic is correctly architected across power domains.
- Provide external control and observation for power gating, retention and isolation signals.
- Manage maximum current and power limitations during test (Hattori et al. 2006a).
- Test the power switching network for correct behavior.
- Test shutdown, isolation, and retention behavior.
- Test the power gating controller.
- Ensure the stability of test modes during test.

Solving these challenges requires a new set of design rules (Keating et al. 2007). These rules are more critical than the traditional rules as, if not complied with, they could lead to untestable designs or even cause damage to designs if test power is considerably more than what is allowed for functional operation. The sub-sections below detail some of the important rules.

11.4.1.1 Stability of Test Modes during Test

The stability of a test mode is a key requirement in the test of power gated designs. If a power domain is powered on (or powered down) in a given test mode, then this state has to be maintained during test. Any nonessential power down (or power up) during scan shift and/or capture must be avoided. Figure 11.4, shows an example where this situation could happen. Scan shift through flip-flop A would power on and down power domain pd1 as the flip-flop directly controls the power switch SW.

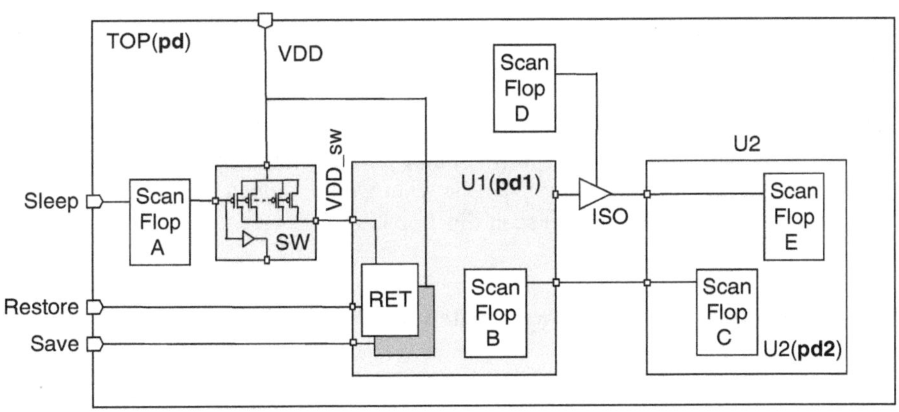

Fig. 11.4 Example of design with test DRC violations

11.4.1.2 Controllability of Isolation Enables

Isolation cells must be disabled (i.e., made transparent) when their associated power domains are powered on, and must be enabled (their outputs forced to 0 or 1) when their associated power domains are powered down. During a scan shift or capture cycle, the enable pin of those cells must not toggle; Otherwise, this would lead to scan chain blockage during scan shift or to serious degradation of fault coverage. Figure 11.4, shows an example of such a violation with scan flip-flop D directly driving the enable signal of the isolation cell at the output of power domain pd1.

Another important rule to watch for is the synchronization between the state of a power domain and its associated isolation cells. That is, an isolation cell must be enabled whenever its corresponding power domain is powered down, and conversely, it must be disabled whenever its corresponding power domain is powered on.

11.4.1.3 Controllability of Retention Signals

Two new issues must be considered when testing switched power domain that has retention capabilities. The first issue may occur when the power domain is powered on. In this case, retention registers operate as regular flip-flops whose control signals, including save and restore, must be controllable for a proper operation. The second issue may occur when the power domain is powered down. In this case, scan shift or capture must not corrupt the retention state. As an example, the save signal that controls the retention latch within register RET in Fig. 11.4 must not toggle during test as it could corrupt the value that is saved on the latch before power domain pd1 is shut down. A corruption could happen when an incorrect logic value is applied on the save signal because of bad test protocol. It could also be caused by an incorrect scan architecture that puts register RET on the same scan chain with other scan cells from live power domains.

11.4.1.4 Scan Architecting across Power Domains

Scan chains must not span power domains that could be independently powered on or down. Any violation of this type will make a scan chain useless because part of the chain is simply not powered on. This is what would happen for a scan chain that goes through scan flip-flop B and scan flip-flop C in Fig. 11.4.

11.4.1.5 Controllability of Power Switches

Switched power domains must be fully controllable in test mode so that the design can be exercised in all valid power states, and illegal transitions between power states must be prevented (Brophy 2008).

11.4.1.6 Power Mode to Test Mode Mapping

To test a power-gated design, user provided power modes (through power state tables) are analyzed and a subset of the modes is then mapped into test modes (Chickermane et al. 2008). Multiple test modes may be required to cover all the logic within the different power domains and between the power domains. The test of logic within a power domain requires a mode where this power domain is on, whereas the test of logic between power domains, referring mainly to isolation logic, would require a mode where source power domains are powered down. The test modes must be a subset of available power modes because the introduction of additional switches during DFT synthesis is generally unacceptable.

11.4.2 Handling of State Retention Registers

Switching off portions of a design during functional mode is a common mechanism to significantly cut down on leakage and dynamic power. The implementation of this technique requires a state retention feature within power domains that could be switched off during functional operation. The state of blocks within such power domains is simply lost when the power supply is switched off. This is why state retention registers are used in practice. These registers are composed of regular state elements that are used during functional operation, with additional latches or flip-flops to store and restore the state of the logic after returning from sleep mode. As show in Fig. 11.5, the regular state element is powered by the primary power supply of the associated power domain, while the retention latch or flip-flop is powered by a backup power supply.

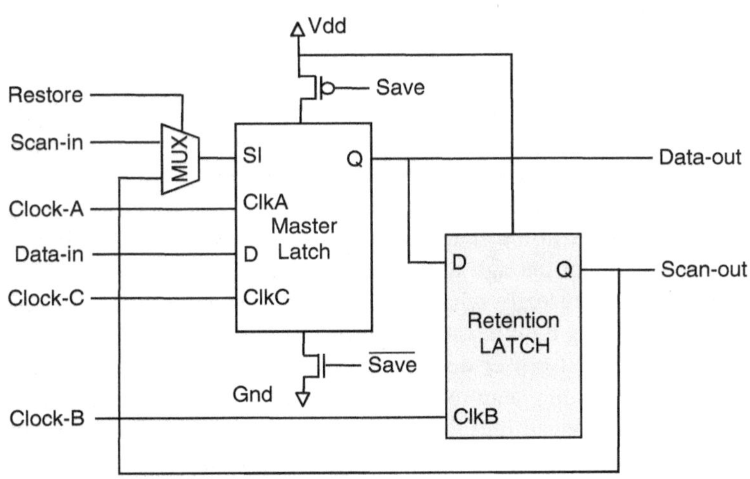

Fig. 11.5 Example of state retention register

However, implementing the scan mechanism separately from the retention mechanism leads to power overhead in the active mode as well as area overhead. The design described in Fig. 11.5 is a low-power, low-area overhead data retention mechanism integrated into a conventional scan flip-flop design. This provides significant area and power efficiency over separate implementations of scan and data retention schemes. The integrated design provides a power-efficient storage mechanism to retain data during power-down (or sleep) mode and an extra path to restore the data from the retention logic to the functional mode flip-flop.

This design enables three modes of operation. During functional mode, the cell operates as a conventional latch. During scan mode (Restore $= 0$), the cell operates as a master–slave flip-flop. While entering the sleep mode of operation (Save $= 1$), clock Clock-B saves data in the retention latch. On returning from the sleep mode, Restore is set to 1 and clock Clock-A restores data from the retention latch to the main latch (Master latch). By integrating the retention mechanism in the scan flip-flop, the additional area and power overhead is greatly minimized due to a reduction in the number of gates that are switching in the active mode (De Colle et al. 2005).

Many possible implementations of retention registers exist (Zyuban and Kosonocky 2002; Ravi 2007; Chakravadhanula et al. 2008). Some cells use the same pin for save and restore operations, while others require two separate pins.

The challenges of supporting retention elements in a one-pass DFT synthesis flow are as follows:

- **Scan replacement**
  During scan replacement, a scan-equivalent register is substituted in place of each regular register. This process is automatically done by synthesis for most registers and scan styles. In the case of state retention registers, this might not always be possible without guidance from the user. The automatic approach would require the user to include some guidance in the library model for the cell. This includes attributes such as power cell type and active state for Save and Restore control pins.
- **Test of state retention**
  The test of state retention happens during two stages. First, the state retention registers are tested as regular scan registers in a mode where the parent power domain is powered on. In this mode, the save and restore signals must be fully controllable just like clocks and reset signals. Any issue found at this level will indicate a problem with the state retention in the power domain. However, this level of testing is not enough as it does not guarantee that the state retention mechanism works correctly when the parent power domain is powered down. State retention testing requires at least three test modes to be tested sequentially. In mode 1, the parent power domain is powered on, and state retention registers are tested as regular scan registers; at the end of the operation, the state of sequential nodes is saved into the retention cells. In mode 2, the parent power domain is switched off. And, in mode 3, the parent power domain is powered up again, and state of the retention cells is inspected and is expected to match the retained state before the domain was switched off. Note that the transition from

sleep mode to power up mode will require a certain number of dummy cycles in the initialization sequence associated with mode 3. The number of dummy cycles is design-specific as it depends on the final implementation of the switch cells, and it determines the latency that is due to the sequential activation of power switches to avoid in-rush current problems.

For the reasons we just mentioned, DFT tools will need to be enhanced by accepting user guidance during test protocol generation.

11.4.3 Impact on DFT Architecture

The support of power gating methodologies has a huge impact on the DFT architecting process. This process must consider power modes to create functional scan chains across power domains. It must allow the user to explore different architecting options and make the best trade-off decisions in term of scan chain budgeting/balancing vs power domain crossings, or to configure compression logic with respect to power domain crossings, etc. At the high level test scheduling has to be power-driven to maximize power savings during test. This section discusses what it takes to make DFT architecture power-aware

11.4.3.1 User Control

A DFT product should be flexible in order to handle different DFT methodologies. It needs to offer orthogonal options regarding mixing of scan structures across different power domains. Depending on how critical the power constraint is, the user could opt to test all power domains on at the same time or decide to test one or a couple of power domains at a time. Figure 11.6 shows a case where scan chains are not mixed across power domains operating at different voltages, while Fig 11.7a shows an example of scan chain mixing across the same power domains The same example could be modified to show scan cell mixing across power domains that can be independently switched on and off.

11.4.3.2 Minimizing Domains Crossing

When DFT structures span different power domains, the tool has to make sure the number of crossings is the minimum possible. In the low-power context, it is very important to minimize domain crossing as each voltage crossing would require a level shifter, and each power domain crossing would require an isolation cell. Both these cells should be used only when necessary. A level shifter has two power rails. Not paying attention to the number of level shifters (as shown in Fig 11.7b) could have a negative impact on routing and area. An isolation cell could have similar impact as it requires a global signal to control the isolation enable. Figure 11.7a shows an example of what is expected when DFT architecture is power-aware.

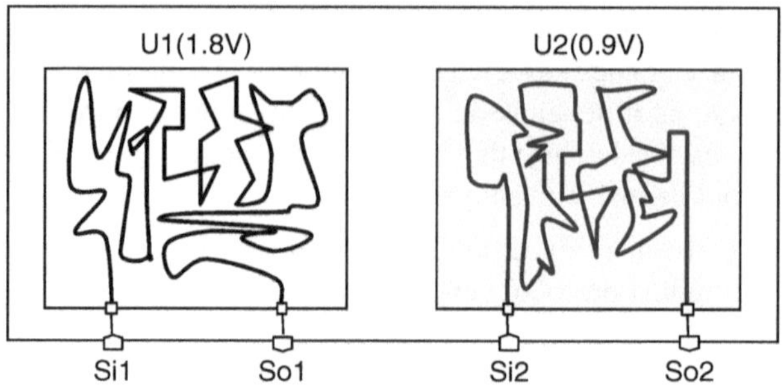

Fig. 11.6 Scan chains not mixed across power domains

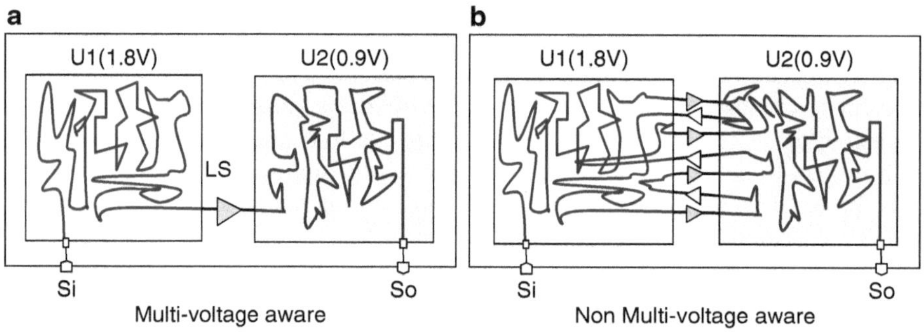

Fig. 11.7 Scan chains are mixed across power domains

DFT methodology should be carefully implemented having in mind the cost in terms of number of level shifters and isolation cells. As an example, if one decides to implement scan compression, it should not be wise to put the decompressor logic, the compressor logic and the related reconfigurable scan chains in different power domains. This would unnecessarily increase the number of isolation cells.

As we said earlier, the tool has to help the user make trade-off decisions between scan chains budgeting/balancing which help in test application time and power domain crossings. If the user, for example, decides not to mix scan chains across power domains, then it becomes very unlikely that scan chain balancing could be achieved.

11.4.3.3 Impact on Scan Chain Reordering

To minimize the area overhead due to level shifter cells, multi-voltage aware scan chain assembly attempts to consider the voltage supply of scan cells while (re)ordering the cells in a scan chain so as to minimize the occurrence of chains that cross voltage regions. Some advantages of multi-voltage aware scan chains assembly include the following: (1) Reduced area overhead due to fewer level shifters, and (2) Reduced wire length and routing congestion since cells in a voltage domain are ordered based on placement information. Another point to note here is that multi-voltage aware scan chain assembly might increase the number of synchronization elements required in case of multiple clock domains. This falls under another trade-off category. The DFT tools should provide options to allow the user decide on which option to implement.

Figures 11.8 and 11.9 show results from an experiment we did earlier to illustrate the benefits of multi-voltage aware scan chain ordering (De Colle et al. 2005). The

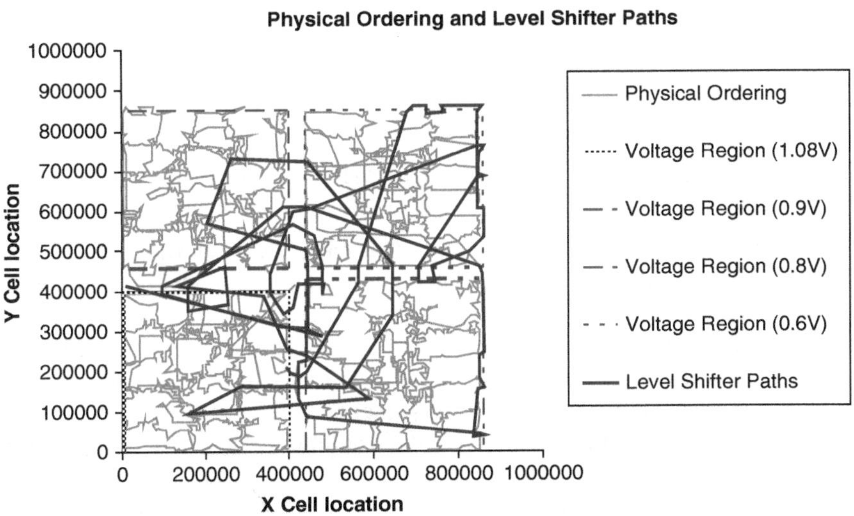

Fig. 11.8 Non-power-aware scan chain reordering

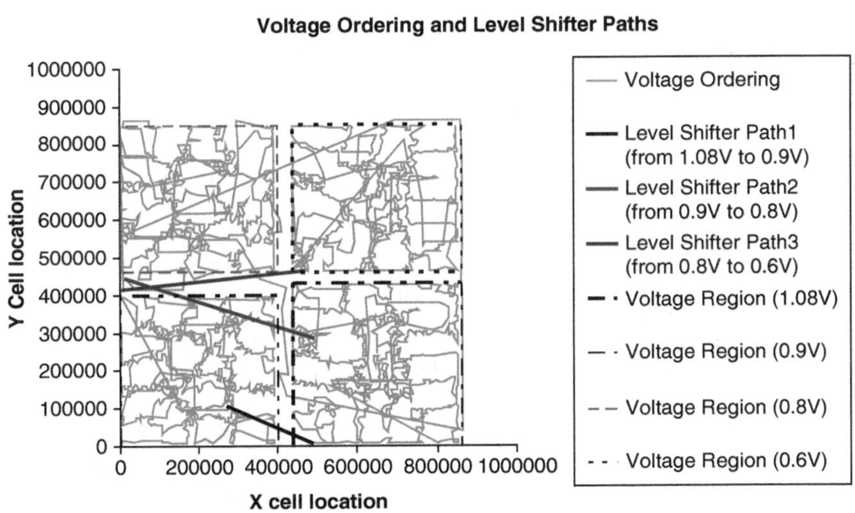

Fig. 11.9 Power-aware scan chain reordering

design under consideration has 3,500 cells; two clock domains and four voltage domains $V1$, $V2$, $V3$, and $V4$. In this case, $V1 = 1.08\,\text{V}$, $V2 = 0.9\,\text{V}$, $V3 = 0.8\,\text{V}$, and $V4 = 0.6\,\text{V}$. We compared the results of physical scan chain assembly and multi-voltage aware scan chain assembly. It can be seen that in the case of physical ordering, the number of level shifters required was 43 and in the case of voltage based scan ordering the number of level shifters needed was 3. We can see a 93% reduction in the need for level shifters. Figure 11.8 plots the scan chain path after physical scan ordering and paths where a level shifter is required. Figure 11.9 shows the scan chain path after multi-voltage-based scan ordering and the paths where a level shifter is required. The figures also show the voltage regions in the design explicitly.

11.4.4 Impact on DFT Implementation

DFT Implementation is a one-step process that modifies the design by realizing a given DFT architecture. It inserts the DFT logic; routes scan chains, and perform logic mapping and local design optimizations. In the low-power environment, it also has to insert the power management structures such as level shifters and isolation cells based on the designer's power intent. To cost-effectively achieve these tasks, DFT implementation has to be tightly integrated with design synthesis so that it relies on the same engines synthesis used. In addition, it has to deal with the following considerations: (1) It needs to re-use existing power management structures whenever it is possible; (2) It inserts DFT logic in order to facilitate test of power management structures; (3) It should produce testable designs and test protocols compliant with the test design rules described earlier; (4) Logic mapping and optimization should not violate voltage regions and/or power domain constraints such as using a non-always-on cell on an always-on path; (5) It should generate test models with power annotation for hierarchical flows support.

11.4.4.1 Re-use of LS and ISO Cells during Scan Stitching

It is critical for DFT insertion to re-use any existing level shifter and isolation cell during scan stitching. The objective is to only create new LS and ISO cells when necessary as the impact on the design could be very expensive. Each LS cell comes with a pair of power supply rails, and each ISO cell needs to be enabled through global control signal. The higher the number of LS/ISO cell is, the higher is the area overhead and the higher is the difficulty to route the design.

To achieve this objective, DFT insertion must re-use any LS/ISO cell unless otherwise instructed by an informed user. This translates into creating the strict minimum number of hierarchical ports during scan stitching. Figure 11.10a shows a design where flop A and flop B are to be stitched on the same scan chain. Figure 11.10b illustrates the case where scan stitching is not power-aware. Here,

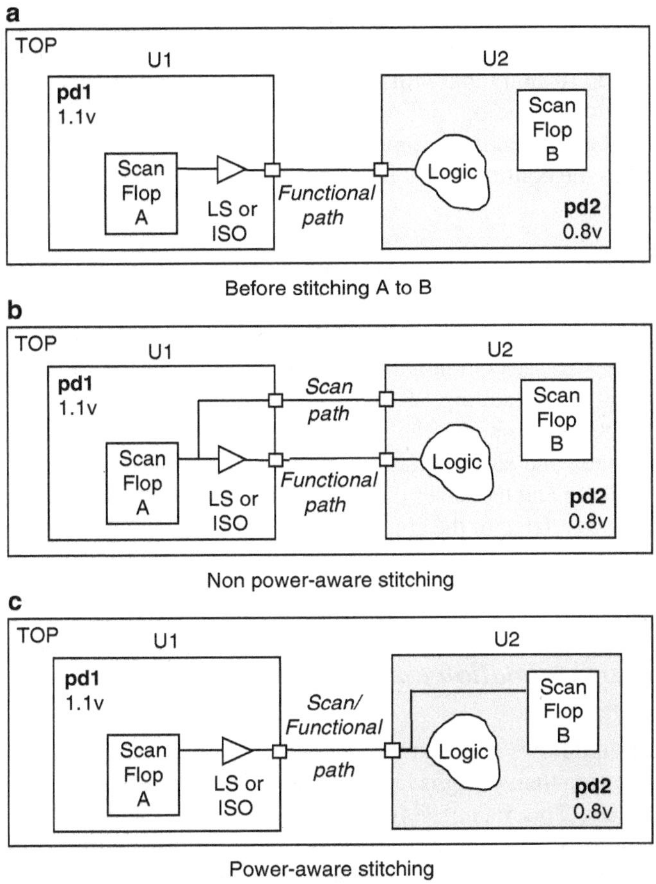

Fig. 11.10 Level shifter/isolation cell re-use during scan stitching

the stitching process ends up creating two additional hierarchical ports and require the insertion of a new LS/ISO cell. When dealing with industrial size designs the issue could be magnified many folds. Finally, with the correct power-aware behavior scan stitching simply ends up re-using the LS/ISO cell without creating any new cell. This is depicted in Fig. 11.10c.

11.4.4.2 Automatic Insertion of LS and ISO Cells

Right after scan stitching, DFT insertion needs to add LS and ISO cells on the nets driven by the newly created hierarchical test ports. This is done according to the power intent as specified by the user. DFT insertion would look at the LS/ISO strategy and the ISO control guidelines. The LS strategy defines the location of any created LS cell with regard to the associated power domain (either inside or outside the power domain). It also defines the type of LS (low to high or high to low), and its applicability to inputs only, outputs only, or both. The ISO strategy defines the type of isolation clamp (0 or 1) that will be used when the ISO cell is enabled. It also defines the location (same options as with LS), the associated power, and ground supply nets, and applicability to inputs, outputs, or both. Finally, the ISO control strategy specifies the control and polarity of the enable signal.

11.4.4.3 Design Synthesis Flow Impact

Having DFT insertion tightly integrated into the design synthesis flows provides a unique opportunity to use common infrastructure, database, and specialized routines and analysis engines. In adding LS and ISO cells, DFT insertion does not need to know a lot about synthesis specific details such as "always on" paths, power and ground supply nets, and so on. It simply populates the hierarchical ports it creates during scan stitching and then uses the same synthesis routines to insert LS and ISO. This way the issues related to the creation of LS/ISO become transparent to the DFT insertion.

11.4.5 Power Annotation and Hierarchical Design Flows

Hierarchical design flows are used to create very large designs and core-based systems. Users partition their designs into modules of manageable size; they build each module separately. This includes synthesis, optimization and DFT insertion, and then, they integrate those modules together at the chip-level. Because of capacity limitation or intellectual property (IP) protection, the integration process usually cannot use gate level representation of modules and/or cores. It relies on small and concise models, instead (Ramnath et al. 2002). In this section, we only look at the models that are used for DFT insertion purpose.

A test model of a design is an equivalent representation of the design that only contains DFT relevant information as described in the IEEE 1450 Core Test Language (CTL 2006).

Test models are side files that are generated as part of the DFT insertion process. Each module that goes through DFT insertion would have an accompanying test model that would describe the implemented scan structures, corresponding clocks with their waveforms, test control signals and their active states, test protocols, etc. The abstracted test information in the model is sufficient for DFT insertion to architect and stitch scan chains when the corresponding module is later integrated into a larger design.

Power details about DFT architecting should also be annotated on the test model, the same way scan chain related clocks are considered (Beausang et al. 1996). Without this critical information, the user will have very limited architecting options. This section defines what power information is important enough to be in the model, and gives illustration examples.

11.4.5.1 Low-Power Annotation

In a hierarchical design flow, a module that has DFT inserted in it is replaced by its test model from the DFT insertion point of view. Each scan chain in that module is treated as a single entity called scan segment (Beausang et al. 1996). Each scan segment is characterized by a set of data that drives a correct integration of the module at a higher-level stage. In a Multiplexed-scan design style, clock domain information is very critical and is annotated on a scan segment. Each scan segment has a capture clock (clock name and capture time associated with the scan cell at the scan input side) and a launch clock (clock name and launch time associated with the scan cell at the scan output side). Along with the clock domain information, all scan segment test access pins and test control signals are also annotated and used in the modeling process. More details on this modeling process can be found in Ramnath et al., (2002).

In the low-power context, a test model alone does not guarantee correct scan architecting in hierarchical flows. For a given module, along with a test model, additional low-power annotation has to be considered as well. This information has to be generated during the module-level DFT insertion process. In addition, any LS/ISO cell that is inserted on DFT signals that are located inside a module has to be known in order to prevent scan integration issues and reduce unnecessary redundancy.

11.4.5.2 Scan Modeling Enhancement

In a similar process to clock domain mixing, the mixing of scan segments across different voltage regions and/or power domains requires voltage and power domain information to be available on a scan segment. Each scan segment will need the following data:

- The scan input pin needs to have associated power domain identification, voltage value, and a Boolean flag saying whether an LS/ISO cell is inserted or not.
- The scan output pin needs to have associated power domain identification, voltage value, and a Boolean flag saying whether an LS/ISO cell is inserted or not.

A given scan segment could have same or different power domains and/or same or different voltage values between its scan input and scan output cells. This depends on power domain mixing and voltage domain mixing options used at the module level DFT insertion.

11.4.5.3 Voltage Annotation for DFT Insertion

This section illustrates how voltage annotation on a scan segment is used during a design integration phase. Figure 11.11 shows a design named "module" that is instantiated in design TOP. When performing DFT insertion on design TOP, only the test model of "module" is used. The DFT architecting process only considers the information annotated on the first and last scan cells of each scan segment in instance U2. Those scan cells are highlighted in Fig. 11.11 by dotted circles.

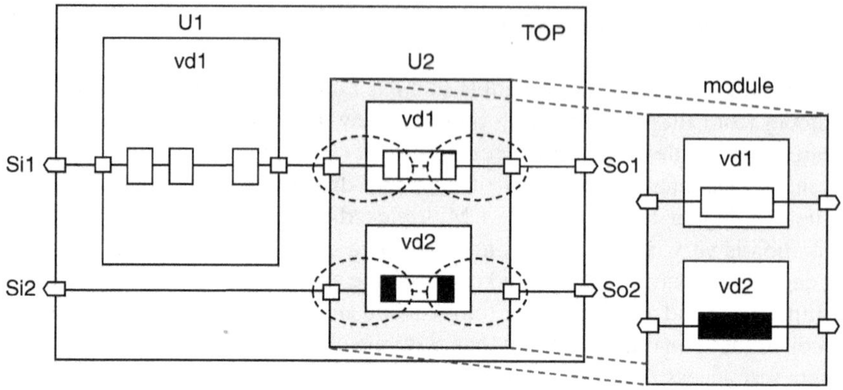

Fig. 11.11 Voltage annotation and scan segment modeling

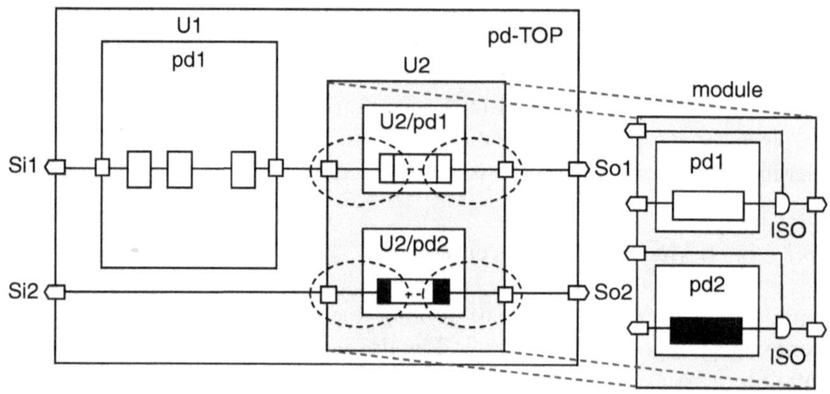

Fig. 11.12 Power domain annotation and scan segment modeling

If the user constraint is not to mix scan cells across voltage regions, then the voltage information that is annotated on the scan segments of instance U2 becomes critical. That's how one can achieve a correct scan architecting. In this example, one would get two scan chains: scan chain 1 (between Si1 and So1) which is comprised of scan cells that operate with voltage supply vd1, and scan chain 2 (between Si2 and So2) which is comprised of scan cells that operate with voltage supply vd2. If the user decides to mix scan cells across voltage regions, the voltage annotation is not that critical as any redundant LS or missing LS cell could be corrected by a post DFT insertion optimization.

11.4.5.4 Power Domain Annotation for DFT Insertion

This section illustrates how power domain annotation on a scan segment is used during a design integration phase. Figure 11.12 shows a design named "module" that is instantiated in design TOP. The DFT architecting process only considers the information annotated on the first and last scan cells of each scan segment in instance U2. Those scan cells are highlighted in Fig. 11.12 by dotted circles.

If the user decides not to mix scan cells across power domains, then the power domain information that is annotated on the scan segments of instance U2 becomes critical. That's how one can achieve a correct scan architecting. In this example, one would get two scan chains: scan chain 1 (between Si1 and So1) which is comprised of scan cells that belong to power domain pd1, and scan chain 2 (between Si2 and So2) which is comprised of scan cells that belong to power domain pd2. If the user opt to mix scan cells across power domains, the power domain annotation is not that critical as any redundant ISO cell could be corrected by a post DFT insertion optimization.

11.5 Test Planning

Test planning is the process by which the test of a design is defined. Based on a multitude of constraints, the challenging exercise is to decide on which DFT techniques to use and how to schedule the test. This process is by and large a manual process today since it is very design specific. The results of test planning can be translated into a series of specifications to implement the DFT structures through automation tools. Accurate test planning helps make the DFT implementation much more predictable and minimizes the need for costly iterative corrections.

Because of all the challenges and tradeoffs we discussed earlier, the test of advanced low-power devices is more than ever in need of intelligent planning. And that planning needs to be done at early stage. Many power-aware DFT techniques do exist but each of them comes with a cost and a set of requirements. The key objective of the planning is to schedule the test of a design and decide on adequate DFT techniques while keeping power consumption under control This section covers the important considerations during test planning.

11.5.1 Predictability of Results

Predictability is the key differentiator in today's EDA flows. Users value design flows based on how quickly they can get to their results, be it timing convergence, routing, or any type of correlation between the logical and physical domain. When it comes to design for low power, many efficient but intrusive techniques exist. Using these techniques will certainly reduce power consumption, but the user never knows by how much until later in the flow. Sometimes the results exceed the power budget; which requires expensive corrections. Sometimes the results are very conservative as power might have been over reduced. Here, the price is paid somewhere else in term of implementation costs. The ideal scenario is to rely on power estimation/analysis and only implement what is necessary to stay close to the power budget.

The power dissipation during test is difficult to predict. Usually, it is discussed in term of switching activity reduction during ATPG (TetraMAX 2008). Even then one cannot predict how much reduction could be achieved. Test experts try to set a budget in term of switching activity threshold of scan flops, for example, but the real power dissipation is only known after the ATPG patterns are generated and analyzed by a power analysis tool.

DFT planning (which DFT techniques to use) and test scheduling (definition of test modes) is the guaranteed way toward keeping power under control. The better the planning at the high level, the better are the chances to quality test within acceptable power margins.

11.5.2 Power Dissipation vs. Test Application Time

During test scheduling, it is important to study the trade-off between power dissipation and test application time (see Fig. 11.13); while the test coverage quality is not negotiable. If power is not an issue, then one could test a design with all power domains powered on at the same time. On the other hand, if the power dissipation is too constrained, then one option could be to only test one power domain at a time. These are the two extreme options that have been used in the industry. When both power dissipation and test application time need to be reduced, the user will need to consider the test of multiple power domains at the same time.

11.5.3 Need for Multi-mode DFT Architecture

Existing techniques such as modified test data de-compressor IP (Mrugalski et al. 2007), scan flop output gating (Gerstendorfer and Wunderlich 1999; ElShoukry et al. 2005), power-friendly test approaches (Czysz et al. 2008), etc., do effectively help reduce power dissipation but might not always yield optimal results

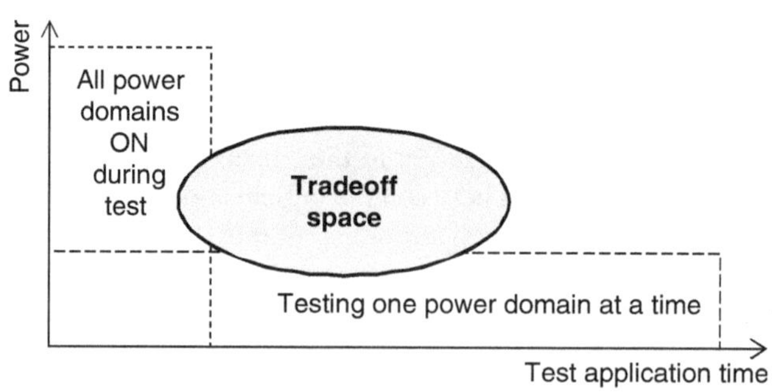

Fig. 11.13 Power dissipation versus test application time

as some of the benefits may be local and not propagated across the system. Substantial system-wide power optimization relies on scheduling multiple test modes. During test, the different test modes are then executed in sequence.

Multi-mode DFT architectures (Ravikumar et al. 2008) provide several modes of operation of a design in which test patterns can be applied. Each test mode targets different portions of the design or different DFT techniques and is associated with its own set of test constraints and procedures. For example, consider an SoC design encapsulating three power domains (not counting the top-level power domain which is always powered on). The design as illustrated in Fig. 11.14 has three different test modes where each power domain is individually tested. The power domains that are not tested in a mode are bypassed as described in Fig. 11.15. This is a major difference when compared to embedded-core testing. Here, the inactive power domains are powered down and cannot be used to implement a bypass functionality. The inactive domains have to be physically bypassed. This new power-aware scheme will reduce power dissipation by reducing switching activity in the inactive power domains. In addition, there could be modes where combinations of power domains are tested in parallel to reduce test application time. The optimal combinations are

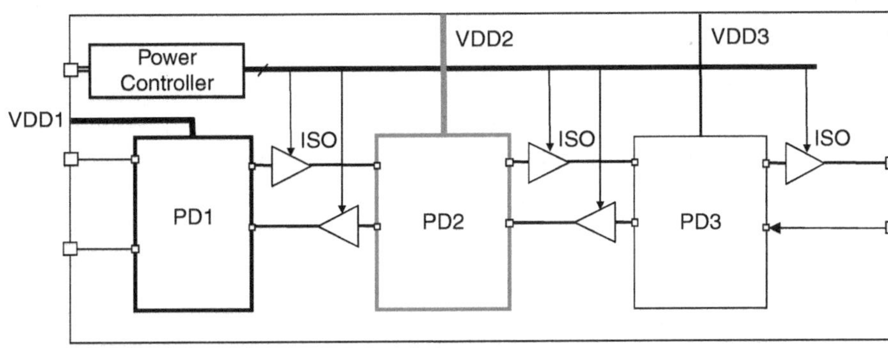

Fig. 11.14 Example of multi-power domain design

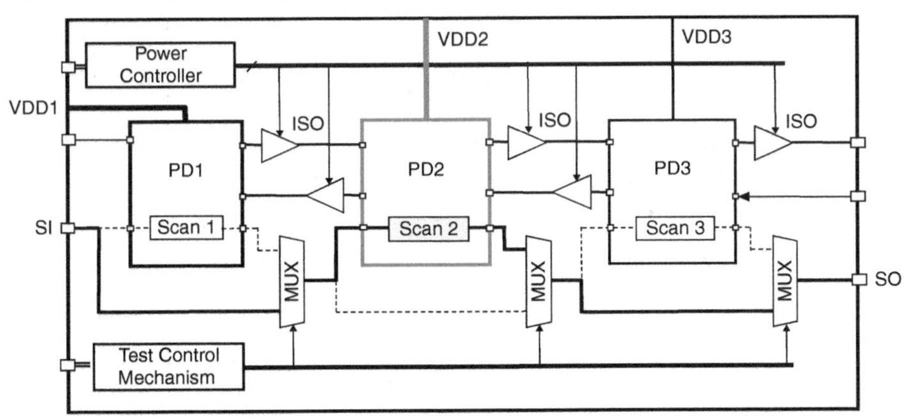

Fig. 11.15 Illustration of multi-mode DFT architecture

determined through intelligent power-aware test scheduling algorithms such as the one described in Chickermane et al., (2008).

11.5.4 Test Scheduling Considerations

Given a multi-mode DFT architecture, test scheduling is the process of defining the different test modes in order to completely test a design. This is where the decision is made regarding the number of test modes and the power domain configurations to be tested in each mode.

For efficient power saving, it is important to be able to test multiple power domains at a time. The idea is to test power domains in a way they were designed to functionally work. The reasons are the following: (1) Having a different partitioning between a test mode and a functional mode might result in higher power dissipation as more power domains could be switched on during test than in functional mode, (2) Usually, there are mode transition restrictions that could limit test scheduling. We don't want to exercise illegal mode transitions during test, and (3) Any deviation from user-specified power modes would require DFT insertion to also insert switch cells for test purpose. Test is not equipped to do such work as this task requires design expertise and information that might only be available later in the physical design flow. Note that DFT could be used to directly control existing power switch cells so that it allows less power domains to be powered on during test.

The following sub-sections describe how user-defined power modes are mapped into test modes, and some of the requirements related to ATPG support.

11.5.4.1 User Power Mode to Test Mode Mapping

As part of the UPF power intent, the user needs to describe the set of legal states of power domains in the design. Each state would tell which power domain is powered off and which one is powered on; and when a power domain is powered on; it shows the voltage state on the power domain corresponding supply network. Figure 11.16 shows an example of UPF power state table for the design of Fig. 11.14.

Given a set of user-defined legal power states, like states s0, s1, and s2 in Fig. 11.16, the objective of the user power modes to test modes mapping is to extract a minimum sub-set of power states in order to achieve the following conflicting goals:

- Cover the test of all the power domains and surrounding logic,
- Minimize the power dissipation during test,
- Minimize the test application time.

To be completely tested, a power domain needs to be tested when it is powered on and also when it is powered down. When a power domain is powered on, the logic inside the power domain and level shifter cells are tested. Isolation cells at the IO

```
• add_port_state VDDR -state {VDDR_on 0.8V} -state {VDDR_off off}
• add_port_state VDDB -state {VDDB_on 0.9V} -state {VDDB_off off}
• add_port_state VDDG -state {VDDG_on 1.1V} -state {VDDG_off off}
• create_pst     my_pst  -supplies     {VDDR     VDDG     VDDB}
• add_pst_state  s0  -pst my_pst  {VDDR_on  VDDG_on  VDDB_off}
• add_pst_state  s1  -pst my_pst  {VDDR_on  VDDG_off VDDB_on}
• add_pst_state  s2  -pst my_pst  {VDDR_off VDDG_on  VDDB_on}
```

Fig. 11.16 Power state table (in UPF format) for example of Fig. 11.14

boundaries of the domain are only tested in their disabled mode. When the power domain is powered down, isolation cells at the IO boundaries of the power domain are tested in their enabled mode. The test of state retention logic inside a power domain PD requires a 3-step process. First, PD needs to be switched on in order to load a pattern into state retention registers. Then, PD is to be switched off to make sure the pattern is saved when all, but the retention registers, are shut down. Finally, PD is powered back on and after some delay (due to power up sequence), the state of the retention registers is unloaded and compared against the initial pattern that was stored before PD was turned off.

This means that defining the test modes will require, for each power domain PD, a mode where PD is powered on and another mode where PD is powered off. New test DRC rules would help analyze the user provided power modes and then alert the user if this basic requirement is not met.

Based on this observation, finding the subset of user power modes suitable for test looks straightforward. But, in reality, the size of the power state table could be too large to allow an exhaustive search for the optimal subset. Indeed, a typical design could have 20–30 power domains (Hattori et al. 2006a, b; Chickermane et al. 2008) and even much more. This makes finding the best subset of power modes an NP complete problem. A practical solution uses a greedy algorithm of set covering. One such example is published in Chickermane et al., (2008).

Now, the questions that need to be raised are the following: how to translate a power state table from power supply-based description into power domain-based description? How to select the power modes? How many modes to select? How to choose between power modes with the same number of active and inactive power domains?

Figure 11.16 shows an example of power state table expressed in UPF. This table is expressed in term of supply nets. This table needs to be translated into power domain-based description where each primary supply net is now replaced by its associated power domain. VDDR is replaced by PD1, VDDG is replaced by PD2, and VDDB is replaced by PD3.

The selection of power modes could be a complete manual process in which the user explicitly specifies power modes for test. This could work for small design with few power domains. In general, the selection process needs to be automatic. It should pick up modes with the right number of active power domains. This number should not be too small as this could lead to a very large test application time that is close to testing one domain at a time. On the other side, this number should not be too high as it could lead to very high power dissipation. The user could impose a limit PD_{MAX} for the selection process so that only modes that have PD_{MAX} or less active power domains will be considered. If no limit is specified, then the selection process will first pick up modes with the most active power domains.

Regarding the number of test modes to be selected, the higher is the number the larger is the test application time as the test modes are activated sequentially. The objective is to define the minimum number of test modes that satisfies the power constraint. Here also, the tool should provide the user a way to guide the process.

To the question on how to choose between equivalent power modes (those having the same number of active power domains), there are two options. The first option is to let the selection algorithm pick up the power mode that provides a better test coverage. The second option is to randomly pick up any of the modes when the logic coverage is roughly the same.

11.5.4.2 ATPG Requirements

In this section we discuss a couple of important ATPG requirements that need to be considered for the test of multiple power domains.

For each test mode, a test protocol needs to be defined. The first issue that needs attention is how to make sure power domains are set to their active and inactive states through the execution of the test protocol. Three options are usually available at this level. The first option is to make this task a user responsibility. The user will need to provide a valid test initialization sequence. The second option is to automatically define the initialization sequence. While this might be possible for simple cases, it could be very difficult to correctly determine the initialization sequence for designs where power switches are controlled through some complex sequential logic. Finally, the third option is for DFT to control the power switches of the power domains directly from PIs. This option is easy to implement but will incur some area overhead. Note that this area overhead is usually offset by the fact that power switches need that same DFT in order to be tested for manufacturing defects.

The second issue is in the identification of powered down regions. The ATPG needs to know which power domains are powered down so that it will not target faults within those regions, and also will not use those regions for simulation and propagations. Fault accounting is yet another area that needs to be managed carefully as a given power domain could be tested several times. One solution here is to require ATPG to understand the user's power intent in term of identification of power domains and power modes. One other solution is to annotate test protocol files with the power domain states.

Finally, there is the important issue of letting the ATPG directly control the switching of power domains during test pattern generation. As we said earlier, switching off a power domain does not cause any problem as the domain is immediately powered down. However, it takes some delay for a power domain to stabilize after it is powered back on. This is because the power switches for a power domain are daisy chained in order to activate them serially to avoid damaging spike with in-rush current (Goel et al. 2006; Souef et al. 2008). For this reason it is not practical for ATPG to switch off and on power domains for each pattern.

11.6 Summary and Conclusions

The test of advanced low-power devices is making designs much more complex and very strictly optimized for power. Design for low power is changing the way designs are built. It is no longer seen as simple, contained incremental changes like gating the clocks, but it rapidly became very intrusive at the point where all of the EDA tools need to be made power aware and must have unified support for users' power intent.

In this chapter, we discussed the expectations, challenges, and practical considerations to making DFT insertion power-aware. The goal was not to insist on comparing DFT techniques or discussing new DFT techniques, but to layout the key considerations in testing power gated designs. We tried to focus on flow integration and impact of implementing any given technique on the flow.

We wanted to highlight the key enhancements that needed to be made to traditional DFT insertion tools in order to support power gated designs, and analyzed some of the new design rules and new trade-offs for which enhancements need to be done in order to help the user make informed decisions.

We talked about the importance of using multi-mode DFT architectures for test planning and test mode scheduling as key to bring predictability into the test of low-power flows.

Since leakage power has also become a challenge, it is important to look for ways to make DFT really transparent in functional mode. DFT should not have any impact when it is not active. One idea that is worth exploring is to group most of the DFT logic in dedicated power domain(s) and then be able to switch these domains off during functional mode.

Looking at the low-power design flow in general, one important work is to enhance power analysis and estimation in terms of accuracy and speed, and have it linked at different stages in the flow where design is being optimized; which, of course, includes DFT synthesis. This will add the needed predictability which helps eliminates very costly design iterations.

Acknowledgments The author wishes to thank James D. Sproch, Senior Director of Research and Development at Synopsys and recognized expert on low-power design and test issues, for his valuable input and detailed review of this chapter; Prof. Xiaoqing Wen of Kyushu Institute of Technology, Japan, Prof. Nicola Nicolici of McMaster University, Canada and Prof. Patrick Girard of LIRMM, France, for their effort and patience in putting together the book and their review of the manuscript.

References

Beausang J, Ellingham C, Robinson M (1996) Integrating scan into hierarchical synthesis methodologies. Proc IEEE Int Test Conf (ITC 1996), 751–756

Brophy D (2008) IEEE P1801 – The unified power format for low power designs. UPF Tutorial at Design Automation and Test in Europe (DATE), http://www.accellera.org/activities/p1801_upf/DATE-UPF-Final_2008.pdf

Chakravadhanula K, Chikermane V, Keller B,Gallagher P, Gregor S (2008) Test generation for state retention logic. Proc IEEE 17th Asian Test Symp (ATS 2008), 237–242

Chickermane V, Gallagher P, Sage J, Yuan P, Chakravadhanula K (2008) A power-aware test methodology for multi-supply multi-voltage designs. Proc IEEE Int Test Conf (ITC 2008), Paper 9.1

CPF (2007) Si2 common power format specifications, http://www.si2.org/, 2007

CTL (2006) IEEE 1450.6 standard test interface language (STIL) for digital test vector data – core test language (CTL), IEEE Computer Society, April 2006

Czysz D, Kassab M, Lin X, Mrugalski G, Rajski J, Tyszer J (2008) Low power scan shift and capture in the EDT environment. Proc IEEE Int Test Conf (ITC 2008), Paper 13.2, 1–10

De Colle A, Ramnath S, Hirech M, Chebiyam S (2005) Power and design for test: a design automation perspective. J Low Power Electronics (JOLPE) 1(1):73–84

DevanathanVR, Ravikumar CP, Mehrotra R, Kamakoti V (2007) PMScan: a power-managed scan for simultaneous reduction of dynamic and leakage power during scan test. Proc IEEE Test Int Conf (ITC 2007), Paper 13.3, 1–9

DFTMAX (2008) DFT Compiler/DFTMAX User Guide, version B-2008.09

ElShoukry M, Tahranipoor M, Ravikumar CP (2005) Partial gating optimization for power reduction during test application. Proc IEEE 14th Asian Test Symp (ATS 2005), 242–247.

Gerstendorfer S, Wunderlich H-J (1999) Minimized power consumption for scan-based BIST. Proc IEEE Int Test Conf (ITC 1999), 77–84

Goel SK, Meijer M, de Gyvez JP (2006) Testing and diagnosis of power switches in SOCs. Proc IEEE 11th Eur Test Symp (ETS 2006), 145–150

Goering R (2008) Automating low power design – a progress report. SCD source, issue 1, http://www.scdsource.com/download.php?=SCDsource_STR_LowPower.pdf

Hattori T, Irita T, Ito M, Kato H, Sado G, Yamada Y, Nishiyama K, Yagi H, Koike T, Tsuchihashi Y, Higashida M, Asano H, Hayashibara I, Tatezawa K, Shimazaki Y, Morino N, Hirose K, Tamaki S, Yoshioka S, Tsuchihashi R, Arai N, Akiyama T, Ohno K (2006a) A power management scheme controlling 20 power domains for a single-chip mobile processor. Digest Tech Papers IEEE Int Solid-State Circuits Conf (ISSCC 2006), Paper 29.5, 2210–2219

Hattori T, Irita T, Ito M, Yamamoto E, Kato H, Yamada T, Nishiyama K, Yagi H, Koike T, Tsuchihashi Y, Higashida M, Asano H, Hayashibara I, Tatezawa K, Shimazaki S, Morino N, Yasu Y, Hoshi T, Miyairi Y, Yanagisawa K, Hirose K, Tamaki S, Yoshioka S, Ishii T, Kanno Y, Mizuno H, Yamada Y, Irie N, Tsuchihashi R, Arai N, Akiyama T, Ohno K (2006b) Hierarchical power distribution and power management scheme for a single chip mobile processor. Proc IEEE Design Automation Conf (DAC 2006), 292–295

Idgunji S (2007) Case study of a low power MTCMOS based ARM926 SoC: design, analysis and test challenges. Proc IEEE Int Test Conf (ITC 2007), Lecture 2.3, 1–10

Keating M, Flynn D, Aitken R, Gibbons A, Shi K (2007) Low power methodology manual: for system-on-chip design. Springer, New York, Edition 2007

Mrugalski G, Rajski J, Czysz D, Tyszer J (2007) New test data decompressor for low power applications. Proc IEEE Design Automation Conf (DAC 2007), 539–544

Ramnath S, Neuveux F, Hirech M, Ng F (2002) Test-model based hierarchical DFT synthesis. Proc IEEE Int Conf Comput Aided Design (ICCAD 2002), 286–293

Ravi S (2007) Power-aware test: challenges and solutions. Proc IEEE Int Test Conf (ITC 2007), Lecture 2.2, 1–10

Ravikumar CP, Hirech M, Wen X (2008) Test strategies for low power devices. Proc IEEE Design Automation Test Eur (DATE 2008), 728–733

Souef L, Eychenne C, Alie E (2008) Architecture for testing multi-voltage domain SOC. Proc IEEE Int Test Conf (ITC 2008), Paper 16.1, 1–10

Std-1801 (2009) 1801 – IEEE standard for design and verification of low power integrated circuits. IEEE Computer Society, March 2009

Synopsys (2007) Synopsys low-power solution. White paper, June 2007, http://www.synopsys.com/lowpower/wp/lp_solution_wp.pdf

TetraMAX (2008) TetraMAX ATPG User Guide, version B-2008.09. Synopsys. Inc., Sept. 2008

UPF (2007) Unified Power Format (UPF) Standard, Version 1.0, Feb. 22, 2007, http://www.accellera.org/apps/group_public/download.php/989/upf.v1.0.pdf

Zyuban V, Kosonocky SV (2002) Low power integrated scan-retention mechanism. Proc IEEE Int Symp Low Power Electronics Design (ISLPED 2002), 98–102

Summary

The topics covered in this book deal with the interrelation between low-power and test of VLSI circuits. The reader has been introduced first to the basic concepts in manufacturing test and power issues during test. In order to avoid destructive testing and overkill, various solutions adopted to reduce power during test have been developed. In the first part of the book, the emphasis was placed on solutions for low-power ATPG, power-aware DFT, BIST and test data compression and power-conscious system-level test planning. The presence of power-management structures, such as clock gating, power gating or multiple supply voltages, introduces additional constraints on the testing process. Therefore, in the second part of the book, the focus was shifted toward the unique test requirements for low-power devices. The book concludes with an overview of the challenges faced by the EDA industry to integrate the constraints and objectives for designing low-power and testable VLSI circuits.

Over the past few decades, the consumers have benefited from Moore's Law by gaining more functionality when shifting from one process node to the next one. As we have already seen in the past few years, the power wall has altered this trend, and innovations in technology, circuits and architectures are necessary to get the most out of the nano-scale process nodes and maintain their performance benefits and hence the added functionality over time. As final remarks for this book, we briefly look at three different directions pursued these days to manage the excessive power requirements and understand their implications on the test technology. These directions can be broadly classified as technology, circuit and architecture oriented.

On the integration technology side, there are major ongoing initiatives for implementing 3D integrated devices. Placing active devices on multiple tiers within the same package can reduce the power consumed by long interconnects and device pins. This, in turn, provides an opportunity to boost the performance without exceeding the power budgets constrained by heat density, packaging and cooling equipment. Nevertheless, regardless how 3D integration is achieved, i.e. through stacked chips, wafer-scale or monolithic integration (each of them with different fabrication cost and size for through-silicon vias), the test technology will need to keep up in order to test these devices cost-effectively. For example, interconnects between layers may require new fault models and devising thermal-aware test plans for 3D circuits will present unique challenges to the test technology.

Future process technologies will introduce even more design variability and predicting circuit performance will become more difficult as the feature size continues to decrease. A worst-case design approach will be impractical either due to performance or yield concerns. As a consequence, there has been a growing interest in using resilient circuits that can tolerate the process parameter variations, temperature gradients or fluctuations in supply voltages, all of which influence power. The use of on-chip temperature and voltage sensors, combined with self-adaptive circuits, has been shown to allow body bias, operating frequency and supply voltage to be dynamically adjusted. Similarly, error detection circuitry can be employed to detect timing errors at runtime, in which case additional clock cycles are required for rollback and recovery. This enables circuit operation at better than worst-case clock period; and as long as the timing errors are not frequent, the improvement in clock frequency outweighs the penalty in clock cycle count. There is little doubt that relaxing the focus from worst-case design will benefit power; however, new test challenges will arise. Screening the fabrication defects in logic blocks in the presence of resilient circuits is not a trivial task and these circuits will also need to be thoroughly characterized and tested for defects, as they are the infrastructure that enables self-adaptation in-field. Besides, guaranteeing their correct operation in-field will require a better understanding how online test and diagnosis can be done in a power-efficient way.

It is well known that multi-core processors are becoming the standard computing architecture. One of the major reasons for their adoption was the power wall faced by the single-core processors, which have relied primarily on scaling the operating frequency for a boost in performance. With the burden shifted to software development, on-chip parallel computing provided by multi-cores is enabling a further improvement in performance. As we gradually move from two to quad to eight and so on to "hundreds-of-core" processors, some of these cores will be used to improve yield and reliability by means of fault tolerance for permanent faults and self-adaptation for transient errors. As it was the case with resilient circuits, the self-adaptive architectural features, which decide at runtime the load on each processor, will pose unique challenges to the test technology. An example is the creation of power-constrained test plans that take into account the temperature gradients for validating the self-adaptive architectural features. On another line of thought, multi-core architectures will provide the opportunity to rethink some of the fundamental EDA algorithms, including ATPG, DFT insertion, test data compression and so on; most of these algorithms will deal with power constraints and hence multi-core architectures will enable a faster and better implementation.

Low-power testing is an active area of research and development that has steadily moved from research labs to practice in the past decade. This book has detailed both the basic and the advanced techniques in the field. It is anticipated that with the growing need for more power efficiency, the low-power testing techniques presented in this book will continue to be widely adopted. With the ongoing advances in technology, circuits and architectures for low-power design, more innovation for low-power testing will happen; in this respect, we believe that this book will serve as an inspiration for future research and development in the field.

Index

Kolibri 360 GmbH
Ferdinand-Jühlke-Straße 7
99095 Erfurt, Deutschland
produktsicherheit@kolibri360.de